COMPUTATIONAL AND EVOLUTIONARY ANALYSIS OF HIV MOLECULAR SEQUENCES

COMPUTATIONAL AND EVOLUTIONARY ANALYSIS OF HIV MOLECULAR SEQUENCES

edited by

Allen G. Rodrigo
University of Auckland

Gerald H. Learn, Jr.
University of Washington

KLUWER ACADEMIC PUBLISHERS
Boston / Dordrecht / London

Distributors for North, Central and South America:
Kluwer Academic Publishers
101 Philip Drive
Assinippi Park
Norwell, Massachusetts 02061 USA
Telephone (781) 871-6600
Fax (781) 681-9045
E-Mail <kluwer@wkap.com>

Distributors for all other countries:
Kluwer Academic Publishers Group
Distribution Centre
Post Office Box 322
3300 AH Dordrecht, THE NETHERLANDS
Telephone 31 78 6392 392
Fax 31 78 6546 474
E-Mail <services@wkap.nl>

Electronic Services <http://www.wkap.nl>

Library of Congress Cataloging-in-Publication Data

Computational and evolutionary analysis of HIV molecular sequences / edited by Allen G. Rodrigo, Gerald H. Learn.

p. cm.

Includes bibliographical references and index.

ISBN 0-7923-7994-2 (alk. paper)

1. HIV (Viruses)—Research—Methodology. 2. Nucleotide sequence. 3. Retroviruses—Genetics—Research—Methodology. I. Rodrigo, Allen G., 1960- II.Learn, Gerald H., 1948-

QR414.6.H58 C66 2000
616.97'92'0072—dc21

00-064314

Printed on acid-free paper. Printed in the United States of America

CONTENTS

PREFACE

HIV research is unusual in that it brings together scientists from quite a range of disciplines: clinicians, pathologists, immunologists, epidemiologists, virologists, computational biologists, structural biologists, evolutionary biologists, statisticians, and mathematicians. This is not a novel insight, but it does have some profound implications. For one thing, there is less likely to be a shared matrix of operational procedures or concepts, experimental or analytic methods than one would expect when researchers are all drawn from the same discipline. An immunologist may abhor the simple dynamical models of HIV infection that mathematicians construct because they relegate to the rather large black box all of the spectacular interactions between our immune systems and the viral invaders. The word "population" means one thing to an evolutionary biologist, something different to a statistician, and something else again to an epidemiologist. Amongst virologists, there is no end of trouble with the word "quasispecies" and its usage.

In Douglas Adams' *The Hitchhiker's Guide to the Galaxy* there was the Babel fish that, when inserted into an ear, fed on incoming sound waves and translated these to sounds that were intelligible to the host. Something like this would not go amiss with HIV research. In part, this book aims to take the place of the Babel fish, at least with respect to the translation of the language of computational and evolutionary biology. Our target audience includes HIV researchers who may have only a nodding acquaintance with computational or evolutionary biology, as well as those who are well versed in some but not all aspects of this very broad field. But this book is not about HIV evolution and as such has little to say about how the virus evolves, or what the biological importance of viral genetic variation is. Instead, the chapters in this book focus on methods so it does have quite a lot to say about the types of analyses researchers can use to address these questions.

The chapters cover a range of analyses that focus largely on HIV genetic variation. In this respect, the content reflects our own research biases. We have not included chapters that deal with the structural biology of HIV, for instance, although such a topic could sit quite comfortably as part of a computational biology text. The contributing authors have tried to make their subject matter accessible to scientists outside their respective disciplines, and we believe they have succeeded. This is not to say that the content is excessively simple. For some of the chapters, a reasonable mathematical foundation is required, and in others, an understanding of basic evolutionary concepts is assumed. Where possible, authors have identified appropriate computer programs that implement the methods they discuss. However, more valuable are the conceptual and theoretical frameworks on which the analytic methods are based and which are described in each chapter. Computer programs are replaced by better ones, algorithms change to suit the computing environment, but concepts and theories persist.

It remains for us to thank the contributing authors. Their contributions made this book (literally). We must also thank the following reviewers for their valuable and helpful comments: Jon Anderson, Peter Beerli, Colombe Chappey,

Keith Crandall, Nicole Mayer, Andrew Rambaut, Mika Salminen, Steven Self, Daniel Shriner, and Christophe Verlinde. We acknowledge the support and encouragement of Jim Mullins. Financial support for JL's work on this book was in part from grants of the US Public Health Service to J. I. Mullins. Mel Norris, our editorial assistant, has our inexpressible gratitude for her painstaking work, doing many of the tasks that make putting together an edited volume so challenging. And finally, to our families, who endured our absences of body or mind with good humor — its time for a holiday!

Allen Rodrigo
University of Auckland

Jerry Learn
University of Washington

SAMPLING AND PROCESSING HIV MOLECULAR SEQUENCES: A COMPUTATIONAL EVOLUTIONARY BIOLOGIST'S PERSPECTIVE

Allen G. Rodrigo*, Edward W. Hanley§, Paul C. Goracke§, and Gerald H. Learn, Jr§.
*School of Biological Sciences, University of Auckland
Auckland, New Zealand
§Department of Microbiology, University of Washington
Seattle, WA 98195-7740 USA

1. INTRODUCTION

In computer science, the acronym GIGO stands for "Garbage In Garbage Out". That an acronym is even required is testimony to the fact that computers are often considered to be infallible, incapable of making numerical errors. However, the correctness and accuracy of even the best computer program is largely dependent on the quality of input data. Computers are as yet only capable of distinguishing good from bad data at the most rudimentary and trivial level. The onus, consequently, is on the user to ensure that the data used in any analysis are suitable to answer the research questions posed and sufficiently free of errors (perhaps down to some acceptable threshold).

Any competent molecular biologist knows the ins-and-outs of molecular experimental design as it relates to sample preparation, amplification, the use of adequate controls, sequencing, error checking and the assembly of sequences into contiguous fragments. It is not our aim, as computational and evolutionary biologists, to teach molecular biologists how to do their jobs. However, many molecular biologists may not be aware of issues that relate to the collection of molecular sequences that are pertinent to the subsequent analysis of the data. Since this book is devoted to the computational and evolutionary analysis of HIV molecular sequences, we believe it appropriate to begin with a computational evolutionary biologist's perspective on sequence collection and experimental design. In particular, as many of the chapters in this book deal with the use of sequence data for evolutionary inferences, we focus this chapter on some aspects of experimental design and the subsequent processing of sequence data in studies where the desired result is to understand the evolutionary processes acting on HIV.

1.1 A very brief molecular biology of HIV

The unit of all analyses described in this book is the HIV molecular sequence. The HIV genome with its approximately 10,000 ribonucleotides, has been well characterized (Figure 1). It has the typical retroviral complement of the *gag*, *pol* and *env* genes coding respectively for the core, matrix and nucleocapsid proteins, viral enzymes (reverse transcriptase, protease and integrase), and the envelope protein. In addition to these genes, there are several accessory genes that are important for replication, integration, viral assembly and intracellular transport (Coffin, 1992). The genome terminates with a long-terminal repeat (LTR) that is involved in reverse transcription. The open-reading frames for different genes overlap in all three reading frames.

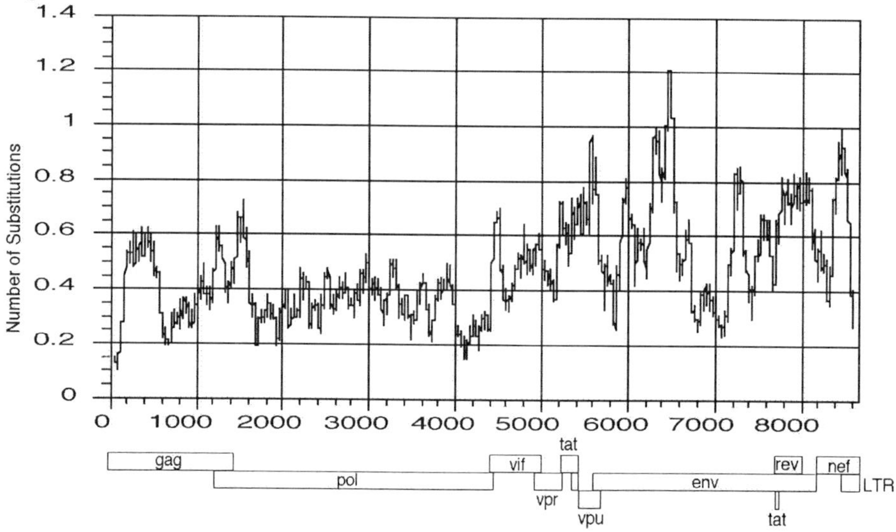

Figure 1. A schematic diagram of the genetic structure of the HIV genome, and a plot of sitewise variation. Note that genes are read in all three reading frames, with some overlap between open reading frames. The plot of sitewise variability was generated by estimating the number of substitutions at each site along the genome using a maximum parsimony tree of a sample of HIV sequences from different subtypes. The plot illustrates the considerable heterogeneity in substitution rates across sites.

The HIV genome in the mature virion is composed of two RNA strands encased in a proteinaceous core that is surrounded by a lipid bilayer impregnated with viral envelope proteins. Infection of target host cells, which include helper T-lymphocytes, monocytes and macrophages, is mediated by the attachment of envelope proteins to the primary cell surface receptor, CD4, and an accessory receptor, e.g., CCR5 or CXCR4. Cell entry involves the fusion of the virion's lipid membrane with the host cell's membrane and the transport of the viral core into the cytoplasm where reverse transcription of the viral RNA to DNA occurs. Reverse transcription is mediated by the viral reverse transcriptase, which has a very high nucleotide misincorporation rate (on the order of 10^{-5} to 10^{-4} per site per replication; Mansky, 1996).

Integration of the viral DNA into the genomes results in a provirus that is available for subsequent transcription and replication whenever the host cell becomes transcriptionally active. Transcription is mediated by the host cell's polymerase. When transcription occurs, viral proteins and genomes are produced, packaged and subsequently released from the host cell by budding, with a part of the host cell's membrane forming the outer membrane of the virion.

When a host cell is infected by two or more virions (cellular superinfection), there is the potential for the transcription and packaging of two different genomes in the same virion. In a subsequent round of infection and reverse transcription, this is likely to result in the production of a proviral genome that is a recombinant of the two viral genomes. The rate of superinfection of host cells, once thought to be low, is obviously sufficiently high to result in an estimated recombination rate of 2 – 3 crossovers per genome per replication cycle, approximately the same order of magnitude as the mutation rate of reverse transcriptase (Jetzt *et al.*, 2000).

The high mutation rate means that many of the proviral genomes have frameshift insertions and deletions, stop-codons, and other mutations that compromise the production of free and mature virions. Nonetheless, enough viruses are produced each day to sustain the infection: it is estimated that as many as 10^{10} viruses are produced in an infected individual daily, with most of these virions being produced by short-lived productively infected cells with a mean generation time of between 1 to 2 days (Perelson *et al.*, 1996). Perhaps 10^7 - 10^8 of these are capable of infecting new host cells. There is therefore the opportunity for the accumulation of a considerable amount of genetic variation within an infected individual (for example see Shankarappa *et al.*, 1999) as well as between infected individuals (Figure 2).

The rate of substitution varies across the genome, with the substitution rate being lowest in *pol*, intermediate in *gag*, and relatively high in *env*. Shankarappa and colleagues (1999) showed that substitutions in a partial *env* fragment accumulated at a rate of approximately 1% per year, a rate that appeared to be remarkably consistent across the 9 patients in their study. This rate heterogeneity is also apparent intragenically. In *env*, for instance, five variable regions (labelled V1 to V5) are separated by conserved regions. Immunogenic epitopes in some of these variable regions have been identified (Carrow *et al.*, 1991; Hwang *et al.*, 1992); in addition, some sites appear to play a role in cell tropism (Chesebro *et al.*, 1996; Peeters *et al.*, 1999).

2. SAMPLING HIV MOLECULAR SEQUENCES

Exactly what gene needs to be sequenced, how long a sequence has to be, and how many sequences are required are determined by the goals of the study itself. If one were interested in assaying the mutations conferring resistance to the nucleoside reverse transcriptase inhibitor AZT, for instance, one would target *pol*, or perhaps, only the RT-encoding domain of *pol*. If, on the other hand, the research focuses on the shift in cell-tropism over the course of infection, sequences of *env* would be the most likely target. In this section, we will not focus on sampling strategies for the entire range of different studies in which HIV sequences are used, but rather concentrate on those strategies that have been used to obtain sequences for the purpose of describing or studying HIV genetic variation. This is partly because of our own research biases, and also because in other areas where molecular sequences need to be sampled, e.g., drug resistance studies and structural biology, there appears to be at least some consensus about how sequences should be sampled.

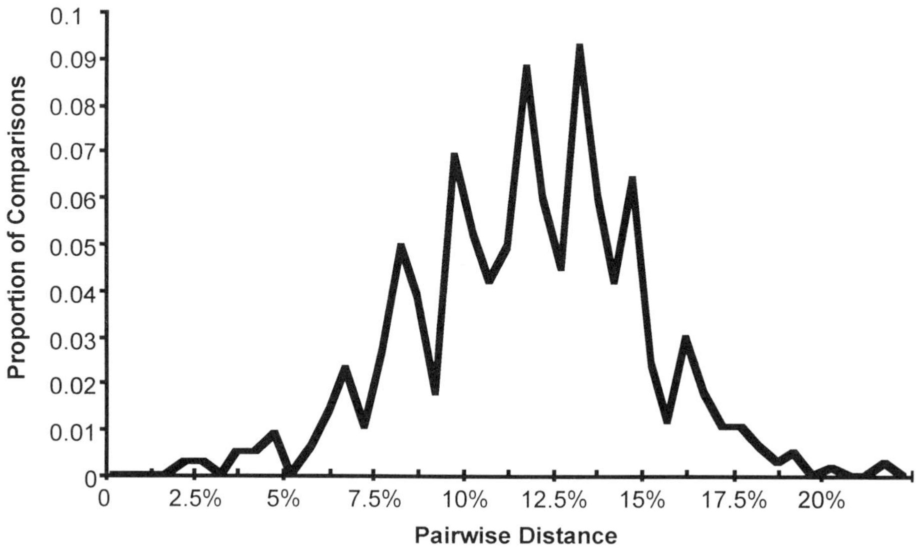

Figure 2 Pairwise genetic distances for *env* gene sequence from different individuals infected with HIV-1 subtype B.

One approach to sampling HIV sequences is consensus sequencing (Leitner *et al.*, 1993). With consensus sequencing, HIV molecules in a sample are extracted either as proviral DNA from infected host-cells or RNA from cell-free virions; if the latter are obtained, there is an preparatory reverse transcription step to produce cDNA. These molecules are then collectively PCR-amplified with primers designed to the target gene or gene fragment. The amplified product, a result of the amplification of all HIV molecules in the original sample, is subsequently sequenced (most often using an automated sequencer). The chromatogram that accompanies the results of an automated sequencing run can distinguish sites at which the gene fragments of all the HIV molecules in the sample have identical nucleotides (conserved sites) from those at which there is some nucleotide heterogeneity (variable sites). In addition, it may be possible to obtain a crude ordinal measure of the relative frequency of each constituent nucleotide at a variable site, although extreme caution is warranted when doing this because of the potential for sequencing artefacts.

If one can be certain that the variable sites are truly the result of genetic variation and not sequencing error or PCR misincorporation, there are some advantages to using consensus sequencing. First, consensus sequencing is a very rapid method for acquiring sequence information. In theory, all that is required is a single PCR per sample, and as many sequencing reactions as will guarantee enough coverage of the fragment of interest and the resolution of any ambiguous automated base-calling. Second, if the sample contains a large number of HIV molecules, consensus sequencing provides us with a snapshot of the level of genetic diversity of the HIV population: the greater the number of variable sites, the higher the level of diversity. In addition, it allows us to pinpoint which sites are variable and which are conserved. This may be important for identifying the emergence of a new tropic variant of HIV, or one that is potentially capable of escaping a specific antibody or cytotoxic T-lymphocyte response, or indeed, one that has acquired a resistance-conferring mutation.

It is sometimes thought that consensus sequencing does not provide us with data that are amenable to any evolutionary analysis. This is not completely true. There are several less commonly applied analyses that do not require knowledge of the sequences themselves, but only of the numbers of variable sites and/or the constituent nucleotides at these sites. Watterson's (1975) estimate of $\Theta = 2N_e\mu$, where N_e is the effective population size and μ is the mutation rate, for instance, requires only the number of variable sites (or segregating sites as they are also called) (see Vasco *et al*, Chapter 9). Tests for neutrality also exist if one can identify the codon-position of each variable site with no more than two constituent nucleotides and determine if the substitution at that site is nonsynonymous or synonymous, e. g., the Hudson-Kreitman-Aguade test (1987) or the McDonald-Kreitman test (1991; also see Nielsen, Chapter 11).

Table 1. Samples of hypothetical sequences showing the same numbers of segregating sites with identical constituent nucleotides at each site, but different patterns of sequence variation. (A) each sequence is different; (B) there are two haplotypes with SI, S2, and S3 having one haplotype and S4 and S5 the other.

A

	Sites						
	1	2	3	4	5	6	7
Con	**C**	**T**	**A**	**T**	**G**	**G**	**T**
S1	A	A	A	T	G	G	T
S2	A	T	A	T	G	G	T
S3	C	T	A	C	G	G	T
S4	C	T	G	T	G	G	C
S5	C	T	A	T	C	G	C

B

	Sites						
	1	2	3	4	5	6	7
Con	**C/A**	**T/A**	**A/G**	**T/C**	**G/C**	**G**	**T/C**
S1	A	T	A	T	G	G	T
S2	A	T	A	T	G	G	T
S3	A	T	A	T	G	G	T
S4	C	A	G	C	C	G	C
S4	C	A	G	C	C	G	C

There are however several drawbacks to consensus sequencing. Putting aside the obvious potential for sequencing artefacts and the difficulty of determining when a minor peak at a given site on the chromatogram is truly above background, consensus sequencing can only work when there are no insertions or deletions in the complement of HIV molecules in the amplified sample. If insertions or deletions are present in some molecules and not in others (or if different insertions or deletions are present in all molecules) then each site on the chromatogram will show peaks that correspond to non-homologous sites on the amplified molecules. In these instances, a "site" on a consensus sequence has no biological meaning. This rules out the use of consensus sequences for some regions of the HIV *env* gene (e.g., V1, V2, V4 and V5), where insertions and deletions accumulate rapidly over the course of infection within an individual.

Sometimes, a consensus sequence is defined by the linear ordering of only the dominant or major nucleotides at each site. In such cases, a consensus sequence may not correspond to any real HIV molecule in the sample (Table 1a). If the consensus sequence does not really exist, one must wonder what it can possibly tell us about the biology of the virus. Nor does the consensus sequence tell us is which specific sequences populate our sample. This is true even if the consensus sequence documents the variation at each site. In Table 1a & b, for instance, the same sites are variable, and share the same constituent nucleotides, in each set of sequences. However, it is apparent that the sequences in sample *B* cluster into two major groups (or haplotypes) whereas those in *A* are each a different haplotype. In our opinion, the fact that consensus sequences may not correspond to real viral sequences, makes their use for phylogenetic or genealogical studies questionable. Instead, for such research, we believe it is a better strategy to obtain individual sequences of HIV; this may be labor-intensive, but the advantages of dealing with real sequences outweigh the costs.

A number of different strategies may be used to obtain a sample of HIV sequences where each is derived from a different molecule. In the simplest case, an aliquot of the amplified product is used in a cloning reaction, and several clonal colonies, each containing a copy of a single PCR product, are picked for sequencing (Figure 3a). There is, however, a critical problem with this approach. If there were only a very few HIV genomes in the original sample to begin with, then all the PCR products are copies of these limited molecules. If, for instance, there are only 10 copies of HIV in the sample, then all the amplified DNA are derived from these 10 copies. Suppose, now, that a researcher desires to obtain 20 clones (and consequently, sequences) from this sample. Obviously, each of these 20 cannot be derived from 20 different HIV genomes, since there were only 10 to start with. The end result is that at least some genomes may be represented more than once, and this in turn will confound any statement one makes about patterns of genetic variation in the sample. This problem of resampling has been discussed by Liu and colleagues (1996). They show that the probability of resampling is given by :

$$P(\text{resampling}) = 1 - \frac{\prod_{i=1}^{n}(N-i+1)}{N^n} \tag{1}$$

where N is the number of target molecules, and n is the number of clones obtained. Plotting the probability of resampling as a function of number of original target molecules (Figure 3b), we see that as more clones are sampled for a given value of N, the probability of resampling at least one molecule increases as well. Surprisingly, even when the number of target molecules is as high as 100, sampling 10

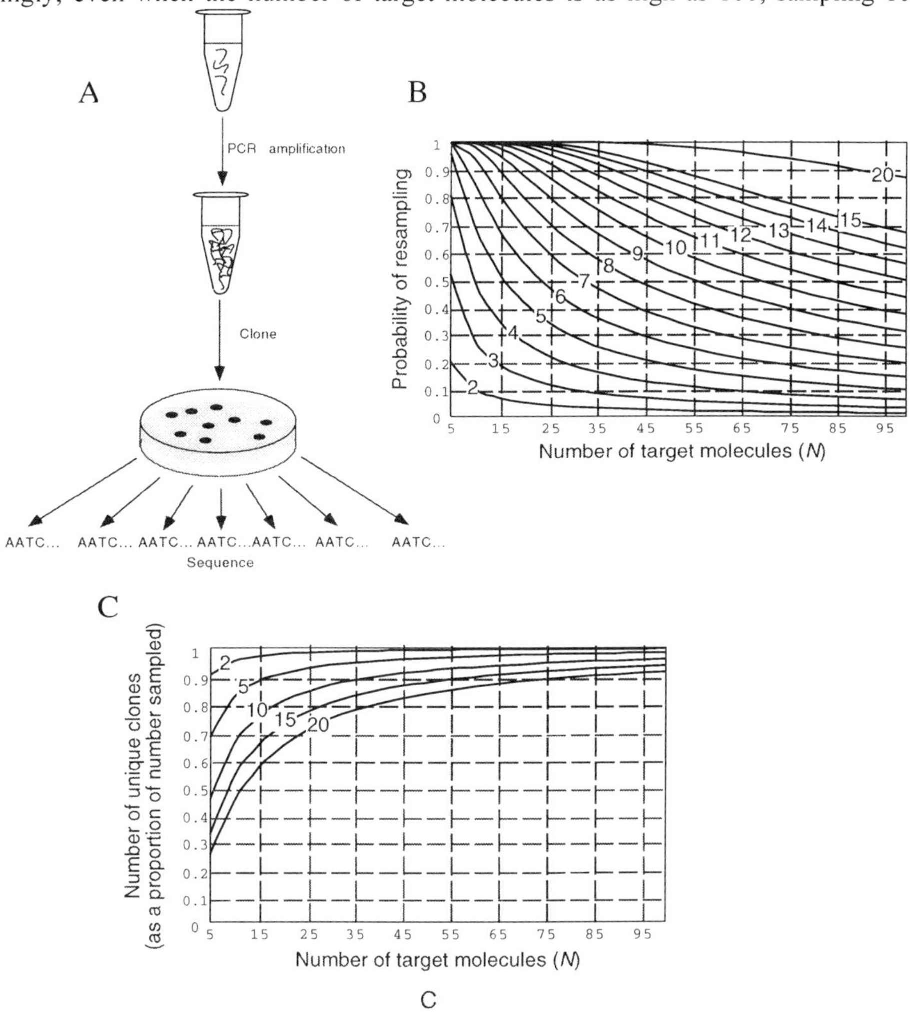

Figure 3. Obtaining molecular sequences. (A) An illustration of a commonly applied procedure for obtaining several molecular sequences of HIV by PCR-amplification, subsequent cloning and sequencing. (B) The probability of resampling when different numbers of clones are sequenced (labelled for each curve) given the number of target molecules in the sample. (C) The expected proportion of sequences derived from unique target molecules in the sample given the number of clones sequenced (labelled on the curves) and the number of target molecules.

clones still gives an appreciable probability of resampling at least one molecule (>40%).

Liu and his co-workers (1996) also provide a formula for the expected number of sequences derived from unique target molecules, when n sequences are sampled from a PCR amplification of N target molecules:

$$E[\text{number of unique sequences}] = N\left[1-\left(1-\frac{1}{N}\right)^n\right] \quad (2)$$

Figure 3c plots this as a function of n and N. Figure 3c reinforces the danger of resampling the same target molecule when several sequences are obtained from an amplification of even moderate numbers of original target molecules. For instance, when 20 clones are sampled from 100 target molecules the expected number of unique molecules these represent is approximately 18.

One obvious way to avoid resampling is to PCR-amplify different aliquots of the original sample, and from each PCR obtain and sequence only a single clone. Since each aliquot of the original sample must necessarily have different target molecules, each clone will be derived from a different viral genome. Alternatively, if one knows the number of viral copies in the original sample, one can use the graph in Figure 3b to work out an upper limit to the value of n, given an acceptable threshold for the probability of resampling. For instance, if it is known that there are 100 target molecules in the original sample, and we are prepared to risk resequencing at least one of these if the probability is 5% or less, then reading off the graph, we will be able to sample as many as 3 clones.

The question of how one estimates the number of copies in the original sample is an important one. It is not sufficient, we believe, to estimate this number using a sensitive quantitation assay that is based on a different fragment of the viral genome than the one targeted. This is because we are really interested in amplifiable copy numbers. For whatever reason, the sensitivity of the PCR may be different with different fragments, and the number of amplifiable copies will differ from region to region. If the intent is to sequence a region with a low number of amplifiable copies, relative to the number of copies estimated some other assay, then we risk resampling only those molecules that are amplifiable. To overcome this problem, a PCR-based quantitation strategy that specifically targets the region of interest can be used (Rodrigo *et al.*, 1997). The method is a variant of the minimum χ^2 (MC) method described by Taswell (1981) for limiting dilution assays.

The MC method, as described by Taswell (1981) and as it applies to PCR-based limiting dilution assays (PLDAs) begins with an estimation of the probabilities of obtaining a negative or positive PCR at each dilution, but assumes that these probabilities are solely due to the absence or presence of a target molecule. In a given homogeneous volume of sample we assume that the probability distribution of the number of molecules in each replicate PCR is Poisson with parameter $\lambda_i = cd_i$ where c is the number of copies per unit volume of sample and d_i is the inverse of the dilution factor of the ith dilution, or the quantity of sample used in the ith dilution. The probability, then, that no target molecules are available in any one replicate at the ith dilution is given by:

$$P_{\lambda_i}(0) = e^{-\lambda_i} \quad (3)$$

The probability that at least one target molecule is present in the PCR is therefore

$$P_{\lambda_i}(>0) = 1 - e^{-\lambda_i} \quad (4)$$

Rodrigo and colleagues (1997) modified Taswell's method to allow for false positive and false negative PCRs. Under this modification, a negative PCR can be the result of two possibilities: 1) the absence of a target molecule, after correcting for the possibility of a false positive PCR; and 2) a false negative PCR, given that a target molecule was indeed present in the PCR. The probability of a false positive PCR is the conditional probability of obtaining a positive result given that a target molecule was absent from the reaction; conversely, the probability of a false negative PCR is the conditional probability of obtaining a negative result given that a target molecule was present in the reaction. The probability of a negative PCR at the ith dilution, $P_{\lambda_i}(-)$, then is given by:

$$P_{\lambda_i}(-) = P_{\lambda_i}(0)(1 - P_f(+)) + P_{\lambda_i}(>0)P_f(-) \tag{5}$$

where $P_f(+)$ and $P_f(-)$ are the probabilities of false positive and false negative results respectively. As can be seen from (5), if $P_f(+)$ and $P_f(-)$ are zero, then the probability of a negative result is solely determined by the absence of a target molecule. The probability of a positive PCR is $P_{\lambda_i}(+) = 1 - P_{\lambda_i}(-)$.

If n_i replicate PCR amplifications are performed at the ith dilution, then the distribution of the number of negative reactions is binomial with mean $n_i P_{\lambda_i}(-)$ and variance $n_i P_{\lambda_i}(-)P_{\lambda_i}(+)$. Similarly, the expected number of positive reactions is $n_i P_{\lambda_i}(+)$ and variance $n_i P_{\lambda_i}(-)P_{\lambda_i}(+)$. A χ^2 statistic (with k-1 degrees of freedom), as a measure of the agreement between observed and expected numbers of positive and negative PCR amplifications, can therefore be calculated as:

$$\chi^2_{k-1} = \sum_i \left[\frac{(r_i - n_i P_{\lambda_i}(-))^2}{n_i P_{\lambda_i}(-)} + \frac{((n_i - r_i) - n_i P_{\lambda_i}(+))^2}{n_i P_{\lambda_i}(+)} \right] \tag{6}$$

where r_i is the number of negative PCR amplifications at the ith dilution, and k is the number of dilutions. Equation (6) reduces to

$$\chi^2_{k-1} = \sum_i \left[\frac{(r_i - n_i P_{\lambda_i}(-))^2}{n_i P_{\lambda_i}(-)P_{\lambda_i}(+)} \right] \tag{7}$$

The number of copies per unit volume of sample is estimated to be the value of c that minimizes the value of χ^2. In addition, Taswell (1981) calculates the standard error of the estimated number of copies as :

$$SE(c) = \sqrt{\frac{2}{F''(c)}} \tag{8}$$

where $F''(c)$ is the second derivative of χ^2 at c. A program for estimating the number of target molecules can be obtained from *http://ubik.microbiol.washington.edu/CBU/quality*.

A limiting dilution assay allows us to kill two birds with one stone. First, we obtain an estimate of the number of amplifiable copies in the original sample, always a valuable estimate particularly if more work is to be done with the sample. Second, we end up with several PCRs that have amplified target molecules. We may now proceed to obtain a single clone from each, thus removing the probability of resampling.

In the experimental strategy outlined above, we have only considered how we may obtain clones from PCRs of our sample. However, it is often preferable to sequence a PCR product directly. This is particularly true if one is concerned about the possibility of polymerase-induced misincorporation of nucleotides during the PCR. Over the 70 or so cycles of a typical nested PCR, the error rate may be as high as 0.7% (unpublished data), so that in a fragment 1000 nucleotides long, 7 "substitutions" may be the result of polymerase error. PCR-induced errors may be seen in each clone; however, if the PCR product is sequenced directly, then any error at a given nucleotide position will be masked by the majority of sequences that have not acquired that substitution. But direct sequencing has its drawbacks. As discussed above, if there is a mixture of target molecules, the direct sequences obtained from the PCR product will be a consensus of these sequences, and may not reflect any real viral sequence. Consequently, it is safest to sequence the PCR directly if one is reasonably certain that the PCR products are copies of only a single amplifiable target molecule.

It would appear, at first glance, that the easiest way to obtain PCR products of a single molecule of starting material is to quantify the number of copies of the target molecule, and dilute appropriately so that the expected number of copies in the PCR reaction is one. However, this procedure is wrong for the following reason. Suppose we follow this procedure and add a diluted aliquot of our sample to ten PCRs. The expected number of molecules in each reaction is one. Using a simple Poisson distribution to obtain the probability of getting a negative reaction:

$$P(\text{negative reaction}) = P(\text{number of molecules} = 0) = e^{-\lambda} \qquad (9)$$

where λ is the expected number of target molecules at the given dilution; in our example, λ =1, and the probability of a negative PCR is 0.37. We expect that approximately six of ten reactions will be positive, and four negative. However, since the average number of molecules over the ten reactions is 1, and we know that four of these PCRs had no target molecules, it must mean that the average number of starting molecules that went into the six positive PCRs is 10/6 = 1.7 molecules. In other words, if we were to sequence the products of these PCRs directly, we would stand a chance of obtaining a consensus sequence of two or more target molecules. The probability that any one of the products of any one of these PCRs is an amplification of a single molecule is:

$$P(\#\,\text{molecules} = 1 \text{ given that PCR is positive}) = \frac{\lambda e^{-\lambda}}{1 - e^{-\lambda}} \qquad (10)$$

Consequently, the probability that any one positive PCR has only a single molecule is 0.58. Therefore, just over 40% of the positive PCRs are expected to have more than one molecule. In Figure 4, we plot the probability that all our amplified copies in a given PCR come from a single target molecule for different values of l. From Figure 4, if we want to be 95% certain that the PCR reaction we plan to direct-sequence has a single molecule, we would need to dilute our sample to an expected 0.1 copies of target molecules per PCR. Therefore, we should expect only one out of 10 PCRs to be positive.

What should we do if more than one out of 10 PCRs are positive at this dilution? By chance alone, we expect that there will be some variation in the number of positive reactions. We can calculate the probability of obtaining an observed number of positive reactions, although it turns out to be mathematically easier to calculate the probability of obtaining the observed number of negative reactions. Returning to our example above, suppose that of 10 reactions, each with an expected number of target molecules = 0.1 per reaction, 2 of 10 are positive (i.e., 8 of 10 negative). The probability of obtaining a single negative reaction is the probability that there are no target molecules available given that $\lambda = 0.1$:

$$P(\text{negative PCR given } \lambda = 0.1) = e^{-\lambda} = 0.90$$

Therefore, using the binomial distribution, we can obtain the probability of seeing 8 of 10 negative reactions:

$$P(\#\text{ negative reactions } = 8) = \binom{10}{8} 0.9^8 (1-0.9)^2 = 0.39$$

There is a good chance, then, of getting 2 of 10 positive reactions when the expected copy number is 0.1 per reaction. However, the probability of getting 3 of 10 positive reactions is 0.04, and since this is less than a 1-in-20 chance of one would be cautious about accepting the result. It may be, for instance, that the original estimate of copy number is sufficiently off the mark to explain this discrepancy. In this case, one is probably best advised to re-estimate the copy number by pooling both old and new dilution series results.

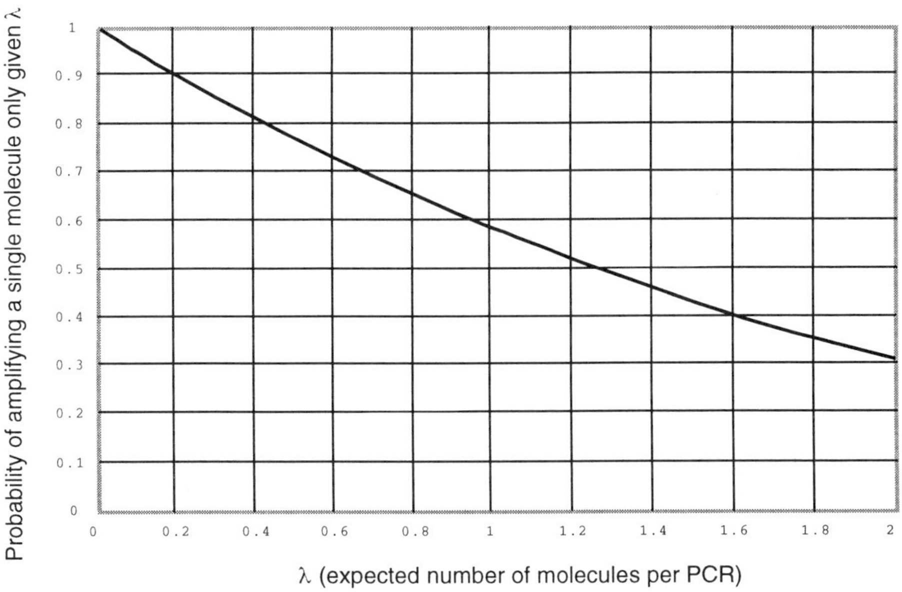

Figure 4. Probability that there is only one target molecule in a PCR reaction, given the expected number of molecules in the dilution.

3. PRESERVING POSITIONAL INFORMATION IN SEQUENCE ALIGNMENTS

Once sequences are obtained, a whole host of other issues arise. Sequences need to be error-checked and assembled into contiguous fragments (or contigs). With HIV sequences it is important to check if any of the sequences are potential contaminants (Korber *et al.*, 1995; Learn *et al.*, 1996; Foley, Chapter 2). In addition, if multiple

HIV sequences have been obtained, these need to be aligned so that homologous sites appear in the same column.

Multiple alignment of HIV sequences, particularly in regions where insertions and deletions occur frequently, can be challenging. Typically, the task of aligning such sequences involves an initial automated alignment followed by manual editing of the resulting alignment. Although appropriate procedures have been developed for aligning coding nucleotide sequences when there are frameshift mutations (Huang and Zhang, 1996) or frameshifts due to sequencing errors (Guan and Uberbacher, 1996), relatively little has been written on how frameshifted regions should be handled once a multiple sequence alignment has been constructed. This problem tends to be exacerbated with HIV sequences, particularly those from highly variable regions, e.g., sequences from the *env*. In fact, the evolutionary analysis of multiple nucleotide sequences in which gaps have been introduced to preserve alignment can be problematic, particularly if the types of analyses that are performed require the correct assignment of first, second, and third codon positions to each nucleotide in a sequence at each column of a multiple alignment (hereafter referred to as "positional information"). Typical examples of analyses that rely on positional information include the estimation of the numbers of synonymous and nonsynonymous substitutions in a sample of sequences (Muse, 1996; Muse and Gaut, 1994; Nei and Gojobori, 1986), as well as any phylogenetic analysis performed on specific codon positions ((Swofford *et al.*, 1996; p.503). Consider, for instance, the following pair of aligned sequences and their translations:

```
                 1111
        1234567890123
Seq1    A-GAGCCCCAGGT →    AGA    GCC    CCA    GGT    → RAPG
Seq2    ACGAGTCCC-TGT →    ACG    AGT    CCC    TGT    → TSPC
```

Because the gap in Seq1 causes a frameshift, nucleotides at columns 3 to 9 are one codon position upstream relative to nucleotides in Seq2. If the aim is to perform a phylogenetic analysis using only third positions, this region presents a dilemma: there is no way that we can extract nucleotides based on their codon positions in this stretch of the alignment. Similarly, consider column 6: the C in Seq1 is at codon position 2, whereas the T is at codon position 3. Does the C↔T substitution represent a synonymous or nonsynonymous substitution? The answer to this question actually depends on whether there was an insertion or a deletion at column 2, and on whether this insertion/deletion occurred before or after the substitution at column 6. In any case, there is no reasonable and easy way as yet to make inferences about the nature of the substitution at column 6.

At this point, one may claim that it is unlikely to see such configurations of insertions and deletions in real and well-characterized sequences. This may be true; however, our own experience indicates that the frequency of such insertions and deletions is sufficiently high to be potentially worrisome. It is, of course, entirely likely that these insertions and deletions are sequencing errors that can be corrected by resequencing the original template. However, this is often not possible – none of the sample may remain, or the sequences are from a database. To see how such insertions and deletions distort the positional alignment consider the following sequences, both of which start at codon position 1.

```
                 1111
        1234567890123
Seq1    A-GAGCCCCAGGT
        1-23123123123  ←  codon position
Seq2    ACGAGTCCC-TGT
        123123123-123  ←  codon position
```

The gap in Seq1 introduces a misalignment in codon positions of the aligned nucleotides—the aligned Gs at column 3 do not share the same codon positions for their respective sequences. Analyzing these bases as if they are correspondent codon positions will yield incorrect results. These sequences remain out of positional alignment until column 11. The gap in Seq2 at column 10 effectively counterbalances the upstream gap in Seq1. Columns 11–13 constitute the only region of the sequence pair properly aligned for positional analysis.

One method of dealing with alignment gaps is to remove from an alignment all columns in which gaps are present ("gap-stripping"; Learn *et al.*, 1996). However, for coding sequences, blind gap-stripping may distort positional information. For example, consider again the alignment above. Removing the gaps at positions 2 and 10, would result in:

```
            1111                        1    11
        1234567890123       134   567   891  23
Seq1    A GAGCCCC GGT→      AGA   GCC   CCG  GT      →RAP
Seq2    A GAGTCCC TGT→      AGA   GTC   CCT  GT      →RVP
```

By removing the C nucleotide at column 2 of Seq2, the codon position of each subsequent nucleotide has been shifted by one, and a translation is created which is not coded for by Seq2.

Another method commonly employed is to strip the entire codons in which a gap occurs ("codon-stripping"; Kumar *et al.*, 1994). This works well if all gaps are arranged as contiguous triplets. However, when gaps are dispersed, codon-stripping will also distort positional information. Consider codon-stripping applied to the same alignment as the gap-stripping example:

```
                    111 1                        1
        123 456 789 012 3         456   789      3
Seq1    A-G AGC CCC AGG T  →      AGC   CCC      T      →SP
Seq2    ACG AGT CCC -TG T  →      AGT   CCC      T      →SP
```

By stripping columns 1–3, a full codon is removed from Seq2, but only a partial (2 of 3 nucleotides) codon is removed from Seq1. Again, codon positions have shifted: the A at column 4 of Seq1 is being analyzed as if it were position 1, when it is really position 3.

To allow easy identification of positional misalignment, the following simple notation may be used. For each sequence, replace all codon positions 1 with a left bracket ("[") and positions 3 with a right bracket ("]"). These readily indicate the beginning and end, respectively, of the codon. Position 2 is indicated by the one-letter identifier for the coded amino acid. All gaps are left as gaps.

```
              1111
     1234567890123
Seq1 A-GAGCCCCAGGT
     1-23123123123  ← codon position
     [-R][A][P][G]
Seq2 ACGAGTCCC-TGT
     123123123-123  ← codon position
     [T][S][P]-[C]
```

It is now possible to compare these sequences and quickly note which codons are properly aligned for positional analysis:

```
              1111
     1234567890123
Seq1 [-R][A][P][G]
Seq2 [T][S][P]-[C]
```

Only codon pairs in which positions 1, 2, and 3 are aligned are appropriate for analysis. At a glance, it is apparent that the last codon pair (columns 11–13) is the only one to meet this requirement. (Note: caution must be exercised with this notation. It is tempting to use it to improve the sequence alignment, but any modifications must be made with reference to the underlying nucleotide alignment. In the above example, a first thought is to move the gap in Seq2 from column 10 to column 7—that would align the Pro codons, and give one more codon to analyze. However, this would result in a A↔C transversion at column 10. One must decide whether it is appropriate to preserve identity of amino acids at this position and invoke a transversion at the nucleotide level. In any case, alignment adjustments should not be made on the strength of the notation alone).

Once a sequence pair becomes positionally misaligned, how does it come back into alignment? As shown above, the simplest way is for the second sequence to "balance" the gaps in the first, as has been used for our examples thus far. It follows that if two gaps are introduced in one sequence before they are in the other, it will take two gaps in the other to offset them.

```
              11111
     12345678901234
Seq1 A-GAG-CCCCAGGT
     [-R][-A][P][G]
Seq2 ACGAGT-CC-ATGT
     [T][S]-[P-][C]
```

If one sequence accumulates three gaps, it will also realign (a triplet grouping of gaps is a simple form of this realignment).

```
              111111
     123456789012345
Seq1 A-GAG-CCCC-GGTA
     [-R][-A][P-][V]
Seq2 ACGAGTCCCCATGTT
     [T][S][P][H][V]
```

In contrast to gap-stripping or codon-stripping, gap-balancing results in the availability of fewer columns in an alignment for any subsequent analysis. However, if the gaps that are introduced in an alignment represent real insertion/deletion events that have occurred in the evolution of the sequences, then columns that are removed by gap-balancing are positionally misinformative and should be excluded. If these insertions-deletions are the result of sequencing errors, but there is no way to go back and obtain more sequence information to correct the error, then it is still advisable to remove positionally misinformative sites. For an alignment of two sequences, gap-balancing and codon-stripping give the same results when gaps are contiguous, and have a length that is a multiple of three. However, in other cases, codon-stripping may include aligned sites that, in reality, do not share the same positional information.

Gap-balancing can be extended to multiple sequences in two ways, depending on the types of analyses for which positional information is required. First, each pair of sequences can be analyzed separately, and comparisons made using only those columns that are positionally identical for the pair of sequences in question. This kind of pairwise-deletion will work if positional information is used in any kind of analysis in which the unit of estimation is based on a pairwise comparison, e.g., the construction of a pairwise matrix of evolutionary distances. Obviously, pairwise gap-balancing would not be appropriate if the goal is to extract columns based on codon positions for site-based analyses, e.g., phylogenetic analyses using maximum parsimony or maximum-likelihood. Also, note that with pairwise gap-balancing, as with any pairwise deletion scheme, the standard errors of the estimated distances for pairs of sequences will differ depending on the effective number of sites compared.

The alternative to pairwise deletion is to extend the method to include only those columns over the entire alignment for which positional information is identical. For example, with the following sequences:

```
                 1111111111
        1234567890123456789
Seq1    AAAA-GAGTCCGAAGTATA    →    [K][-R][V][P][S][I]
Seq2    AAAACGAGTCCCAA-CATA    →    [K][T][S][P][N-][I]
Seq3    AAGACGAGTCCGA-GTATA    →    [K][T][S][P][-S][I]
```

the positional information of the first three and the last three columns is the same for all three sequences. Note, however, that by using only these columns, we are discarding sites from Seq2 and Seq3 that share the same pairwise positional information, and, as with the C↔G substitution at column 12, indicate differences between Seq1 and Seq2 that are not apparent when these sites are excluded. Therefore, we recommend that if it is appropriate to do so, users apply gap-balancing to pairs of sequences instead of to the entire alignment.

3. SUMMARY

In this chapter we have touched on two aspects of experimental design and data processing as these relate to HIV molecular sequences. We have focused on the collection of sequence data and strategies that prevent the sampling of amplified products that are derived from the same target molecule. This is particularly pertinent in any study that aims to describe HIV genetic variation (Korber, Chapter 4, Pinheiro and Seillier-Moisewitsch, Chapter 6), or that uses several HIV sequences to make inferences about the evolutionary history (Posada *et al*, Chapter 7) or evolutionary dynamics of the HIV population (Vasco *et al*, Chapter 9, Beerli *et al*, Chapter 10). All of these require that the sequences obtained have a one-to-one correspondence with sequences in the pre-amplified sample. This ensures that there is no disproportionate representation of a haplotype that is due solely to artefacts of post-amplification sampling. The importance of this cannot be overstated: inferences about the population can change drastically depending on the extent of sequence resampling in a study.

In studies for which it is important to identify the codon positions of each nucleotide correctly, it is important to have a method that can quickly allow a researcher to weed out those nucleotides that may be positionally misaligned. In particular, studies on selection (Nielsen, Chapter 11) require information about the numbers of synonymous and nonsynonymous substitutions that have occurred in the evolutionary history of a sample of sequences, and this in turn requires that sequences are aligned correctly with respect to their codon positions. We explore two available methods for dealing with insertions and deletions, and describe a third – gap-balancing – that allows a researcher to determine which sites in a multiple alignment share the same positional information.

ACKNOWLEDGMENTS

We thank Drs. Bette Korber and James Mullins for comments on parts of this manuscript. Support for this research was made available through US Public Health Service and National Institutes of Health grants.

REFERENCES

Carrow, E. W., Vujcic, L. K., Glass, W. L., Seamon, K. B., Rastogi, S. C., Hendry, R. M., Boulos, R., Nzila, N. and Quinnan, G. V., Jr. 1991. High prevalence of antibodies to the gp120 V3 region principal neutralizing determinant of HIV-1mn in sera from Africa and the Americas. *AIDS Res. Hum. Retovir.* 7: 831-838.

Chesebro, B., Wehrly, K., Nishio, J. and Perryman, S. 1996. Mapping of independent V3 envelope determinants of HIV-1 macropage tropism and syncytium formation in lymphocytes. *J. Virol.* **70**: 9055-9059.

Coffin, J. M. 1992. Structure and classification of retroviruses, In *The Retroviridae, Vol. 1* (Levy, J. A., ed.) Plenum Press, New York.

Hudson, R. R., Kreitman, M. and Aguade, M. 1987. A test for neutral molecular evolution based on nucleotide data. *Genetics* **116**: 153-159.

Hwang, S. S., Boyle, T. J., Lyerly, H. K. and Cullen, B. R. 1992. Identification of envelope V3 loop as the major determinant of CD4 neutralization sensitivity of HIV-1. *Science* **257**: 535-537.

Jetzt, A. E., Yu, H., Klarmann, G. J., Ron, Y., Preston, B. D. and Dougherty, J. P. 2000. High rate of recombination throughout the human immunodeficiency virus type 1 genome. *J Virol.* **74**: 1234-1240.

Korber, B. T. M., Learn, G., Mullins, J. I., Hahn, B. H. and Wolinsky, S. 1995. Protecting HIV sequence databases. *Nature*, **378**: 242-243.

Kumar, S., Tamura, K. and Nei, M. 1994. MEGA: Molecular Evolutionary Genetics Analysis software for microcomputers. *Comput. Appl. Biosci.* **10**: 189–191.

Learn, G. H., Jr., Korber, B. T. M., Foley, B., Hahn, B. H., Wolinsky, S. M. and Mullins, J. I. 1996. Maintaining the integrity of HIV sequence databases. *J. Virol.* **70**: 5720-5730.

Leitner, T., Halapi, E., Scarlatti, G., Rossi, P., Albert, J., Fenyo, E.-M. and Uhlen, M. 1993. Analysis of heterogeneous viral populations by direct DNA sequencing. *BioTechniques* **15**: 120-127.

Liu, S.-L., Rodrigo, A. G., Shankarappa, R., Learn, G. H., Hsu, L., Davidov, O., Zhao, L. P. and Mullins, J. I. 1996. HIV quasispecies and resampling. *Science* **273**: 415-416.

Mansky, L. M. 1996. Forward mutation rate of Human Immunodeficiency Virus Type 1 in a T lymphoid cell line. *AIDS Res. Hum. Retrovir.* **12**: 307-314.

McDonald, J. H. and Kreitman, M. 1991. Adaptive protein evolution at the Adh locus in Drosophila. *Nature* **351**: 652-654.

Muse, S. 1996. Estimating synonymous and nonsynonymous substitution rates. *Mol. Biol. Evol.* **13**: 105-114.

Muse, S. and Gaut, B. S. 1994. A likelihood approach for comparing synonymous and nonsynonymous nucleotide substitution rates with application to the chloroplast genome. *Mol. Biol. Evol.* **11**: 715-724.

Myers, G., Korber, B., Hahn, B.H., Jeang, K.-T., Mellors, J.W., McCutchan, F.E., Henderson, L.E. and Pavlakis, G.N. 1995. *Human Retroviruses and AIDS 1995. A Compilation of Nucleic Acid and Amino Acid Sequences.* Los Alamos National Laboratory, Los Alamos, NM.

Nei, M. and Gojobori, T. 1986. Simple methods for estimating the numbers of synonymous and nonsynonymous nucleotide substitutions. *Mol. Biol. Evol.* **3**: 418-426.

Peeters, M., Vincent, R., Perret, J. L., Lasky, M., Patrel, D., Liegeois, F., Courgnaud, V., Seng, R., Matton, T., Molinier, S. and Delaporte, E. 1999. Evidence for differences in MT2 cell tropism according to genetic subtypes of HIV-1: syncytium-inducing variants seem rare among subtype C HIV-1 viruses. *J. Acquir. Immune Defic. Syndr. Hum. Retrovirol.* **20**: 115-121.

Perelson, A. S., Neumann, A. U., Markowitz, M., Leonard, J. M. and Ho, D. D. 1996. HIV-1 dynamics in vivo: virion clearance rate, infected cell life-span, and viral generation time. *Science* **271**: 1582-1586.

Rodrigo, A. G., Goracke, P. C., Rowhanian, K. and Mullins, J. I. 1997. Quantitation of target molecules from Polymerase Chain Reaction-based limiting dilution assays. *AIDS Res. Hum. Retrovir.* **13**: 737-742.

Shankarappa, R., Margolick, J. B., Gange, S. J., Rodrigo, A. G., Upchurch, D., Farzadegan, H., Gupta, P., Rinaldo, C. R., Learn, G. H., He, X., Huang, X.-L. and Mullins, J. I. 1999. Consistent viral evolutionary dynamics associated with the progression of HIV-1 infection. *J. Virol.* **73**: 10489-10502.

Swofford, D. L., Olsen, G. J., Waddell, P. J. and Hillis, D. M. 1996. Phylogenetic inference, In *Molecular Systematics* (Hillis, D. M., Moritz C. and Mable, B. K. eds.) Sinauer Associates, Sunderland, MA.

Taswell, C. 1981. Limiting dilution assays for the determination of immunocompetent cell frequencies. *J. Immunol.* **126**: 1614-1619.

Watterson, G. A. 1975. On the number of segregating sites in genetical models without recombination. *Theor. Popul. Biol.*, **7**: 256-276.

ACCESSING HIV MOLECULAR INFORMATION

Brian T. Foley

Theoretical Biology and Biophysics Group,
Los Alamos National Laboratory, Los Alamos, NM 87545 USA

1. INTRODUCTION

1.1 History of the HIV sequence database at LANL

The HIV Sequence Database and Analysis Project began in late 1986 under the aegis of the AIDS Program (now the AIDS Division) of the United States National Institute of Allergy and Infectious Diseases (NIAID) through an interagency agreement with the Department of Energy (DOE). Dr. Gerald Myers was the principal investigator of the project from 1986 through 1997. Drs. Bette Korber and Carla Kuiken currently share the principal investigator role. The goals of the project have centered upon compilation and analysis of HIV and HIV-related gene and protein sequences, and the publication of annotated sequences and their analyses. Beginning in 1987, the database project assumed an additional role of tracking viral genotypes for molecular epidemiological studies, and in 1995 an immunology database was added to integrate sequence and immunological information.

1.2 Differences and similarities between HIV-DB and GenBank/EMBL

The HIV Database is a specialized molecular-sequence database that provides services, including annual hard-copy Compendia of the sequence and immunology databases, to the HIV research community at no cost to the individual researcher. Unlike the general sequence databases (GenBank, EMBL, GSDB, DDBJ), which must gather macromolecular sequences from a wide variety of sources, the specialized HIV sequence database concentrates only on primate lentiviral sequences and a few primate genes and proteins which are known to interact with the immunodeficiency viruses. The HIV database shares data with the general databases, so that all sequences appearing in the HIV database also appear in the GenBank, EMBL, GSDB and DDBJ databases. However, the HIV sequence database also stores data elements which do not fit in the general databases, such as multiple sequence alignments, detailed patient information, and information resulting from analyses done by the HIV database staff. In addition, the HIV database performs some screening of data, on top of that done by the general databases. Erroneous or suspicious HIV sequences frequently make it through the peer review system into publications, and into the general databases. Although the HIV database staff does not have authority to label entries in GenBank as erroneous, the sequences can be so labeled in the HIV database. A discussion of the types of errors frequently encountered in HIV sequence data has been presented elsewhere (Learn *et al.*, 1996). We have recently added www pages (*http://hiv-web.lanl.gov/ HTML/Contam/contam_main.html* and *http://hiv-web.lanl.gov/HTML/Contam/ contam_conserved.html*) to our web site (*http://hiv-web.lanl.gov*) which assist researchers in examining sequence data for common problems.

Being a specialized sequence database, and working in concert with a relatively small and focussed research community, allows the HIV sequence database to be flexible and responsive to the needs of the user community. It can also act as a guide in pointing out new areas of research that are likely to yield useful results, or to fill gaps in the existing data.

2 HIV-DB SEARCH AND RETRIEVAL

2.1 By region of genome

A recent addition to the HIV Database sequence search and retrieval tools located at *http://hiv-web.lanl.gov/hivDB_search/index.html*, is a tool for retrieving sequences by region of the genome (*http://hiv-web.lanl.gov/MAP/hivmap.html*). In the past, such retrieval relied upon annotation of each gene, which made it impossible to retrieve all sequences if the only annotation was "complete genome", or for example if the *vif* gene was annotated as "viral infectivity factor". A second complication was that it was difficult to obtain just the region of interest from complete genome entries. This new tool, MAP, is based on the addition of genomic map coordinates for the beginning and end of each sequence, compared to a model reference sequence of the HIV genome. As each sequence enters the database, it is aligned with the model HIV sequence and both the map coordinates and the alignment itself are stored with the entry. This information is then used to retrieve and (if desired) build a global multiple sequence alignment of any or all entries containing the desired region of the genome. The user can either retrieve the entire sequence of each entry, or just the desired region clipped to any beginning and ending point. There is an option for including only those sequences that span the entire desired region, or all entries that contain any portion of the desired region.

The search tool includes a map of the genome with the start and endpoints of each gene enumerated. A pull-down menu of genes and gene regions, such as *gag*-P17 or *env*-V3, is also available to assist researchers in selecting the desired region. The sequences can be obtained in GenBank format (not aligned) or in FASTA or Intelligenetics multiple sequence alignment formats. In the alignment formats, the sequences are each aligned to the model sequence and then gaps are introduced where needed to produce a multiple sequence alignment. While this method does not attempt to produce an "optimal" global alignment, it does produce a global alignment that in our experience is often better than the results of some programs that do attempt optimal alignment. The speed of this process is its major advantage over other multiple alignment programs such as CLUSTALW or HMMER.

The proviral LTR region remains problematic for this tool. Many complete genomes include one LTR at either the 5' end or the 3' end. Some include the R and U5 region on the 5' end and the U3 and R regions on the 3' end, so that a complete LTR can only be produced by splicing the two ends together. The *nef* gene is usually but not always complete at the 3' end of complete genome entries.

2.2 By subtype or country of origin

The HIV database search page at *http://hiv-web.lanl.gov/hivDB_search/index.html* allows users to search for sequences on several criteria, including genetic subtype and country of origin. In the subtype field, recombinant genomes are stored with all subtypes listed in alphabetical order. A recombinant between subtypes A and G will be listed as "AG" regardless of which regions of the genome are derived from subtypes A and G. In cases where only one gene is sequenced and this gene does not contain a recombination site, the sequence will be listed as recombinant only if it can be inferred to be from a recombinant genome based upon other data. For example the envelope gp120 region from the AE recombinant form found in Thailand is subtype E, but there is ample evidence that all genomes from Thailand that are subtype E in *env* are subtype A in *gag* so these genes are listed as circulating recombinant form CRF01_AE. In some cases the subtype cannot be determined, either because the sequence is too short for accurate analyses, or because the sequence does not cluster with the existing subtypes that have been previously defined. These sequences are listed as subtype "U" for unclassified. Most sequences enter the HIV database without any subtype information included. The subtype field remains blank until the database staff can either determine the subtype via phylogenetic analysis, or find accurate subtype information in primary publications. Phylogenetic analysis is primarily done with the PHYLIP programs (Felsenstein, 1989) and including reference sequences for each subtype taken from the HIV database. When more extensive sequencing or analysis is performed on a given isolate, efforts are made to update all sequence records pertaining to that isolate. For example, a short *gag* region sequence might initially be classified as subtype A, and later be found to be contained within an A/G recombinant complete genome.

The country of origin field is defined as the country in which the patient was living at the time his/her blood was drawn. If the patient was a recent immigrant from another country, the DEFINITION line should state both countries. For example it may say "BFP90 from Australia (infected in Burkina Faso)." The HIV database has added a field for country of infection, but few entries have yet had this field filled in, and we foresee filling it only in those entries for which the country of infection differs from the country of residence at time of sampling.

2.3 By year of sample

Researchers rarely provide information on sampling dates with their sequence submissions or publications. In many cases where dates are provided in the materials and methods sections, they are not specific to each sample. When a publication states that 30 samples were obtained from patients between June, 1995 and September, 1997 the date of sample field in the database is left blank, rather than taking an average date, or attempting to store a range of dates in this field. The World

Health Organization is a notable exception. They have collected samples globally and noted date, patient health, and other information with each sample. In (Korber *et al.*, 1994) a standard nomenclature system for HIV isolates was proposed, which included date and country of sample collection. However few researchers have used this nomenclature system in practice.

A related date/time type of information that would be useful is time elapsed between infection and sampling for sequencing. This would be useful in tracking intra-patient evolution, and for correlating viral genotype with length of survival. In most cases, especially outside the USA, the date of seroconversion is unknown.

Date of sample information is useful for tracking the rate of HIV evolution, and using that to back calculate the date of the beginning of the HIV pandemic. An example of this, as well as a discussion of the current limits of this analysis is found in (Korber *et al.*, 1998). The rate of HIV evolution is also useful in predictions of the growth in viral diversity over time, in order to estimate vaccine efficacy. Rates of evolution in intra-patient samples can be compared to rates of evolution in inter-patient samples to provide clues about population sizes, transmission bottlenecks, and other epidemiological factors.

Because most sequence entries have a blank date field, it might be tempting to use date of publication or date of entry into the database as indicators of sampling date. This should not be done, because many researchers use stored blood or serum samples for sequence analysis, and in some cases the time lag between sampling and publication is very large (e.g., Zhu *et al.*, 1998).

2.4 By health of patient

Efforts to correlate viral genotype with pathogenesis or rate of disease progression are laudable. While HIV-2 has been shown to be both less transmissible and less pathogenic than HIV-1 (Marlink *et al.*, 1994), there is a lack of evidence for different genotypes within HIV type 1 correlating with differences in pathogenicity or transmissibility. An abstract (*http://acoma.santafe.edu/hiv5/hoelscher.html*) at a recent meeting presents the results of an analysis that used seroprevalence methods to imply that individuals infected with subtype C progress to AIDS and death more quickly than those infected with subtype A. Long-term cohort studies would more rigorously address this issue.

Health information alone would not allow a correlation of viral genotype with rate of progression, because in those countries where HIV-1 subtype B is most prevalent, patients may tend to get health care before they become ill, while in other countries the sampling may be biased toward those who are already very ill at time of sampling. However, if both health of patient and date of seroconversion were reported, these statistics would become available. One nontrivial problem is encoding the health information in a standard format. The CDC has codes to standardize recording of patient health data that can be viewed at *http://www.cdc.gov/data/index.htm*.

Efforts have also been made to correlate patient factors that can contribute to disease progression. These include patient age, HLA types, CCR5 gene mutations, antiretroviral therapies, CD4 cell counts, and other factors. At this time, some of these data are so variable and so rarely reported, that they are included in the free text of the COMMENT field if possible, but not in separate database fields.

As more data become available in standardized formats, database fields can be added.

2.5 By viral phenotype

The descriptions of HIV-1 phenotypes have changed over the years. Replication kinetics were classified as rapid/high or slow/low; cell tropism was classified as macrophage tropic or T-cell tropic; and cytopathic effect was classified as syncytium inducing (SI) or non-syncytium inducing (NSI). More recently it has been shown that these characteristics are related to a large degree, and new nomenclature schemes have been suggested (Fenyo *et al.*, 1997; Doms and Moore, 1997).

Assigning a phenotype that was often determined for a population of virions to a sequence that is most often derived from a single molecular clone, is uncertain. In quite a few cases, the phenotype has actually been determined from the same complete genome molecular clone that was sequenced, or the envelope V3 region from an SI viral clone has been shown to confer the SI phenotype on an NSI clone when this domain is swapped. However, in many more cases, it cannot be determined if the viral sequence is from the same clone used for phenotype determination.

The HIV database has data fields for phenotype (SI, NSI, etc.) as reported in the publication, and also coreceptor usage (R5, X4, R5X4, etc.) as reported. When the sequence is reported in a publication separate from the publication describing the phenotype, an effort is made to determine whether the same clone or isolate was used prior to entering the data. If a paper reports phenotype based only on sequence data, assuming that changes in the V3 loop are predictive of coreceptor usage, we do not enter this in the database. The phenotype must be determined in a biological assay.

2.6 By similarity to a given sequence

The National Center for Biotechnology Information offers a WEB-based BLAST search at *http://www.ncbi.nlm.nih.gov/BLAST/*. The National Center for Genomic Research also offers a WEB-based BLAST server at *http://seqsim.ncgr.org/newBlast.html*. These two services are roughly equivalent. The HIV database offers a WEB-based BLAST search at *http://hiv-web.lanl.gov/BASIC_BLAST/basic_blast.html*. The advantage of using the HIV database search is that subtype and country of origin are included in the output, so the results can give a quick first estimate of the subtype of the query sequence. The output from most of these searches, including the HIV-DB BLAST, is in HTML format, linking each match to its database entry. Several sites also offer e-mail based BLAST searches.

A BLAST search is useful for ascertaining that a newly sequenced PCR product is truly from a new isolate, and not a PCR contaminant from a lab strain of HIV or a previously sequenced sample. This search will quickly identify any sequence in the database with 99% to 100% identity to the query sequence. While 98% identity in small conserved regions of the *gag* and *pol* genes is not unexpected, identity of greater than 97% over the V3 region of *env* is prima facie evidence that the sequences are derived from very closely linked samples. BLAST is also useful for gathering sets of related sequences, sorted by similarity to the query. However, BLAST scores do not accurately reflect overall similarity when the sequences have insertions or deletions relative to the query. Sequences are scored only over the length of ungapped alignment, and raw scores are based on length as well as percent identity within the ungapped segment.

2.7 By sets of sequences from longitudinal studies

Sets of sequences from longitudinal studies are invaluable for studying the evolution of HIV. Several types of sets are available, such as a set of transmission events traced over a twenty-plus year period from Haiti to Sweden (Leitner *et al.*, 1996), sets of mother-infant transmission events (Pasquier *et al.*, 1998), (Wolinsky *et al.*, 1992), and sets of sequences from patients followed over time (Wolinsky *et al.*, 1996). Each individual set is easily retrieved by a search for the entries that contain the author and publication information. To obtain many similar sets of data, such as all mother-infant transmission studies, is more difficult. A separate data filed for "Type of study" is not yet available. Text searches of the publication titles, or of comprehensive reviews of data sets can be used to obtain collections of related sets.

2.8 Formats for retrieval

Complete sequence records with annotation can be retrieved in GenBank flat file format. Most sequence analysis tools either use this format directly, or provide a program to convert from GenBank flat file format into the format they use. Multiple sequence alignments can be retrieved in FASTA format, Intelligenetics format, or printable text format. Other formats such as the GCG MSF format are not supplied, but again conversion programs are readily available.

2.9 Limitations of searches

Researchers should keep in mind that not all data fields are filled out for all entries. For example if a search for "sample date = 1994" turns up 93 entries, it is not because only 93 samples from that year were sequenced, but because most sequences from that year did not specify the sample date and the date field is blank. Likewise searching for "phenotype = NSI" will not pull up all non syncytium-inducing isolates, but only those for which the phenotype has been rigorously determined for the same clone from which the sequence was determined. In many cases it is advisable to call on the expertise of the HIV database staff, by sending e-mail to *seq-info@t10.lanl.gov*, for help in obtaining data sets that are as complete as possible.

3. PRE-BUILT SEQUENCE ALIGNMENTS

3.1 DNA or protein, by gene or complete genomes

In addition to storing individual sequence entries, the HIV database stores many different pre-built multiple sequence alignments. The more sequences that are included in an alignment, the more difficult it becomes to produce an optimal alignment, and the more gaps are necessary. For this reason, and to keep file sizes manageable, many of the alignments stored for retrieval from the database are not comprehensive alignments of all available sequences, but rather are composed of just a few representative sequences from each genetic subtype. The annual printed compendium provides alignments with representative sequences, and more extensive alignments are available from the HIV web site. Comprehensive alignments can be built on the fly using the MAP tool described above. An extensive alignment of the V3 region of *env* gp120 is compiled each year including one sequence per patient. Limiting the alignment to one sequence per patient is important, when assaying the global variability of HIV, because data sets from the USA and Europe often contain more than 100 sequences per patient, but this is rare for data sets from third world countries.

Both DNA and protein sequence alignments are stored in pre-built form. Only DNA alignments are produced by the MAP program. DNA alignments do not always translate nicely into protein alignments, because gaps may have been introduced which interrupt reading frames for protein translation. Optimal DNA alignments do not always result in optimal protein alignments, and vice versa. If the purpose of the analysis is to study protein properties such as antigenicity or hydrophobicity, it is more informative to work directly from protein alignments, than to translate DNA alignments into protein. For phylogenetic analyses, DNA alignments are more informative. Columns in a sequence alignment that contain gaps are usually excluded from phylogenetic analysis, due to uncertainty of how to score an insertion or deletion, relative to a point mutation.

For each gene in the HIV genome, both DNA and protein alignments are available in pre-built form at *http://hiv-web.lanl.gov/* under the alignments link. The exact URL path will change each year, as the alignments are updated, but previous years' alignments will remain accessible, so that future research can be carried out on consistent data sets. Included with each alignment is a table of sequence descriptions, including the accession numbers that can be used to obtain the full database entry for each sequence.

Not all sequence alignments are found under the single table of alignments organized by genes. Some specialized alignments, such as the alignment of HIV-1 subtype reference genomes, are obtained from other sections of the database. In addition to the HIV Sequence Database, the HIV Immunology Database stores alignments of epitopes known to react with cytotoxic T lymphocytes or monoclonal antibodies. If the particular alignment you are interested in cannot be found, send e-mail to *hiv-info@t10.lanl.gov* explaining exactly what type of sequence alignment you need.

3.2 Subtype Reference Set

In order to facilitate the classification of newly sequenced HIV isolates, the database provides an alignment of a set of complete genomes, which are chosen as representatives of each subtype or circulating recombinant form of HIV-1. Four sequences of each subtype are chosen for this alignment.

3.3 Envelope V3 Region

Because it was once thought to be the "principal neutralizing domain" and therefor a very important region of the HIV-1 genome for vaccine design, the V3 loop region of the *env* gene is more frequently sequenced than any other region of the genome. This region of *env* evolves quite rapidly, and is apparently under positive selection to change and thus avoid the human immune response. This region of the genome has thus become most useful for tracking the global diversity of HIV-1 and for studying the molecular epidemiology of HIV-1 infection. For purposes of assessing the global variability of this region of the *env* gene, the HIV Database compiles a list of V3 region sequences such that only one sequence per patient (or group of very closely related patients) is included. Including all sequences would heavily bias the sampling toward those sets of patients from which hundreds of clones have been sequenced.

3.4 Consensus sequences and how are they calculated

Included with most multiple sequence alignments are "consensus" sequences for each subtype of HIV-1 or HIV-2. These consensus sequences are calculated from the data set in the alignment, not from the broader set of all sequences. The nucleotide or amino acid listed at each position is that which is found in the greatest number of sequences.

REFERENCES:

Doms, R. W. and Moore, J. P. 1997. HIV-1 coreceptor use: A molecular window into viral tropism. In *Human Retroviruses and AIDS 1997* (Korber, B., Hahn, B., Foley, B., Mellors, J. W., Leitner, T., Myers, G., McCutchan, F. and Kuiken, C. L., eds). Los Alamos National Laboratory, Los Alamos, NM.

Felsenstein, J. 1989. PHYLIP -- Phylogeny Inference Package (Version 3.2). *Cladistics* **5**: 164-166.

Fenyo, E. M., Schuitemaker, H., Asjo, B. and McKeating, J. 1997. The history of HIV-1 biological phenotypes, past, present and future. In *Human Retroviruses and AIDS 1997* (Korber, B., Hahn, B., Foley, B., Mellors, J. W., Leitner, T., Myers, G., McCutchan, F. and Kuiken, C. L., eds). Los Alamos National Laboratory, Los Alamos, NM.

Korber, B. T. M., Osmanov, S., Esparza, J. and Myers, G. 1994. The WHO Network for HIV Isolation and Characterization. The World Health Organization Global Programme on AIDS proposal for standardization of HIV sequence nomenclature. *AIDS Res. Human Retroviruses* **10**: 1355-1358.

Korber, B., Theiler, J. and Wolinsky, S. 1998. Limitations of a molecular clock applied to considerations of the origin of HIV-1. *Science* **280**: 1868-1871.

Leitner,T., Escanilla, D., Franzen, C., Uhlen, M. and Albert, J. 1996. Accurate reconstruction of a known transmission history by phylogenetic tree analysis. *Proc. Natl. Acad. Sci. USA.* **20**: 10864-10869.

Learn, G. H., Korber, B. T. M., Foley, B., Hahn, B. H., Wolinsky, S. M. and Mullins, J. I. 1996. Maintaining the integrity of human immunodeficiency virus databases. *J. Virol.* **70**: 5720-5730.

Marlink, R., Kanki, P., Thior, I., Travers, K., Eisen, G., Siby, T., Traore, I., Hsieh, C. C., Dia, M. C., Gueye, E. H., Hellinger, J., Gueye-Ndiaya, A., Sankale, J.-L., Ndoye, L., M'boup, S. and Essex, M. Reduced rate of disease development after HIV-2 infection as compared to HIV-1. *Science* **265**: 1587-1590.

Pasquier, C., Cayrou, C., Blancher, A., Tourne-Petheil, C., Berrebi, A., Tricoire, J., Puel, J. and Izopet, J.J. 1998. Molecular evidence for mother-to-child transmission of multiple variants by analysis of RNA and DNA sequences of human immunodeficiency virus type 1. *Virology* **72**: 8493-8501.

Wolinsky, S. M., Wike, C. M., Korber, B. T., Hutto, C., Parks, W. P., Rosenblum, L. L., Kunstman, K. J., Furtado, M. R. and Munoz, J. L. 1992. Selective transmission of human immunodeficiency virus type-1 variants from mothers to infants. *Science* **255**: 1134-1137.

Wolinsky, S. M., Korber, B. T., Neumann, A. U., Daniels, M., Kunstman, K. J., Whetsell, A. J., Furtado, M. R., Cao, Y., Ho, D. D. and Safrit, J. T. 1996. Adaptive evolution of human immunodeficiency virus-type 1 during the natural course of infection. *Science* **272**: 537-542

Zhu, T., Korber, B. T., Nahmias, A. J., Hooper, E., Sharp, P. M. and Ho, D. D. 1998. An African HIV-1 sequence from 1959 and implications for the origin of the epidemic. *Nature* **391**: 594-597.

HIV-1 SUBTYPING

Carla L. Kuiken* and Thomas Leitner*§

*Theoretical Biology and Biophysics Group, Los Alamos National Laboratory
Los Alamos, NM 87545 USA
§Department of Virology, Swedish Institute for Infectious Disease Control,
17182 Solna, Sweden

1. INTRODUCTION

The family of human immunodeficiency viruses (HIV) consists of groups of clearly related but genetically distinct retroviruses. The most basic division is between the two distantly related types of viruses that cause AIDS in humans, human immunodeficiency virus type 1 (HIV-1) and type 2 (HIV-2). Early sequencing efforts have led to an appreciation of the extraordinary variability of both HIV-1 (Hahn *et al.*, 1984; Srinivasan *et al.*, 1987) and HIV-2 (Clavel *et al.*, 1986; Zagury *et al.*, 1988), and from those earliest observations, researchers have attempted to organize the viruses according to patterns observed in phylogenetic reconstructions. Within HIV-1, three major groups are now defined, designated as HIV-1 group M (for Main), group O (for Outlier), and group N (for Neither, Next, or New) (De Leys *et al.*, 1990; Gurtler *et al.*, 1994; Simon *et al.*, 1998). Phylogenetically defined subtypes have been recognized for both HIV-1 group M (Myers, 1992; Louwagie *et al.*, 1993; 1995; Gao *et al.*, 1996a) and HIV-2 (Sharp *et al.*, 1995; Korber *et al.*, 1995; Gao *et al.*, 1994). There are now indications that a subtype pattern can also be seen within group O sequences (Quinones-Mateu *et al.*, 1998; Janssens *et al.*, 1999). The variability of group O sequences, coupled with the fact that a cluster of HIV-1 group O infections dating back to the 1960s has been identified in Norway (Jonassen *et al.*, 1997), suggests that the group O epidemic is not new; it is not known why group O has not spread to the same extent as group M (Loussert-Ajaka *et al.*, 1995).

2. CLASSIFICATION METHODS

The subtype classification that is in use for HIV is a phylogenetic system. This means that the groups are based on their inferred evolutionary relationship, rather than on characteristics such as co-receptor usage, serological reactivity, phenotype and many other possible biological characteristics. It is important to stress that the classification is a phylogenetic one, as the issue can easily become confused if this fact is overlooked. A case in point is the recurrent designation of a Brazilian variant of subtype B as subtype B". This HIV variant has a distinct tetramer (GWGR) at the tip of the V3 loop. It is serologically distinct (i.e. plasma from people infected with this variant reacts with different V3 peptides) from variants that that have other V3 loop tetramers like GPGR or GPGQ. However, the distinction is limited to a few amino acids in the V3 loop, and the phylogenetic classification of sequences that have the GWGR is not distinct from other subtype B sequences (Figure 1); a Brazilian sequence not having the GWGR tetramer clusters in the 'GWGR' cluster.

Therefore, unlike the Thai B' variant that does cluster separately from the 'main' subtype B, this distinction, although it may be biologically relevant, has no place in the subtype classification system. It should be noted that the concordance of (V3 peptide) serological classification and genetic subtyping is generally not very good (Cheingsong-Popov *et al.*, 1998).

Even on the basis of sequence data alone, there are two conceptually different classification methods available, namely evolutionary as opposed to genotypic classification. In most sequence-based classification methods the aim of the analysis is to group together sequences that have similar evolutionary origins. For this reason, the methods frequently attempt to correct for different rates and modes of evolution: these factors might cause two sequences to diverge differently, even though they share a common origin; or conversely, two sequences from very different evolutionary backgrounds to become almost identical (convergent evolution). This last possibility might for example affect the protease and RT genes, which contain little genetic variation and therefore little phylogenetic information, and are subject to strong pressure to evolve drug-resistant phenotypes. It does not seem likely that this pressure to converge will overshadow the original subtype distinction.

It should be noted that at present, the results of genotypic and evolutionary classification rarely conflict in the case of HIV nomenclature. When they do, there is very good reason to study the offending sequence, as it is very likely that there is something interesting or strange going on (recombination, hypermutation, etc).

3. WHAT CONSTITUTES A SUBTYPE?

The phylogenetically based classification system of HIV-1 subtypes has existed since 1992. It arose as a consequence of the clustering patterns that form when viral sequences are organized through phylogenetic analysis (Myers, 1992; Louwagie *et al.*, 1993). In 1992, five subtypes were known for *env* (A through E) and four for *gag* (A through D). Since then, the classification system has been continually updated as new viral isolates were sequenced and new data became available; the yearly publication of the "Human Retroviruses and AIDS" Compendium, published in Los Alamos, provides a historical overview of the changes. In 1993, subtypes F, G and H were added (as well as group O, which falls outside the framework of this chapter) (Myers, 1993). Subtypes G and H were not yet defined for *env*, only *gag* sequences were available. In this edition, too, the first attempt was made to define what constituted a subtype. Three criteria were proposed:

1) subtypes are approximately equidistant from one another in *env* (a "star phylogeny");

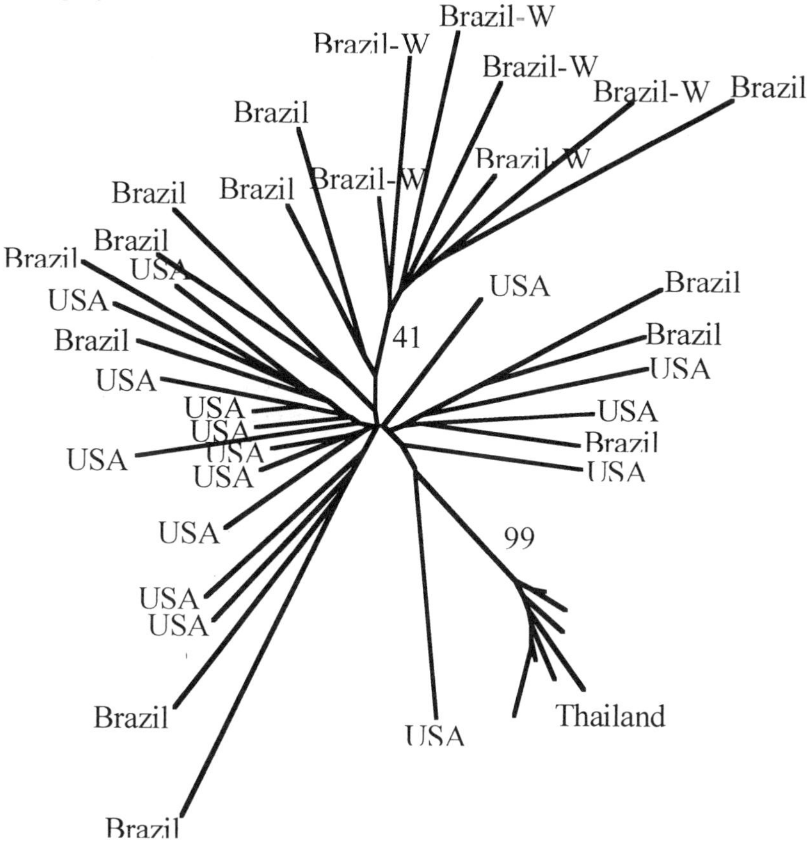

Figure 1. Neighbor-joining tree (Kimura 2-parameter distances) of the V3 region showing the distribution of American, Brazilian, and Thai subtype B sequences. Although more sophisticated methods for estimating genetic distances exist, Kimura 2-parameter distance estimation was used here because it does a fairly good job and is widely available. When the sequences in a tree are fairly similar (roughly, up to 10% difference), the distance measure used has little effect on the accuracy of tree reconstruction; only when differences are large (over 20%) the effect becomes important (Leitner *et al.*, 1996). It should be noted that when the exact branching order or the branch length of trees are important, a more sophisticated distance estimation method (or a non-distance-based tree reconstruction method) would be called for. The codon encoding the 'W' in the Brazil-associated GWGR motif in the V3 loop was deleted from the sequences to create this tree. The GWGR sequences intermingle with other Brazilian sequences, while the Thai B cluster is clearly separate.

2) the *env* phylogenetic tree is for the most part congruent with the *gag* phylogenetic tree;
3) two or more samples are required to define a sequence subtype.

These are both sensible and practical criteria for the subdivision of sequences based on phylogenetic trees; criterion 1 excludes new groups (N and O) from the subtype category, criterion 2 is a first attempt at excluding recombinants, and criterion 3 excludes 'oddballs' that do not fit in the classification. However, many sequences were found to be unclassifiable because they clustered differently using different phylogenetic reconstruction methods or in different reference data sets.

With recent knowledge and methods of detecting intersubtype recombinants, a significant proportion of these unclassified sequences have been explained. In fact, ambiguous classification typifies chimeric sequences. Therefore, Salminen and coworkers (1995a) suggested that the criteria for new subtypes should include:

4) the appearance of at least two isolates that are not directly epidemiologically related, that cluster together and are distinct from established genotypes;
5) the availability of at least 1.5 kilobases of contiguous sequence from each;
6) the absence of any subsegment that can join established genotypes (no recombinants).

The combination of these criteria forms a sound basis for establishing a new subtype. However, when attempting to classify a sequence as an established subtype, less strict criteria may be used, as long as this is made clear. For instance, if only the *env* V3 region (typically 300 bp) was used to classify a virus one might propose that the *env* V3 region belongs to a given subtype. It would be better, of course, if more of the above criteria were fulfilled. Should the sequence fall outside any of the established clusters, it is desirable that further sequences be generated and that additional analyses be performed, as described below.

The extension of the subtype classification has not always proceeded smoothly. In 1995, a paper appeared that described the discovery of a new HIV-1 subtype, called subtype I (Kostrikis *et al.*, 1995). However, the sequences on which the announcement was based were from two epidemiologically related infections in Cyprus, thus violating requirement 4, and they encompassed only the V3 region, in contradiction to requirement 5. Recently, more apparently unrelated samples were discovered that are very similar to this isolate (Nasioulas *et al.*, 1998). However, when the entire genome of this new form was sequenced, "Subtype I" turned out to be a complex mosaic of known and unknown subtypes (see below).

Recently, Triques and colleagues (1999) suggested that subtype F be split into three sub-subtypes labeled F1-F3, because the subtype F sequences formed three subclusters that were clearly distinct, but not far enough apart to be called separate subtypes. In a later publication, sub-subtype F3 was re-classified as a subtype, K. The sub-subtypes are as distant from each other as subtypes B and D, which in retrospect should also have been labeled sub-subtypes. Interestingly, many fragments that had initially been assigned to subtype I and later designated as unknown appear to be very similar to subtype K. However the CY032 genome, which was the prototype for "subtype I", still contains some unclassifiable areas aside from regions that are subtype A, G and K.

Because of all these changes in the classification of subtypes, sub-subtypes and recombinants, in September 1999, a meeting was held to once again look at the HIV nomenclature and to try to define objective, unambiguous criteria for assigning a sequence to an existing subtype and for erecting new subtypes or sub-subtypes (see Robertson, 2000). The most important decisions made at that meeting are listed here.

1) The previously defined groups, the most genetically distant divisions of HIV-1, will continue to be called M, N, and O. If new groups are discovered, they should be named by continuing through the alphabet: P, Q, R etc.
2) Some epidemic strains are derived from viruses that have genomes that are recombinants of different subtypes and are responsible for large numbers of infections. These strains are as epidemiologically important as subtypes. They have been called circulating recombinant forms (CRFs) and numbered sequentially, with the first fully sequenced virus of a CRF serving as prototype (Carr *et al.*, 1998a). There are currently four defined CRFs. If a CRF is recombinant of two or three subtypes, the subtypes will be listed in the name alphabetically. For example, CRF02_AG(IbNG) refers to a CRF that is very common in parts of Africa (McCutchan *et al.*, 1999) and contains regions which resemble A and G subtypes, with a prototype sequence called IbNG. Mosaic viruses containing regions that resemble four or more subtypes will be called complex, designated "cpx".
3) Before a new subtype or CRF is named, three full-length genomes should be obtained from individuals who do not have direct epidemiologically linked infections. Unusual variants should be designated "U" until this criterion is met.
4) Most M group subtype designations will be retained. Subtypes A-K are currently defined, subtype K being the most recent designation (Triques *et al.*, 2000). The two exceptions to this are subtypes I and E.
   - Subtype I was originally given a subtype designation based on partial gene sequences (Kostrikis *et al.*, 1995). Subsequent analysis of full-length genomes revealed that the genomes were very complex mosaics of various HIV-1 subtypes, and regions designated I could not be classified consistently. Based on subsequent analyses, it was decided that the name "subtype I" should be retracted from the nomenclature. The strains previously designated "I" are now grouped in CRF04_cpx(CY032). The former subtype I has been discussed at length by Robertson and coworkers (1997).
   - Subtype E is a very well known subtype, with a high prevalence in many parts of Asia. It is considered a recombinant of an A subtype virus with a parental E subtype virus (Carr *et al.*, 1996; Gao *et al.*, 1996b). The *env* gene region is clearly distinct from other subtypes. Other portions of the genome show a clear affinity to sequences from subtype A viruses. There is extensive literature referring to the designation of the E subtype, while there has been much confusion about what to call regions outside of the envelope. We propose that sequences formerly assigned to subtype E be called CRF01_AE(CM240), retaining the E subtype designation in the name to refer to the regions from an inferred parental subtype E, but now naming it as a CRF.
5) As new subtypes are discovered, they will be named by continuing through the alphabet, so that there could eventually be a group M, subtype M virus (written M:M when a distinction is required).

6) The B and D subtypes are more similar genetically than any of the other subtypes, but are more distant than within-subtype genetic distances. For historical consistency, the B and D designations will remain. However, if new lineages with similar intermediate relationships are found, they should be given sub-subtype designations (for example, the distinct lineages F1 and F2 were designated as sub-subtypes within subtype F, Triques *et al.*, 1999).
7) It would be generally helpful if a simplified version of the WHO nomenclature to name HIV sequences (Korber *et al.*, 1994a) could be more generally adopted. In particular, sequences should be named with a two-letter country code designation that refers to the country the person was living when sampled, and the year of sampling. The laboratory identifier and clone number should also be indicated in the name.

4. USES OF THE SUBTYPE CLASSIFICATION

The classification scheme of HIV-1 into subtypes has proven useful in the phylogenetic analysis of new sequences and in clarifying epidemiological relationships and possible ancestry of HIV. The main role that subtyping has had is to aid in understanding the epidemic on different levels. On the highest level it has given a picture of the worldwide spread. On the next level it has provided important information on how HIV has entered different countries, where patterns differ depending on epidemiological factors. For instance, several European countries have recently displayed many, even all, known subtypes while in other parts of the world only one subtype dominates. Furthermore, it appeared that in several countries there was an association between risk group and subtype, homosexual men usually carrying subtype B virus and heterosexually acquired infections usually caused by non-B subtypes (Lukashov *et al.*, 1995; van Harmelen *et al.*, 1997; Ubolyam *et al.*, 1994). This suggests that the epidemic among homosexuals has spread through international travel, in addition to local acquisition.

In Thailand, the B subtype was originally associated with intravenous drug use (IVDU), while CRF01_AE (formerly known as subtype E) with heterosexual transmission (Ou *et al.*, 1993). This distinction is no longer true; virtually all recent infections in Thailand, both heterosexually acquired and IVDU- related infections, are now caused by CRF01_AE virus (Kalish *et al.*, 1995). Most subtype B viruses from Thailand are distinct from the subtype B strains prevalent worldwide and are often designated B'. There have also been a small number of infections with the main subtype B in Thailand. A mini-epidemic was reported among prisoners (Kalish *et al.*, 1994). It has been found that a large number of infections among homosexual men are caused by subtype B (Ubolyam *et al.*, 1994), suggesting that this branch of the epidemic results from import rather than local transmissions.

Recently, a similar subdivision within a subtype has also been reported for subtype A (Carr *et al.*, 1998b). There appear to be three main lineages of subtype A at present. The first is composed of isolates from Central and East Africa. The second one is found mainly in West Africa (Ivory Coast and Liberia) and in Djibouti, where it was presumably introduced by soldiers of the French Foreign Legion who had contracted the infection in West Africa. The subtype A portions of CRF01_AE constitute the third lineage which always clusters tightly together separate from the other subtype A sequences. The three lineages are particularly visible in a *gag* tree (Figure 2). The clusters formed by these last two lineages are more compact than the first one. In *gag*, the mean difference (Hamming distance) between sequences within each of the two compact clusters is around 6%, while the difference between them is around 14%.

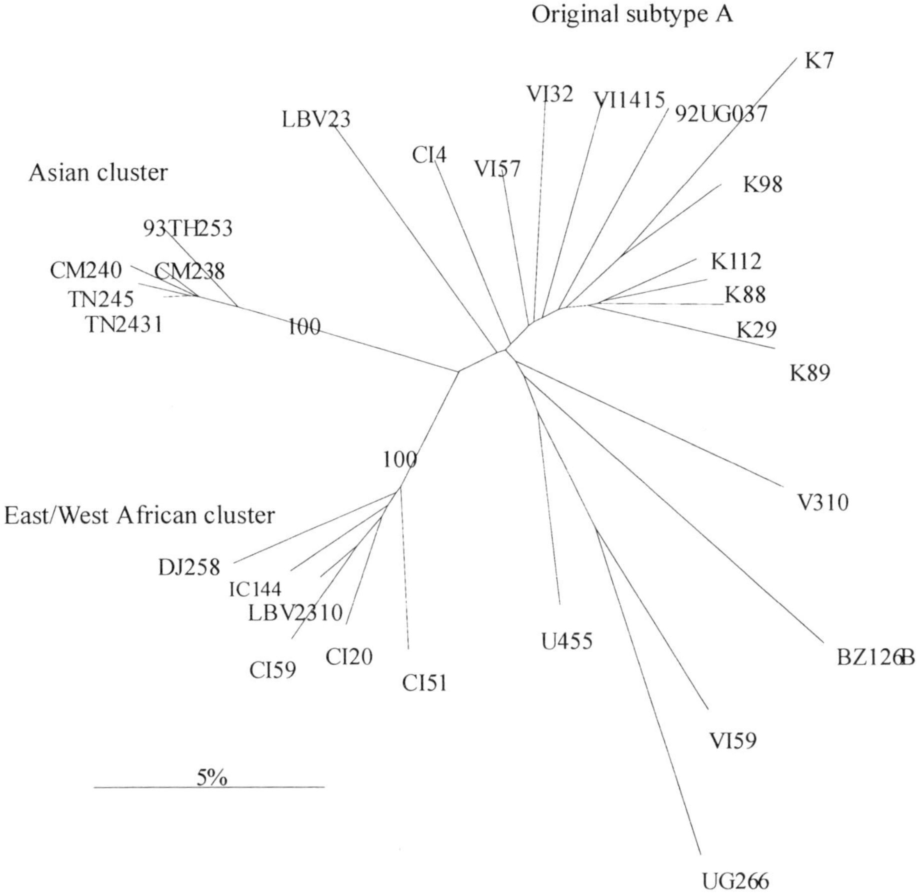

Figure 2. Neighbor-joining tree (Kimura 2-parameter distances) showing the sub-clustering of subtype A sequences in three groups. The tree is based on the entire gag gene. The two distinct subclusters have been indicated. Bootstrap values in % based on 500 replicates.

5. INTERSUBTYPE RECOMBINATION

There have always been isolates that did not fit the subtype classification very well. Some of these have turned out to be the newly discovered representatives of a new subtype, such as the first mention of subtype F (Potts *et al.*, 1993). Others were later found to be recombinants that could not be assigned to a subtype unambiguously.

The subtype classification yielded a convenient handle to detect and label recombinants. Two former subtypes that were named on the basis of partial genome sequences, CRF01_AE and CRF04_cpx, now appear to be recombinant in their own right (Robertson *et al.*, 1997; Gao *et al.*, 1996b). "Subtype I"-like fragments have been found in various A/G/I and A/D/I recombinants, some of which have been found in multiple epidemiologically unlinked infections (Robertson *et al.*, 1997), but despite extensive searching no parental subtype E or subtype I genome has been found yet.

Some recombinants (most notably the MAL isolate) were identified even before the subtypes were introduced. In the 1987 Human retroviruses and AIDS compendium, Myers mentioned that "the MAL genes display a disjunctive pattern of divergence relative to other isolates... A plausible explanation for this effect is that the MAL genome has been involved in recombination with a more divergent virus" (Myers, 1987). Li and colleagues (1988) also noticed this. It is likely that recombination research would have proceeded much more slowly without the possibility for unequivocal classification that the subtype designation provided. To date, some within-subtype recombinants have been found (Zhu *et al.*, 1995; Diaz *et al.*, 1995), but their number is far smaller than that of between-subtype recombinants. The reason for this is that detection of intra-subtype recombinants is much more difficult due to the greater similarity of the 'parent' variants.

It should be emphasized that the present subtype classification system is to some extent a historical accident. For example, if the subtype I mosaic had been sequenced before subtypes A and G, it would now be regarded as a 'pure' subtype, and subtypes A and G would now both be termed 'recombinants', since they contain parts of subtype I and parts that do not resemble the subtype I genome. Thus, the distinction between pure subtypes and recombinants is artificial. The reason subtype I is called a recombinant is that it is important that the nomenclature indicates that regions in this mosaic closely resemble another subtype, while other regions are very diverged.

6. THE ORIGIN OF SUBTYPES

The question about the origin of the subtypes is closely related to that of the origin of HIV. It is clear that HIV has its roots in African primates. The origin of HIV-1 is appears to be different than that of HIV-2. HIV-2 subtypes are interspersed among viral sequences from sooty mangabey virus (SIVsm), suggesting several cross-species transmissions from sooty mangabeys to humans (Gao *et al.*, 1992; Sharp *et al.*, 1994). In the HIV-1 M group subtypes, no such evidence has been reported: they are not divided by any sequences from nonhuman prmates. Hence, it seems like all the subtypes of HIV-1 group M have a common origin. However, groups M and O are separated by the chimpanzee lentivirus (SIVcpz). These data strongly suggest that HIV-1 was introduced into humans from different chimpanzee sources on several occasions (Gao *et al.*, 1999).

However, with recombination operating throughout the evolution of HIV, multiple cross-species transmissions may obscure each other, perhaps leading us to conclude mistakenly that they occurred only once. Thus, one can speculate whether the subtypes in the HIV-1 M group represent more than one introduction. The extreme view would be that each subtype represents an introduction, and thus that the basal structure of the subtype tree displays evolution in the primate source. Recent evidence, however, contradicts that theory. Sequence material from a virus from the former Zaire, dated 1959, was shown to be close to the common ancestor of subtypes B and D (Zhu *et al.*, 1999), suggesting that at least these two subtypes are not independent introductions. Much remains to be discovered about lentiviruses in nonhuman primates, however. About twenty nonhuman primate species or subspecies are presently known to be hosts of lentiviruses, while about 200 primate species are known (of which more than a third are Old World primates). Wild chimpanzees have not been extensively sampled to date, and it is possible that more clues about the origin of the HIV epidemics emerge later.

Since we still do not have a full understanding of the origins and relationships of HIV and the subtypes, the timing of HIV's history is made even more difficult. Many suggestions on the age of HIV have been proposed, ranging from 40 years to several million years (Li *et al.*, 1988; Sharp *et al.*, 1994; Fukasawa *et al.*, 1988; Smith *et al.*, 1988; Sharp and Li, 1988). Today the common view is that the origin of SIV may date back millions of years, but that the current HIV-1 and HIV-2 epidemics were caused by several more recent introductions of SIV strains into humans. While there has been some debate whether a molecular clock is applicable to HIV or not, it has recently been shown that HIV-1 does evolve in a clock-like fashion (Leitner and Albert,1999). A recent attempt to date several nodes in the HIV-1 *env* gene phylogeny using maximum likelihood calculations under the general reversible substitution model and variable rates across nucleotide sites (Korber *et al.*, 2000) estimates the origin of the group M viruses at about 1930 (± 20 years). This result seems to be reliable based on both internal and external validation of other predictions using this method (using the known date of origin of the Thai E group, and a simple regression method that dated the early 1959 sequence very accurately and arrived at the same estimate for the group M node, respectively).

It is interesting to note the surprising evolutionary space the subtypes occupy. Figure 3 shows typical trees displaying the relationships between the subtypes, using three different phylogenetic methods for three different genes (*env*, *gag* and *pol*). Any sequence form of the virus within the space of the subtypes is capable of establishing infection and epidemic spread, whether it has a common or distinct origin. Moreover, even the space between the subtypes is possible, clearly demonstrated by the increasing amount of detected recombinant forms that are successfully spreading. However, in phylogenetic analyses recombinant sequences cause problems since the methods assume that recombination does not occur. Therefore, the location of a recombinant sequence does not tell about its evolutionary origin but rather about the fraction of sequence it derived from its parents, and time estimates from histories that involve recombination will thus be confounded.

7. BIOLOGICAL DIFFERENCES BETWEEN SUBTYPES

There are two possible causes for the existence of viral subtypes. First, they may be the result of a 'founder event': certain variants of the virus become founders of a

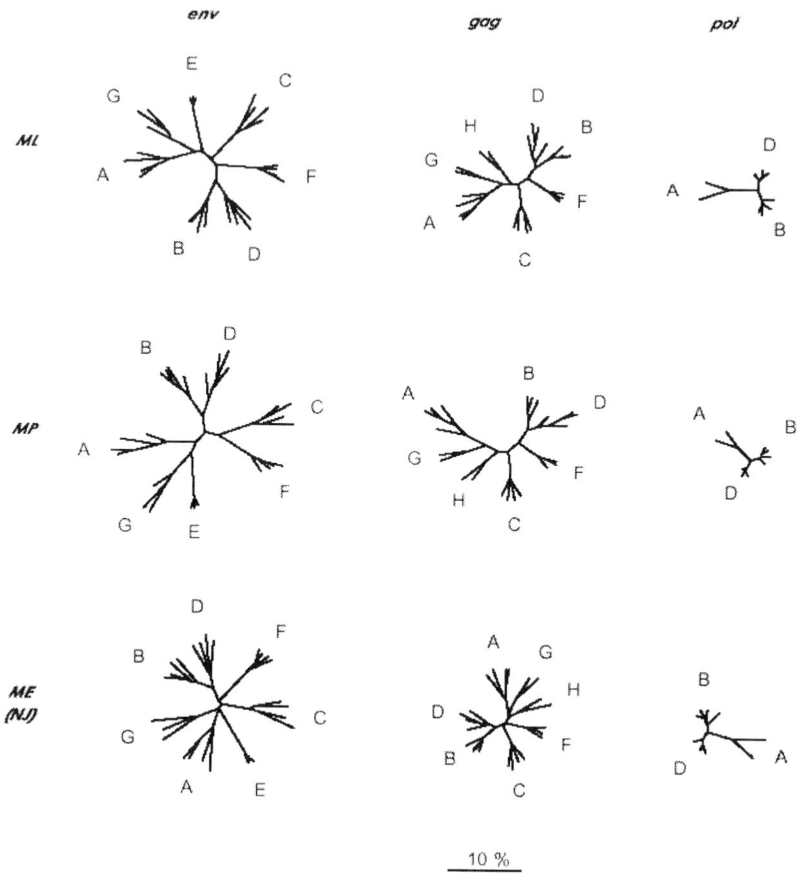

Figure 3. Unrooted trees calculated according to the maximum likelihood (ML) criteria, the maximum parsimony (MP) criteria, and the minimum evolution (ME) criteria. The ML search was conducted using the PHYLIP package (Felsenstein, 1984) with the F84 substitution model, the MP search was conducted with a heuristic search using the PAUP program (Swofford, 1991) under uniform weighting, and the ME search was conducted with the neighbor joining approximation using the PHYLIP package with the F84 substitution model. Each method was applied to the complete *env*, *gag*, and *pol* gene alignments, gapstripped to 2301, 1380, and 3000 nucleotide sites, respectively. Branch lengths in ML and ME trees were adjusted to the indicated scale bar, and MP trees were drawn arbitrarily to reflect the information content in a comparable scale.

epidemic because they happen to be involved in an extensive transmission chain. In this scenario, the subtypes can be biologically equivalent even though they are genetically very different. The alternative explanation is that the subtypes had certain characteristics that allowed them to out-compete less fit viral variants. This pre-supposes biological differences between the subtypes. These two explanations are not mutually exclusive.

Ever since the existence of subtypes was established, the hunt has been on for biological differences between the subtypes. In spite of intensive efforts to explore this issue, little conclusive evidence exists as yet to suggest that biologically important differences exist. In 1996, a biological distinction was reported in susceptibility of Langerhans' cells to infection, which might make viruses of Thai CRF01_AE more easily transmissible via heterosexual intercourse than Thai subtype B. Langerhans' cells are found in the vaginal walls in the cervix and in the penile foreskin. This biological difference could explain why CRF01_AE is outcompeting subtype B in Thailand at present (Soto-Ramirez *et al.*, 1996; Essex *et al.*, 1997). The report created a turmoil in the popular press and the WHO convened a meeting of experts devoted to discussion of the possible implications and a course of action (anonymous, 1997). Recent studies have not been able to reproduce this result; while isolates did differ in their ability to infect Langerhans' cells, no clear correlation with viral subtype was (Dittmar *et al.*, 1997; Pope *et al.*, 1997). Ongoing studies are attempting to examine the rate of T-cell decline, an indicator of immune dysfunction and predictor of the onset of AIDS, in individuals infected with different subtypes. Some African cohorts with two (or more) subtypes co-circulating in the same population and well documented incident cases, are ideal study groups for attempting to answer this kind of question. So far, however, the data from these cohorts are too sparse to draw conclusions about subtype differences in pathogenicity and infectivity.

Despite the lack of clear correlation between subtypes and biological properties, more subtle phenotypic distinctions have been reported, which may have biological importance if confirmed. An example is a recent observation concerning patterns of co-receptor usage. Some background information is necessary to explain this phenomenon. HIV-1 usually requires two receptors to enter a target cell, the CD4 protein and one of the chemokine receptors CCR5 and CXCR4. The co-receptor functions of these proteins are inhibited by their natural chemokine ligands (for a review see Paxton *et al.*, 1996), and a homozygous deletion in the CCR5 gene confers some resistance to HIV-1 infections (Huang *et al*, 1996). Examples of viruses capable of using both of these co-receptors have been found in all group M subtypes tested, as well as among group O viruses (Zhang *et al.*, 1996). Closely associated with co-receptor usage is another important biological property of HIV, the syncytium-inducing (SI) capacity and rapid/high replication kinetics. Although the formation of syncytia (clumps of fused cells) in vivo has not been demonstrated, the in vitro SI capacity of HIV-1 (usually measured in a standard assays using MT-2 cell culture) is associated with disease progression. Viruses that use CCR5 are generally not syncytium inducing, and are referred to as macrophage-tropic viruses. Viruses that can use CXCR4 for viral entry usually retain their ability to use CCR5, and thus are dual tropic (Lu *et al.*, 1997), are syncytium inducing, and can grow in MT2 cells, which lack the CCR5 receptor. It appears that subtype D isolates, if they use the CXCR4 co-receptor, tend to use that co-receptor exclusively, while viruses of other subtypes that use CXCR4 are usually also capable of using the CCR5 receptor (Tscherning *et al.*, 1998). Furthermore, there are indications to suggest that usage of the CXCR4 receptor by subtype C viruses may be rare (Tscherning *et al.*, 1998; Abebe *et al.*, 1999).

A second property associated with subtypes was noted by Gao and colleagues (1996b) who found an RNA secondary structure difference in an important transcriptional regulatory domain, TAR, of subtype A and CRF01_AE viruses. The TAR element is a conserved, stable stem-loop structure required for Tat-mediated transactivation of HIV-1 gene expression, and the level of activity of this system may influence the rate of disease progression (Kirchhoff *et al.*, 1997). In addition, the number of binding sites for NF-κB, a cis-acting transcriptional activator which may also influence the level of viral gene expression (Gao *et al.*, 1996a; Carr *et al.*, 1996; Montano *et al.*, 1997), has been observed to differ among the long terminal repeats (LTRs) of different subtypes.

Another difference between subtypes that is of potentially biological significance is reflected in subtype specific patterns of genetic variation. There is an elevated rate of nonsynonymous (amino acid altering) substitutions in the third variable region of *env*-gp120 (V3 loop) of subtype D viruses (Korber *et al.*, 1994b), and apparently in a subset of CRF01_AE viruses, compared with other subtypes. Serological studies have further indicated that the V3 loop of the D subtype may be the most diverse (Barin *et al.*, 1996). The V3 loop is a functionally important domain of the viral envelope. Positively charged amino acids in certain positions in the V3 loop correlate with a syncytium-inducing phenotype in culture (De Jong *et al.*, 1992; De Wolf *et al.*, 1994), and the V3 loop influences co-receptor usage (Cocchi *et al.*, 1996; Bieniasz *et al.*, 1997). Although the V3 loop is one of the more variable regions of the Env protein, a highly conserved form of the V3 loop is found in many C-subtype viruses and a subset of A-subtype viruses (Korber *et al.*, 1994b). This common, conserved form of the V3 loop is probably the reason that A- and C-subtype viruses are difficult to distinguish by V3 peptide serology, which can be used with some success to discriminate between other subtypes (Barin *et al.*, 1996; Sherefa *et al.*, 1994; Hoelscher *et al.*, 1998). Recently, the analysis presented by Korber in 1994 was repeated using the vastly increased number of V3 sequences that are now available in the Los Alamos database. The results of this analysis strongly suggested that the V3 loop, but not the flanking regions (outside the loop-defining cysteines), of subtype D virus was much more variable than the V3 loop of other subtypes, while that of subtype C was more conserved. Although the biological meaning of these differences is unknown, they suggest that there may be differences at least in the restrictions to which the different viruses are subject, which in turn may be indicative of biological differences between the subtypes (see Kuiken *et al.*, 1998 for details).

8. SUBTYPING A SEQUENCE

This section is a practical cookbook-style guide on how to go about subtyping a new sequence. We have taken the HIV-1 isolate MAL as an example. MAL is a mosaic containing subtypes A, D, and K fragments as well as sections that are as yet unclassified.

The first step in finding the subtype of your sequence (and in any type of sequence analysis) should be a contamination check. Contamination is a common and serious problem in HIV research. The conclusions drawn from bad data can be misleading, and can cause fundamental misconceptions in our understanding of HIV biology. Publications based on contaminated sequences have led to scattered erroneous reports regarding virtually all aspects of HIV biology that involve sequences: viral clearance, transmission patterns, rapid and slow progression, drug resistance, central nervous system tropism, immune escape, and variability in populations. Contamination can happen in anyone's laboratory, and is not a sign of sloppy work. It is a fact of life with HIV culture, PCR amplification, and sequencing. For a more detailed discussion of contamination in a number of real data sets, see Learn *et al.*, 1996 and the Los Alamos website (*http://hiv-web.lanl.gov*). There is no method that can prove that a sequence is not a contaminant, but there are ways to detect the most obvious contamination problems.

The two main sources for contamination are lab strain DNA and DNA from virus that was worked on simultaneously with the sample in question. Lab strain contamination can often be detected by comparing your sequence to the sequences in the Los Alamos HIV database or in GenBank. Doing a BLAST search (Altschul *et al.*, 1996) is the easiest way to do this. BLAST very rapidly compares a query sequence to all sequences in the database and returns the most similar ones. Recently, gapped BLAST was introduced, which allows insertions and deletions (indels) in the homology searches (Altschul *et al.*, 1997); for HIV, where indels are quite common, this is a major improvement. If your sequence, or part of it, is unexpectedly similar to a sequence in the database, this is a reason to be suspicious.

The BLAST search on the Los Alamos web site uses the same gapped BLAST program, but gives slightly different output than GenBank's BLAST, because the output from the Los Alamos web site is in order of percent similarity, rather than by the BLAST score. Since BLAST also identifies small fragments that are very similar, we have added a lower limit on the length of the fragments that are retrieved. This was done because for example in running BLAST on a whole genome, a multitude of fragments will be identical to the query sequence that are too short to be of any use. Figure 4a shows the BLAST output of a comparison of MAL to all sequences in the HIV database. The sequence that resembles MAL most still is 8% different from it. The highest-scoring sequence in the GenBank BLAST is the NDK isolate, which is 91% similar to MAL over its entire length (this cannot be seen in the figure, but is reported in the alignment that BLAST also produces). These scores are low enough that contamination with an HIV strain that is present in GenBank can be excluded.

The second most common source of contamination, viral sequences from other samples that were analyzed simultaneously, can frequently be detected when one creates a phylogenetic tree that contains all sequences in the set (or even better, all sequences obtained from samples that were processed in the same laboratory in the same time period). Sample mix-ups, sequences that fall in an unexpected cluster, and sequences from two patients that cluster together very closely are all warning signs. If no explanation can be found for the unexpected sequence(s), the only solution is to try to find another sample from the same patient, sequence again, and see if the results are congruent.

The next step in the analysis is to find a reference alignment to compare your sequence to. Which alignment is good for subtyping depends on many things. The Los Alamos web site provides background alignments that are specifically designed for subtyping new sequences. These are available for the whole genome and for the *gag*, *pol*, *env* and *nef* subregions. If you have a sequence from a different region, a good starting point is the extensive gene alignments that are also on the Los Alamos web site. By weeding out the recombinants (these are clearly marked in the sequence names) and retaining a few representatives for each subtype, it is usually easy to create a good background alignment.

A

```
BLASTN 2.0.6 [Sept-16-1998]
Reference: Altschul, Stephen F., Thomas L. Madden, Alejandro A. Schaffer, Jinghui Zhang
, Zheng Zhang, Webb Miller, and David J. Lipman (1997), "Gapped BLAST and PSI-
BLAST: a new generation of protein database search programs", Nucleic Acids Res. 25:338
9-3402.

Query= (9229 letters)
Database: public.fasta 38,700 sequences; 18,884,188 total letters

Searching..............................................done          Score E
Sequences producing significant alignments: (bits) Value
seid|13275|HIV-1|ADK|CD|X04415|HIVMALCG|HIV-1 isolate MAL from Za... .830e+040.0
seid|2022|HIV-1|D|CD|M27323|HIVNDK|HIV-1 isolate NDK from France...  6946 0.0
seid|2414|HIV-1|D|CD|M22639|HIVZ2Z6|HIV-1 isolate Z2 (originally...  6839 0.0
seid|1221|HIV-1|D|CD|K03454|HIVELICG|HIV-1 isolate ELI from Zaire... 6807 0.0
seid|15269|HIV-1|D|CD|U88822|84ZR085|HIV-1 isolate 84ZR085 from Z... 6794 0.0
seid|12336|HIV-1|D|UG|U88824|94UG114|HIV-1 isolate 94UG114 clone...  6605 0.0
seid|58084|HIV-1|D|KE|AF133821|MB2059|HIV-1 isolate MB2059 from K... 6570 0.0
seid|4703|HIV-1|B|US|M93259|HIVYU10X|HIV-1 isolate YU-10 from USA... 6556 0.0
seid|47600|HIV-1|B|US|U69588|WCIPR855|HIV-1 patient WCIPR sample...  6544 0.0
seid|2234|HIV-1|B|US|M17451|HIVRF|HIV-1 isolate RF (HAT-3) from U... 6540 0.0
seid|47598|HIV-1|B|US|U69586|WCIPR8546|HIV-1 patient WCIPR sample... 6520 0.0
```

B

```
BLASTN 2.0.11 [Jan-20-2000]

Reference: Altschul, Stephen F., Thomas L. Madden, Alejandro A. Schäffer, Jinghui Zhang
, Zheng Zhang, Webb Miller, and David J. Lipman (1997), "Gapped BLAST and PSI-
BLAST: a new generation of protein database search programs", Nucleic Acids Res. 25:338
9-3402. RID: 956087922-12952-19679

Query= K03456|HIVMALCG|gi|| (9229 letters)
Database: nt 614,744 sequences; 1,848,934,927 total letters

                                                                 Score E
Sequences producing significant alignments:                      (bits) Value
emb|X04415.1|HIVMALCG Human lymphadenopathy virus (MAL isol...   9607 0.0
emb|AJ237565.1|HIM237565 Human immunodeficiency virus type...    3358 0.0
gb|AF064699.1|AF064699 HIV-1 isolate BFP90 from Burkina Fas...   3013 0.0
emb|AJ245481.1|HIM245481 Human immunodeficiency virus type...    3005 0.0
emb|AJ245482.1|HIM245482 Human immunodeficiency virus type...    2813 0.0
emb|AJ249235.1|HIM249235 Human immunodeficiency virus type...    2809 0.0
emb|AJ249239.1|HIM249239 Human immunodeficiency virus type...    2728 0.0
gb|AF076475.1|AF076475 HIV-1 isolate VI1126 from Democratic...   2656 0.0
gb|AF049337.1|AF049337 HIV-1 clone 94CY032-3 from Cyprus, c...   2617 0.0
gb|AF119819.1|AF119819 HIV-1 strain GR84 complete genome         2597 0.0
```

Figure 4 HIV-BLAST output (**A**) vs. the output from an NCBI BLAST search (**B**). The original BLAST output orders the sequences by BLAST score (a similarity score weighted by sequence length), the HIV-BLAST output is sorted by % similarity. Short fragments (less than 20% of the query sequence length, or less than 500 nucleotides) are discarded. "..." indicates a deletion of part of the output.

Aligning your sequence to the reference alignment is not very difficult, especially if the reference alignment already contains very different variants. There is usually no need to insert more gaps into the existing alignment, and it is sufficient to just copy (some of) the gaps into the new sequence to align it properly.

There are many automatic multiple sequence alignment programs, but aligning many sequences simultaneously is still a very difficult computational problem that has not been solved in a satisfactory way. Some programs (CLUSTAL (Higgins *et al.*, 1996) for example) do a reasonable job, but all computer-generated alignments must be manually checked. Recently, new alignment programs based on the Hidden Markov Model approach have been doing improved alignments (Eddy *et al.*, 1995; Eddy, 1995). These programs build a model of the data based on a training set. The model can be improved manually, and then used to generate new alignments. However, even these programs sometimes make glaring mistakes in multiple sequence alignments. In addition, few programs are able to codon-align, i.e. to keep the reading frame intact in the alignment. Since this often is the biologically most meaningful alignment for coding regions, it makes sense to manually correct the alignments.

For manual checking and correction of an alignment, a sequence editor is indispensable. For UNIX machines, GDE (*ftp://megasun.BCH.UMontreal.CA/pub/gde/*) is a good choice; for Macintoshes, the program SeAl (*http://evolve.zoo.ox.ac.uk/Se-Al/Se-Al.html*) is very convenient, because it is able to work with codon alignments. For PC's running Windows, Tom Hall's Bioedit (*http://www.mbio.ncsu.edu/RNaseP/info/programs/BIOEDIT/bioedit.html*) is a wonderfully versatile alignment editor, which can do on-the-fly amino translations, a feature that makes correctly aligning coding sequences much easier. If one has a choice of platforms, Bioedit is probably the user-friendliest and most all-round editor. GDE and Bioedit also allow phylogenetic and other sequence analysis by providing an interface to other programs (PHYLIP, CLUSTAL, BLAST etc.).

Most analyses in this chapter were done on the basis of the Los Alamos HIV-1 whole genome alignment, available on the web site.

Once the alignment is finished, the next step is to build a phylogenetic tree. Note that all tree building methods assume that the alignment is correct; errors in the alignment can lead to a very misleading tree. There are many different methods to create phylogenies. While an extensive discussion of the differences between all these methods is beyond the scope of this chapter, we will provide an overview of them. Roughly, the methods can be divided into distance-based and character-based methods. Character-based methods use the individual substitutions among the sequences to determine the most likely ancestral relationships, while distance-based methods first calculate the overall distance between the all pairs of sequences, and then calculate a tree based on those distances. Maximum parsimony (MP) and maximum likelihood (ML) are the most important character-based methods. In most comparative studies, ML seems to be the method that yields the best trees (Leitner *et al.*, 1996).

The most important drawback of ML is that it is very computationally intensive; it is almost unusable with more than a few dozen sequences. MP is much faster than ML, but still slow compared to most distance-based methods. Both these methods calculate large numbers of trees and compare them by either the likelihood or the parsimony score. There are programs available that speed up the process considerably, such as FastDNAml (Olsen *et al.*, 1994) but for large numbers of sequences the computational burden is still a major hurdle. In this group, Neighbor Joining (NJ) is by far the most popular method. It is very fast and generally quite good, although there are conditions under which it is biased (i. e., it systematically produces a wrong tree). When the number of sequences is large (>500), NJ is the only viable option. The performance of a distance-based method depends to some extent on the way the distances are calculated; there are many such methods, from a simple proportion difference between the sequences (Hamming distance) to methods that weigh each of the transitions and transversions differently and compensate for different base frequencies. Generally the more sophisticated methods tend to work slightly better. A more elaborate discussion of phylogeny reconstruction methods is found in the chapter on phylogeny elsewhere in this book.

For the purpose of subtyping sequences, an NJ tree based on a matrix of genetic distances is generally good enough. It is not vital that the tree is correct in every branch, only that the sequence of interest is clustered with the right subtype. Almost any method will solve this problem correctly (see Figure 3). However, when more detailed information on the evolutionary relationships is important (for instance in forensic analyses, or when studying the origin of lentiviruses), more realistic models of evolution and unbiased tree reconstruction methods should be used (Leitner *et al.*, 1996; 1997).

The interpretation of the tree you have obtained can be very simple. If the new sequence clearly clusters with an established non-recombinant subtype, the problem is solved. To finish the analysis, a bootstrap value is usually calculated on the cluster of interest. The criteria for "high" bootstrap values are vague, but >70% over 500 or more replicates is generally deemed to indicate a stable cluster (Hillis and Bull, 1993). Figure 5 shows the ML tree for a small set of sequences from different subtypes. The placement of MAL is striking, and very typical for a recombinant sequence: outside all the main clusters, with a branch length that takes it about as far outside the center of the tree as the other sequences that do cluster with a subtype. This is very distinct from old sequences, for example, that also tend to branch off from or close to an interior branch, but are expected to have very short branch lengths (for an example, see Zhu *et al.*, 1998). Figure 6 shows ML trees for the three structural genes of sequences – it jumps from one place to another in the three trees, while the other (reference) sequences stay in well defined clusters. Note that although this analysis already suggests a recombinant origin of MAL (it clusters either close to subtype A, D, or K), the division of the genome into gag, pol, and env regions are arbitrary from a recombinational point of view, i.e., recombination can occur virtually anywhere in the genome.

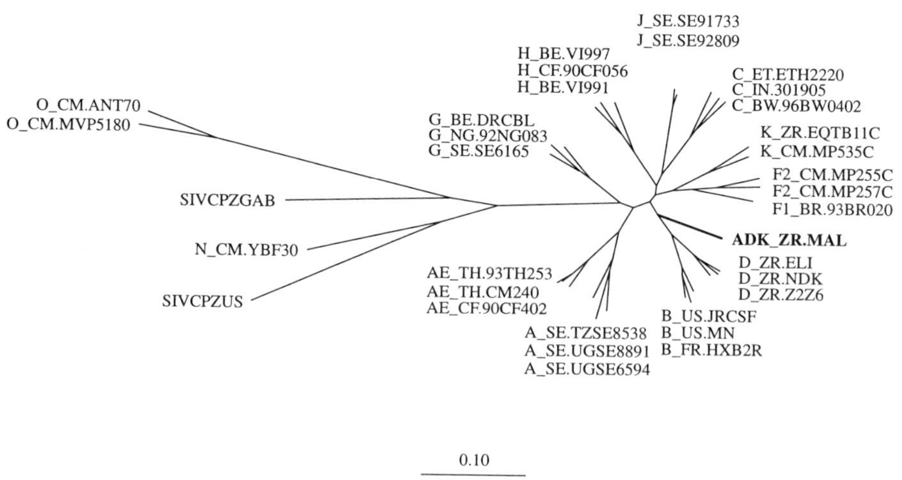

Figure 5. Maximum likelihood tree of complete genome HIV-1 and SIVcpz sequences. Subtypes of the M-group are designated with their subtype letter as a prefix to the sequence name, and groups O and N are indicated in the same way. MAL does not seem to relate to any of the established subtypes. Distances (F84) are according to the indicated scale bar.

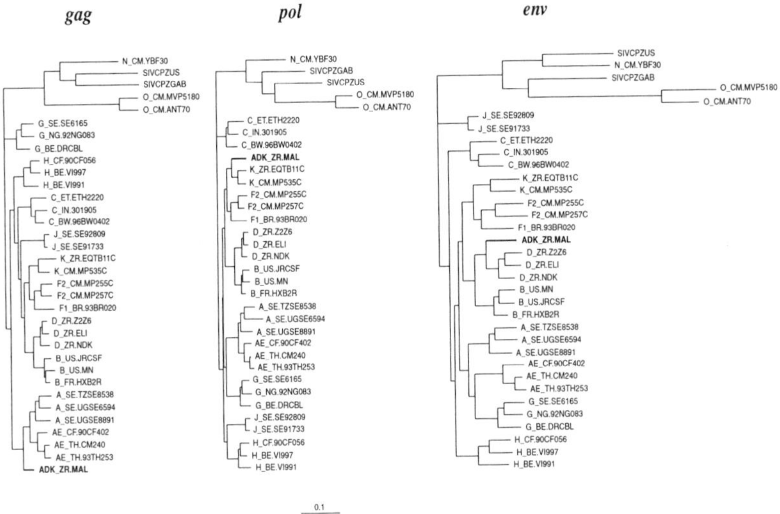

Figure 6. Maximum likelihood trees of the three structural genes of HIV. The trees contain the same sequences as figure 5. In these trees the MAL sequence is located outside subtype A in pol, outside subtype K in gag, and outside subtype D in env. The reason for its different clustering patter is because it is a recombinant sequence. However, the recombination points do not exactly follow the genetic organization of HIV, why it does not cluster within any subtype in these plots. Once the recombination points have been established (see text and figures 7-9), trees of more adequate regions should be inferred.

One important and often overlooked factor in the interpretation of the tree is the length of the sequences, or more accurately, the amount of information it is based on. Many tree-building programs use the concept of "informative site", but unfortunately its meaning is different for different methods. In distance-based methods, all positions that vary in a data set give information on the relationships between the sequences. In parsimony methods on the other hand, a site is only considered informative if it partitions the sequences into at least two groups, any two of which must have at least two members. In maximum likelihood all sites are considered informative (though they may not actually contain any variation). Many tree-building programs report either the number of variable or (in the case of parsimony programs) the number of informative positions, which is usually lower.

The number of positions that vary or are informative needed for a reliable tree also depends on the agreement between the information derived from the informative sites, but 100 sites that vary are generally enough to allow conclusions about the subtype classification. It is possible that a much longer sequence yields less than 100 informative positions. This can happen if there are a lot of gaps in the alignment (most tree building programs discard the whole column if one sequence has a gap at a position) or because the sequence is from a very conserved region (such as *gag*-p24 or *pol* and especially the protease region of this gene). With fewer than 100 informative positions in the sequence, the results of the analysis can still be valid, but they may give misleading information; furthermore, detection of recombination (see below) may be all but impossible in short sequences.

If the query sequence does not clearly cluster with any of the subtypes, there are a number of possible causes. The first thing to check is the alignment. Misalignment of the query sequence can cause it to branch off by itself. There are several ways to check for misalignment. The best way to do that is by eye. A good alignment editor (see above), especially one that shows different bases in different colors, makes it very easy to visually inspect the alignment. If it is too large or too complicated to evaluate by eye, another method is to use different alignment programs, and to build a tree based on each resulting alignment, using the same tree construction method. If the trees show important differences, the alignment clearly is problematic. It is not easy to then decide which alignment is best. Alignments and alignment programs are discussed more extensively elsewhere in this book.

Another possibility is that the sequence has somehow evolved in a strange way. Hypermutated sequences are one example of this. Hypermutation is a rare phenomenon where a very large percentage of the Guanines in the genome are replaced by Adenines (Vartanian *et al.*, 1991). This phenomenon has also been shown to exist in other lentiviruses (Wain-Hobson *et al.*, 1995) and in hepatitis B virus (Gunther *et al.*, 1997). Hypermutation can result in extremely long branches in a phylogenetic tree, but it usually does not affect the subtype classification of the sequence. The extreme preponderance of A's in the alignment can easily be spotted on visual inspection. In the GenBank entry for MAL (accession numbers X04415 and K03456), the number of A's over the entire genome is 3355, against 1627 C, 2204 G, 2043 T. For HIV-1, this is a very typical distribution. Visual scanning doesn't show an extreme density of A's in any section of the genome.

If neither misalignment nor hypermutation can explain the behavior of the sequence, the next step is to check if it may be a recombinant. The likelihood of isolating a recombinant sequence depends to a large extent on where your sample came from: samples from the US or Europe are much less likely to be recombinant (at least inter-subtype recombinants) than samples from regions with high prevalence and co-circulation of multiple subtypes.

The first step in checking for recombination is usually the Recombinant Identification Program (RIP) (Siepel *et al.*, 1995). A web interface for this program is available (*http://hiv-web.lanl.gov/RIP/RIPsubmit.html*). Like most tools to investigate recombination, it uses a sliding window approach: over the whole length of the sequence, the similarity to a number of subtype reference sequences is calculated for short stretches at a time. If there is recombination, it will show in a sudden lowered similarity to one subtype. The analysis can be based on all positions or only on informative sites. An example of a RIP plot that shows strong evidence of recombination is shown in figure 7. All subtypes were originally present in this analysis, but those that were never the most similar to the query sequence were deleted from the figure. RIP does a simple significance test (z-test). It should be noted that one of the main assumptions of this test, independent evolution in all positions, is certain to be violated, so the test results should be interpreted with care. This assumption is, however, also a cornerstone in most phylogenetic reconstruction programs.

The Simplot program (*http://www.med.jhu.edu/deptmed/sray/download/*) is a very user-friendly program that is available only for PC's running Windows. It creates similarity plots, which are very similar to RIP output. An example of a similarity plot is shown in Figure 8. As expected, this plot is very similar to the one produced by RIP, although Simplot does not include any kind of significance test. Results from both RIP and Simplot suggest that the MAL genome contains regions

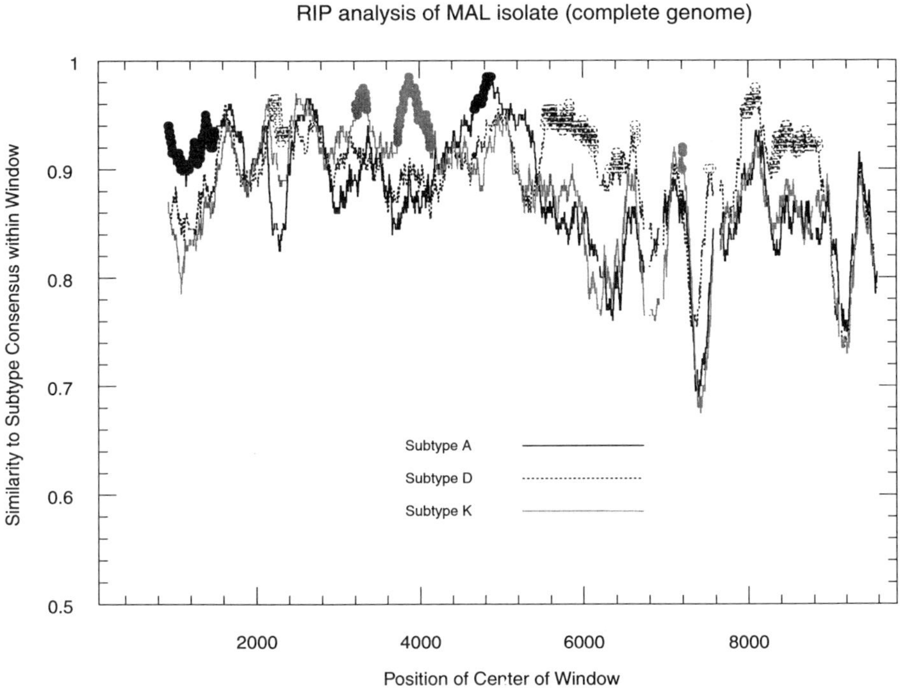

Figure 7. RIP graphical output. Lines represent the similarity of each of the subtypes to the query sequence, MAL. For simplicity, lines representing subtypes that never were significantly more similar to MAL than any of the others have been omitted from this graph. The thick lines indicate significantly more similarity to MAL than any other subtype, according to a Fisher's Z-test (see (Siepel *et al.*, 1995) for details).

of high homology to subtypes A, D and K, as well as a small segment that do not resemble any sequence of the comparison set.

The next step is to verify this using a more robust method, such as bootscanning (Salminen *et al.*, 1995b). Bootscanning also uses a sliding window approach, but instead of just calculating the distance to each reference sequence, a tree is constructed for every window, and bootstrap values are then calculated for the clusters. In the region of the crossover point, the bootstrap value first drops for the cluster that contains the recombinant. At some point, the sequence will move from its cluster into a new one, and the bootstrap value for this cluster should then increase as the sliding window moves past the crossover point. Simplot can also be used to do bootscanning (it uses the PHYLIP package to do the phylogenetic trees), and is presently the only publicly available bootscanning program. Unlike distance plotting, bootscanning results can be heavily influenced by the sequences included in the background set. It is usually safe to omit sequences that have low similarity to the query sequence over the entire length of the genome (RIP and/or similarity plotting can be used to ascertain this) but representatives of all other sequences need to be included in the background set. In addition, it is recommended that results obtained with one set of representatives of the relevant subtypes are always checked

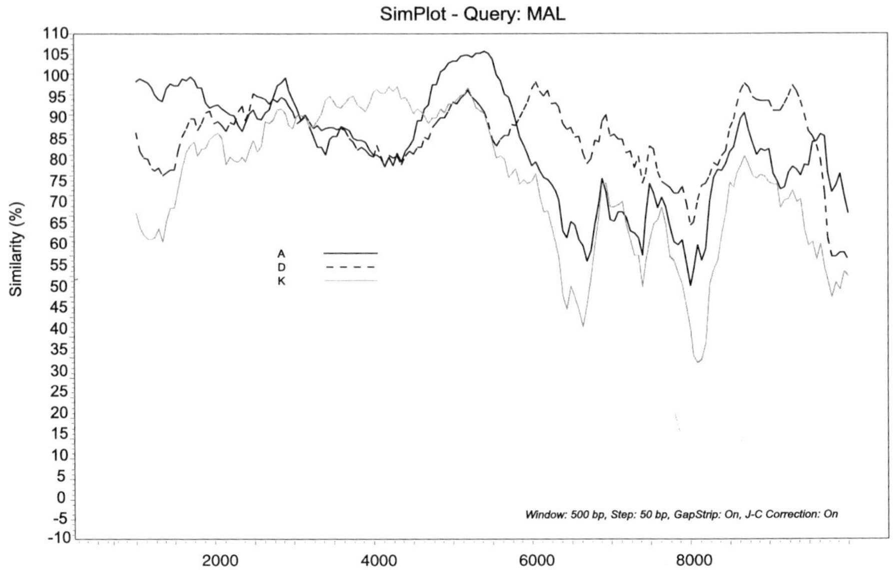

Figure 8. Similarity plot of MAL against consensus sequences of subtypes A, D and K. Lines represent the similarity scores of Mal against the three consensus sequences, corrected for multiple hits (Jukes-Cantor correction). The window size was set at 500.

against a second set, because random (i.e. subtype-independent) similarities to individual sequences can distort the results.

Figure 9 shows the bootscanning results of MAL against subtypes A, D and K (bootscanning was performed against all subtypes but for the reason of clarity only comparisons against subtypes A, D and K are shown). The mosaic nature of the MAL genome can be seen clearly in this graph. One short stretch remains (position 2200-3200 in this alignment) in which none of the subtypes resemble MAL. This stretch can now be subjected to another BLAST search, to see if anything similar is pulled out of the sequence database. In this case, the recently published sequence NOGIL3, from a Norwegian transmission cluster, is the one BLAST pulls up as most similar to this stretch of MAL (not shown). Indeed, as the authors report (Jonassen *et al.*, 2000), the entire 5' end of this genome closely resembles MAL, while the 3' end is most related to subtype H sequences. Interestingly, there is a short stretch around position 7200 in the env gene where MAL clusters with subtype A rather than with D, the subtype that it is most similar to according to the similarity plot. The similarity to subtype D does dip to below 70% here, which is low even for env, and indicates that the region doesn't fit well with this subtype. When we BLAST this region against the HIV database, it turns out to be most similar to subtype B. However, when subtype B is then included in the analysis, it does not come out as most similar to MAL in this region, nor does MAL ever cluster with it in the bootscan analysis. The conclusion must be that this is either an accidental similarity to subtype B (subtypes B and D are more similar than any other

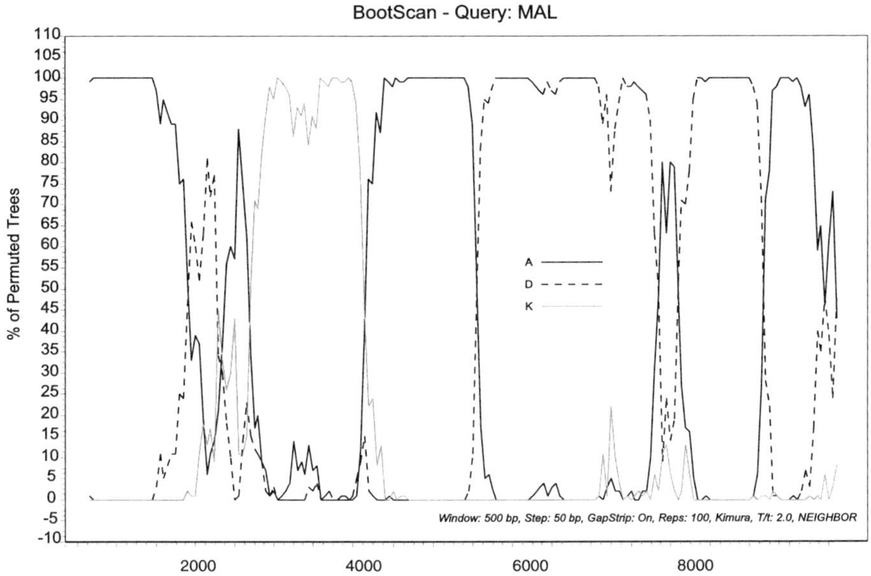

Figure 9. Results of a bootscanning analysis of the query sequence against subtype A, D and K. Scores on the Y axis represent the percentage of times the query sequence clusters with a representative sample of subtype A, D and K sequences. As in figure 8, the window size was set at 500.

subtype) or, much less likely, another real recombination event involving a region that is too short to show up in the analysis. Either way, there is insufficient reason to decide that this region is not subtype D. Figure 10 shows the probable mosaic pattern of HIV-1 MAL on the basis of the analyses carried out here.

Once the recombination points have been established, one should confirm that the regions suggested by the recombination tool are derived from different subtypes by reconstruction of phylogenetic trees of each such distinct region.

A recent addition to the subtyping arsenal is a web site at NCBI (*http://www.ncbi.nlm.nih.gov/retroviruses/HIV1/subtypeHV1.html).* Again, it has an

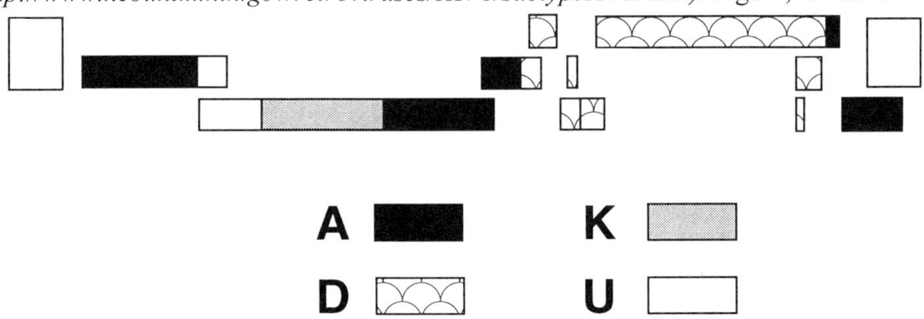

Figure 10. Mosaic structure of MAL, as determined on the basis of the analyses performed in this chapter.

easy to use cut-and-paste interface. It uses a sliding window technique as well, but rather than calculating the similarity to a set of reference sequences, it calculates BLAST scores against a small number of reference sequences of various subtypes. These scores are then plotted in a way similar to RIP. This tool has the advantage that the query sequence doesn't have to be aligned; BLAST finds the region with the highest homology and returns the score for that. An example of the output from this web site is presented in figure 11. It should be noted that this tool might give very misleading results in cases where the query sequence has large inserts or deletions; it should only be used for exploratory work, and should always be followed up by analyses based on aligned sequences.

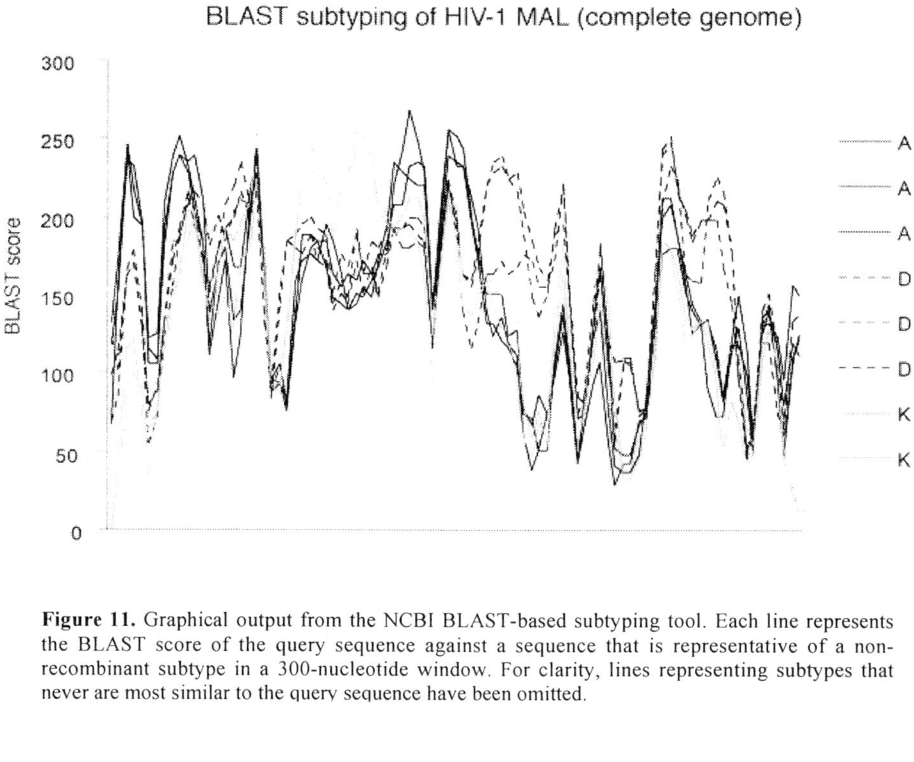

Figure 11. Graphical output from the NCBI BLAST-based subtyping tool. Each line represents the BLAST score of the query sequence against a sequence that is representative of a non-recombinant subtype in a 300-nucleotide window. For clarity, lines representing subtypes that never are most similar to the query sequence have been omitted.

REFERENCES

Abebe, A., Demissie, D., Goudsmit, J., Brouwer, M., Kuiken, C. L., Pollakis, G., Schuitemaker, H., Fontanet, A. L. and Rinke de Wit, T. F. 1999. HIV-1 subtype C syncytium- and non-syncytium-inducing phenotypes and coreceptor usage among Ethiopian patients with AIDS. *AIDS* **13**: 1305-1311.

Altschul, S. F., Gish, W., Miller, W., Myers, E. W. and Lipman, D. J. 1990. Basic local alignment search tool. *J. Mol. Biol.* **215**: 403-410.

Altschul, S. F., Madden, T. L., Schaffer, A. A., Zhang, J., Zhang, Z., Miller, W. and Lipman, D. J. 1997 Gapped BLAST and PSI-BLAST: a new generation of protein database search programs. *Nucleic Acids Res.* **25**: 3389-3402.

anonymous 1997. HIV-1 subtypes: implications for epidemiology, pathogenicity, vaccines and diagnostics. Workshop Report from the European Commission (DG XII, INCO-DC) and the Joint United Nations Programme on HIV/AIDS. *AIDS* **11**: 17-36.

Barin, F., Lahbabi, Y., Buzelay, L., Lejeune, B., Baillou-Beaufils, A., Denis, F., Mathiot, C., M'Boup, S., Vithayasai, V., Dietrich, U. and Goudeau, A. 1996. Diversity of antibody binding to V3 peptides representing consensus sequences of HIV type 1 genotypes A to E: an approach for HIV type 1 serological subtyping. *AIDS Res. Hum. Retroviruses* **12**: 1279-1289.

Bieniasz , P. D., Fridell, R. A., Aramori, I., Ferguson, S. S., Caron, M. G. and Cullen, B. R. 1997. HIV-1-induced cell fusion is mediated by multiple regions within both the viral envelope and the CCR-5 co-receptor. *EMBO J.* **16**: 2599-2609.

Carr, J. K., Salminen, M. O., Koch, C., Gotte, D., Artenstein, A. W., Hegerich, P. A., St Louis, D., Burke, D. S. and McCutchan, F. E. 1996. Full-length sequence and mosaic structure of a human immunodeficiency virus type 1 isolate from Thailand. *J. Virol.* **70**: 5935-5943.

Carr, J. K., Foley, B. T. F., Leitner, T., Salminen, M., Korber, B. T. M. and McCutchan, F. 1998a. Reference sequences representing the principal genetic diversity of HIV-1 in the pandemic, In *Human Retroviruses and AIDS 1998: A Compilation and Analysis of Nucleic Acid and Amino Acid Sequences* (Korber, B., Kuiken, C. L., Foley, B., Hahn, B., McCutchan, F., Mellors, J. and Sodroski, J., eds), Los Alamos National Laboratory, Los Alamo, NM.

Carr, J. K., Salminen, M. O., Albert, J., Sanders-Buell, E., Gotte, D., Birx, D. L. and McCutchan, F. E. 1998b. Full genome sequences of Human Immunodeficiency Virus type 1 subtypes G and A/G intersubtype recombinants. *Virology* **249**: 22-31.

Cheingsong-Popov, R., Osmanov, S., Pau, C. P., Schochetman, G., Barin, F., Holmes, H., Francis, G., Ruppach, H., Dietrich, U., Lister, S. and Weber, J. 1998. Serotyping of HIV type 1 infections: definition, relationship to viral genetic subtypes, and assay evaluation. UNAIDS Network for HIV-1 Isolation and Characterization. *AIDS Res. Hum. Retroviruses* **14**: 311-318.

Clavel, F., Guyader, M., Guetard, D., Salle, M., Montagnier, L., Alizon, M. 1986. Molecular cloning and polymorphism of the human immune deficiency virus type 2. *Nature* **324**: 691-695.

Cocchi, F., DeVico, A. L., Garzino-Demo, A., Cara, A., Gallo, R. C. and Lusso, P. 1996. The V3 domain of the HIV-1 gp120 envelope glycoprotein is critical for chemokine-mediated blockade of infection. *Nat. Med.* **2**: 1244-1247.

De Jong, J. J., De Ronde, A., Keulen, W., Tersmette, M. and Goudsmit, J. 1992. Minimal requirements for the human immunodeficiency virus type 1 V3 domain to support the syncytium-inducing phenotype: analysis by single amino acid substitution. *J.Virol.* **66**: 6777-6780.

De Leys, R., Vanderborght, B., Vanden Haesevelde, M., Heyndrickx, L., van Geel, A., Wauters, C., Bernaerts, R., Saman, E., Nijs, P., Willems, B., Taelman, H., Van der Groen, G., Piot, P., Tersmette, T., Huisman, J.G. and Van Heuverswyn, H. 1990. Isolation and partial characterization of an unusual human immunodeficiency retrovirus from two persons of west-central African origin. *J. Virol.* **64**: 1207-1216.

De Wolf, F., Hogervorst, E., Goudsmit, J., Fenyo, E.M., Rubsamen-Waigmann, H., Holmes, H., Galvao-Castro, B., Karita, E., Wasi, C., Sempala, S.D., Baan, E., Zorgdrager, F., Lukashov, V., Osmanov, S., Kuiken, C., Cornelissen, M. and the WHO Network for HIV Isolation and Characterization. 1994. Syncytium-inducing and non-syncytium-inducing capacity of human immunodeficiency virus type 1 subtypes other than B: phenotypic and genotypic characteristics. WHO Network for HIV Isolation and Characterization. *AIDS Res. Hum. Retroviruses* **10**: 1387-1400.

Diaz, R. S., Sabino, E. C., Mayer, A., Mosley, J. W. and Busch, M. P. 1995. Dual human immunodeficiency virus type 1 infection and recombination in a dually exposed transfusion recipient. The Transfusion Safety Study Group. *J. Virol.* **69**: 3273-3281.

Dittmar, M. T., Simmons, G., Hibbitts, S., O'Hare, M., Louisirirotchanakul, S., Beddows, S., Weber, J., Clapham, P. R.and Weiss, R. A. 1997. Langerhans cell tropism of human immunodeficiency virus type 1 subtype A through F isolates derived from different transmission groups. *J. Virol.* **71**: 8008-8013.

Eddy, S. R., Mitchison, G. and Durbin, R. 1995. Maximum discrimination hidden Markov models of sequence consensus. *J. Comput. Biol.* **2**: 9-23.

Eddy, S. R. 1995. Multiple alignment using hidden Markov models. *ISMB* **3**: 114-120.

Essex, M., Soto-Ramirez, L. E., Renjifo, E., Wang, W. K. and Lee, T.H. 1997. Genetic variation within human immunodeficiency viruses generates rapid changes in tropism, virulence, and transmission. *Leukemia* **11** Suppl 3: 93-94.

Felsenstein, J. 1984. PHYLIP: Phylogeny Inference Package, V3.52c Seattle, Wa., University of Washington, 1984.

Fukasawa, M., Miura, T., Hasegawa, A., Morikawa, S., Tsujimoto, H., Miki, K., Kitamura, T. and Hayami, M. 1988. Sequence of simian immunodeficiency virus from African green monkey, a new member of the HIV/SIV group. *Nature* **333**: 457-461.

Gao, F., Yue, L., White, A. T., Pappas, P. G., Barchue, J., Hanson, A. P., Greene, B. M., Sharp, P. M., Shaw, G. M. and Hahn, B. H. 1992. Human infection by genetically diverse SIVsm-related HIV-2 in West Africa. *Nature* **358**: 495-499.

Gao, F., Yue, L., Robertson, D. L., Hill, S. C., Hui, H., Biggar, R. J., Neequaye, A. E., Whelan, T. M., Ho, D. D., Shaw, G. M., Beddows, S. Weber, J., Sharp, P. N., Shaw, G. M., Hahn, B.H. and the WHO and NIAID Networks for HIV Isolation and Characterization. 1994. Genetic diversity of human immunodeficiency virus type 2: evidence ,for distinct sequence subtypes with differences in virus biology. *J. Virol.* **68**: 7433-7447.

Gao, F., Morrison, S.G., Robertson, D.L., Thornton, C.L., Craig, S., Karlsson, G., Sodroski, J., Morgado, M., Galvao-Castro, B., von Briesen, H., Beddows, S, Weber, J, Sharp, P, Shaw, G, Hahn, B, and the WHO and NIAID Networks for HIV Isolation and Characterization. 1996a. Molecular cloning and analysis of functional envelope genes from human immunodeficiency virus type 1 sequence subtypes A through G. *J. Virol.* **70**: 1651-2667.

Gao, F., Robertson, D.L., Morrison, S.G., Hui, H., Craig, S., Decker, J., Fultz, P.N., Girard, M., Shaw, G.M., Hahn, B.H. and Sharp, P.M. 1996b. The heterosexual human immunodeficiency virus type 1 epidemic in Thailand is caused by an intersubtype (A/E) recombinant of African origin. *J. Virol.* **70**: 7013-7029.

Gao, F., Bailes, E., Robertson, D. L., Chen, Y., Rodenburg, C. M., Michael, S. F., Cummins, L. B., Arthur, L. O., Peeters, M., Shaw, G. M., Sharp, P. M. and Hahn, B. H. 1999. Origin of HIV-1 in the chimpanzee *Pan troglodytes troglodytes*. *Nature* **397**: 436-441.

Gunther, S., Sommer, G., Plikat, U., Iwanska, A., Wain-Hobson, S., Will, H. and Meyerhans, A. 1997. Naturally occurring hepatitis B virus genomes bearing the hallmarks of retroviral G-->A hypermutation. *Virology* **235**: 104-108.

Gurtler, L. G., Hauser, P. H., Eberle, J., von Brunn, A., Knapp, S., Zekeng, L., Tsague, J. M. and Kaptue, L. 1994. A new subtype of human immunodeficiency virus type 1 (MVP-5180) from Cameroon. *J. Virol.* **68**: 1581-1585.

Hahn, B. H., Shaw, G. M., Arya, S. K., Popovic, M., Gallo, R. C. and Wong-Staal, F. 1984. Molecular cloning and characterization of the HTLV-III virus associated with AIDS. *Nature* **312**: 166-169.

Higgins, D. G., Thompson, J. D. and Gibson, T. J. 1996. Using CLUSTAL for multiple sequence alignments. *Methods Enzymol* .**266**: 383-402.

Hillis, D. M. and Bull, J. J. 1993. An empirical test of bootstrapping as a method of assessing confidence in phylogenetic analysis. *Syst. Biol.* **42**: 182-189.

Hoelscher, M., Hanker, S., Barin, F., Cheingsong-Popov, R., Dietrich, U., Jordan-Harder, B., Olaleye, D., Nagele, E., Markuzzi, A., Mwakagile, D., Minja, F., Weber, J., Gurtler, L. and Von Sonnenburg, F. 1998. HIV type 1 V3 serotyping of Tanzanian samples: probable reasons for mismatching with genetic subtyping. *AIDS Res. Hum. Retroviruses* **14**: 139-149.

Huang, Y., Paxton, W. A., Wolinsky, S. M., Neumann, A. U., Zhang, L., He, T., Kang, S., Ceradini, D. , Jin, Z., Yazdanbakhsh, K., Kunstman, K., Erickson, D., Dragon, E., Landau, N. R., Phair, J., Ho, D. D. and Koup, R. A. 1996. The role of a mutant CCR5 allele in HIV-1 transmission and disease progression. *Nat. Med.* **2**: 1240-1243.

Janssens, W., Heyndrickx, L., Van der Auwera, G., Nkengasong, J., Beirnaert, E., Vereecken, K., Coppens, S., Willems, B., Fransen, K., Peeters, M., Ndumbe, P., Delaporte, E. and van der Groen, G. 1999. Interpatient genetic variability of HIV-1 group O. *AIDS* **13**: 41-48.

Jonassen, T. O., Stene-Johansen, K., Berg, E. S., Hungnes, O., Lindboe, C. F., Froland, S. S. and Grinde, B. 1997. Sequence analysis of HIV-1 group O from Norwegian patients infected in the 1960s. *Virology* **231**: 43-47.

Jonassen , T. O., Grinde, B., Asjo, B., Hasle, G. and Hungnes, O. 2000. Intersubtype recombinant HIV type 1 involving HIV-MAL-like and subtype H-like sequence in four Norwegian cases. *AIDS Research Hum. Retrovir.* **16**: 49-58.

Kalish, M. L., Luo, C. C., Weniger, B. G., Limpakarnjanarat, K., Young, N., Ou, C. Y. and Schochetman, G. 1994. Early HIV type 1 strains in Thailand were not responsible for the current epidemic. *AIDS Res. Hum. Retroviruses* **10**: 1573-1575.

Kalish, M. L., Baldwin, A., Raktham, S., Wasi, C., Luo, C. C., Schochetman, G., Mastro, T. D., Young, N., Vanichseni, S., Rubsamen-Waigmann, H., von Briesen, H., Mullins, J. I., Delwart, E., Herring, B., Esparza, J., Heyward, W. L. and Osmanov, S. 1995. The evolving molecular epidemiology of HIV-1 envelope subtypes in injecting drug users in Bangkok, Thailand: implications for HIV vaccine trials. *AIDS* **9**: 851-857.

Kirchhoff, F., Greenough, T. C., Hamacher, M., Sullivan, J. L. and Desrosiers, R. C. 1997. Activity of human immunodeficiency virus type 1 promoter/TAR regions and tat1 genes derived from individuals with different rates of disease progression. *Virology* **232**: 319-331.

Korber, B. T., Osmanov, S., Esparza, J. and Myers, G. 1994a. The World Health Organization Global Programme on AIDS proposal for standardization of HIV sequence nomenclature. WHO Network for HIV Isolation and Characterization. *AIDS Res. Hum. Retroviruses* **10**: 1355-1358.

Korber, B. T., MacInnes, K., Smith, R. F. and Myers, G. 1994b. Mutational trends in V3 loop protein sequences observed in different genetic lineages of human immunodeficiency virus type 1. *J. Virol.* **68**: 6730-6744.

Korber, B. T., Allen, E. E., Farmer, A. D. and Myers, G. L.1995. Heterogeneity of HIV-1 and HIV-2. *AIDS* **9** Suppl A: S5-18.

Korber, B., Muldoon, M., Theiler, J., Gao, F., Gupta, R., Lapedes, A., Hahn, B. H., Wolinsky, S., and Bhattacharya, T. 2000. Timing the ancestor of the HIV-1 pandemic strains. *Science* **288**: 1789-1796.

Kostrikis, L. G., Bagdades, E., Cao, Y., Zhang, L., Dimitriou, D. and Ho, D. D. 1995. Genetic analysis of human immunodeficiency virus type 1 strains from patients in Cyprus: identification of a new subtype designated subtype I. *J. Virol.* **69**: 6122-6130.

Kuiken, C. L., Foley, B. T. and Korber, B. T. 1998. Determinants of HIV-1 protein evolution, In *Molecular Evolution of HIV.* (Crandall, K., ed), Johns Hopkins University Press, Baltimore, MD.

Learn, G. H., Jr.., Korber, B. T., Foley, B., Hahn, B. H., Wolinsky, S. M. and Mullins, J. I. 1996. Maintaining the integrity of human immunodeficiency virus sequence databases. *J. Virol.* **70**: 5720-5730.

Leitner, T., Escanilla, D., Franzen, C., Uhlen, M. and Albert, J. 1996. Accurate reconstruction of a known HIV-1 transmission history by phylogenetic tree analysis. *Proc. Natl Acad. Sci. USA* **93**: 10864-10869.

Leitner, T., Kumar, S. and Albert, J. 1997. Tempo and mode of nucleotide substitutions in gag and env gene fragments in human immunodeficiency virus type 1 populations with a known transmission history. *J. Virol.* **71**: 4761-4770.

Leitner, T., and Albert, J. 1999. The molecular clock of HIV-1 unveiled through analysis of a known transmission history. *Proc. Natl. Acad. Sci. USA* **96**:10752-10757.

Li, W. H., Tanimura, M. and Sharp, P. M. 1988. Rates and dates of divergence between AIDS virus nucleotide sequences. *Mol. Biol. Evol.* **5**: 313-330.

Loussert-Ajaka, I., Chaix, M. L., Korber, B., Letourneur, F., Gomas, E., Allen, E., Ly, T. D., Brun-Vezinet, F., Simon, F. and Saragosti, S. 1995. Variability of human immunodeficiency virus type 1 group O strains isolated from Cameroonian patients living in France. *J. Virol.* **69**: 5640-5649.

Louwagie, J., McCutchan, F. E., Peeters, M., Brennan, T. P., Sanders-Buell, E., Eddy, G. A., van der Groen, G., Fransen, K., Gershy-Damet, G. M., Deleys, R. and Burke, D. S. 1993. Phylogenetic analysis of gag genes from 70 international HIV-1 isolates provides evidence for multiple genotypes. *AIDS* **7**: 769-780.

Louwagie, J., Janssens, W., Mascola, J., Heyndrickx, L., Hegerich, P., van der Groen, G., McCutchan, F. E. and Burke, D. S. 1995. Genetic diversity of the envelope glycoprotein from human immunodeficiency virus type 1 isolates of African origin. *J. Virol.* **69**: 263-271.

Lu, Z., Berson, J. F., Chen, Y., Turner, J. D., Zhang, T., Sharron, M., Jenks, M. H., Wang, Z., Kim, J., Rucker, J., Hoxie, J. A., Peiper, S. C. and Doms, R. W. 1997. Evolution of HIV-1 coreceptor usage through interactions with distinct CCR5 and CXCR4 domains. *Proc. Natl Acad. Sci. USA* **94**: 6426-6431.

Lukashov, V. V., Cornelissen, M. T., Goudsmit, J., Papuashvilli, M. N., Rytik, P. G., Khaitov, R. M., Karamov, E. V. and de Wolf, F. 1995. Simultaneous introduction of distinct HIV-1 subtypes into different risk groups in Russia, Byelorussia and Lithuania. *AIDS* **9**: 435-439.

McCutchan, F. E., Carr, J. K., Bajani, M., Sanders-Buell, E., Harry, T. O., Stoeckli, T. C., Robbins, K. E., Gashau, W., Nasidi, A., Janssens, W. and Kalish, M. L. 1999. Subtype G and multiple forms of A/G intersubtype recombinant human immunodeficiency virus type 1 in Nigeria. *Virology* **254**: 226-234.

Montano, M. A., Novitsky, V. A., Blackard, J. T., Cho, N. L., Katzenstein, D. A. and Essex, M. 1997. Divergent transcriptional regulation among expanding human immunodeficiency virus type 1 subtypes. *J. Virol.* **71**: 8657-8665.

Myers, G. 1987. Consensus trees for complete HIV-1 genomic sequences, In *Human Retroviruses and AIDS* (Myers, G., Josephs, S. F., Rabson, A. B. and Smith, T. F., eds), Los Alamos National Laboratory, Los Alamos, NM..

Myers, G. 1992. HIV-1 and HIV-2 sequence subtypes, In *Human Retroviruses and AIDS 1992*. (Myers, G., Korber, B., Berzofsky, J. A. and Smith, R. A., eds) Los Alamos National Laboratory, Los Alamos, NM.

Myers, G. 1993. HIV-1 sequence subtypes and phylogenetic trees, In *Human Retroviruses and AIDS 1993* (Myers, G., Korber, B., Wain-Hobson, S. and Smith, R. A., eds), Los Alamos National Laboratory, Los Alamos, NM.

Nasioulas, G., Paraskevis, D., Paparizos, V., Lazanas, M., Karafoulidou, A. and Hatzakis, A. 1998. Genotypic characterization of human immunodeficiency virus type 1 in Greece. Multicentre Study on HIV-1 Heterogeneity. *AIDS Res. Hum. Retroviruses* **14**: 685-690.

Olsen, G. J., Matsuda, H., Hagstrom, R. and Overbeek, R. 1994. fastDnamL: a tool for construction of phylogenetic trees of DNA sequences using maximum likelihood. *Comput. Appl. Biosci.* **10**: 41-48.

Ou, C. Y., Takebe, Y., Weniger, B. G., Luo, C. C., Kalish, M. L., Auwanit, W., Yamazaki, S., Gayle, H. D., Young, N. L. and Schochetman, G. 1993. Independent introduction of two major HIV-1 genotypes into distinct high-risk populations in Thailand. *Lancet* **341**: 1171-1174.

Paxton, W. A., Dragic, T., Koup, R. A. and Moore, J. P. 1996. The beta-chemokines, HIV type 1 second receptors, and exposed uninfected persons. *AIDS Res. Hum. Retroviruses* **12**: 1203-1207.

Pope, M., Ho, D. D., Moore, J. P., Weber, J., Dittmar, M. T. and Weiss, R. A. 1997. Different subtypes of HIV-1 and cutaneous dendritic cells. *Science* **278**: 786-788.

Potts, K. E., Kalish, M. L., Bandea, C. I., Orloff, G. M., St. Louis, M., Brown, C., Malanda, N., Kavuka, M., Schochetman, G., Ou, C.-Y. and Heyward, W. L. 1993. Genetic diversity of human immunodeficiency virus type 1 strains in Kinshasa, Zaire. *AIDS Res. Hum. Retroviruses* **9**: 613-618.

Quinones-Mateu, M. E., Albright, J. L., Mas, A., Soriano, V. and Arts, E. J. 1998. Analysis of pol gene heterogeneity, viral quasispecies, and drug resistance in individuals infected with group O strains of human immunodeficiency virus type 1. *J. Virol.* **72**: 9002-9015.

Robertson, D. L., Gao, F., Hahn, H. B. and Sharp, P. M. 1997. Intersubtype recombinant HIV-1 sequences, In *Human Retroviruses and AIDS 1997* (Korber, B. T. M., Hahn ,B. H., Foley, B. T. F., Mellors, J. W., Leitner, T. K., Myers, G., McCutchan, F. E. and Kuiken, C. L., eds), Los Alamos National Laboratory, Los Alamos, NM.

Robertson, D. L., Anderson, J. P., Bradac, J. A., Carr, J. K., Foley, B., Funkhouser, R. K., Gao, F., Hahn, B. H., Kalish, M. L., Kuiken, C., Learn, G. H., Leitner, T., McCutchan, F., Osmanov, S., Peeters, M., Pieniazek, D., Salminen, M., Sharp, P. M., Wolinsky, S. and Korber, B. 2000. HIV-1 nomenclature proposal. Science. **288**:55-56.

Salminen, M. O., Carr, J. K., Burke, D. S. and McCutchan, F. E. 1995. Genotyping of HIV-1, In *Human Retroviruses and AIDS 1995* (Myers, G., Korber, B., Hahn, B. H., Jeang, K.-T., Mellors, J. W., McCutchan, F. E., Henderson, L. E. and Pavlakis, G. N., eds), Los Alamos National Laboratory, Los Alamos, NM.

Salminen, M. O., Carr, J. K., Burke, D. S. and McCutchan, F. E. 1995b. Identification of breakpoints in intergenotypic recombinants of HIV type 1 by bootscanning. *AIDS Res Hum Retroviruses* **11**: 1423-1425.

Sharp, P. M. and Li, W. H. 1988. Understanding the origins of AIDS viruses. *Nature* **336**: 315.

Sharp, P. M., Robertson, D. L., Gao, F., Hahn, B. H. 1994. Origins and diversity of human immunodeficiency viruses. *AIDS* **8** (suppl. 1): S27-S42.

Sharp, P. M., Robertson, D. L. and Hahn, B. H. 1995. Cross-species transmission and recombination of 'AIDS' viruses. *Philos. Trans. R. Soc. Lond. B. Biol. Sci.* **349**: 41-47.

Sherefa, K., Sonnerborg, A., Steinbergs, J. and Sallberg, M. 1994. Rapid grouping of HIV-1 infection in subtypes A to E by V3 peptide serotyping and its relation to sequence analysis. *Biochem. Biophys. Res. Commun.* **205**: 1658-1664.

Siepel, A. C., Halpern, A. L., Macken, C. and Korber, B. T. 1995. A computer program designed to screen rapidly for HIV type 1 intersubtype recombinant sequences. *AIDS Res. Hum. Retroviruses* **11**: 1413-1416.

Simon, F., Mauclere, P., Roques, P., Loussert-Ajaka, I., Müller-Trutwin, M. C., Saragosti, S., Georges-Courbot, M. C., Barre-Sinoussi, F. and Brun-Vezinet, F. 1998. Identification of a new human immunodeficiency virus type 1 distinct from group M and group O. *Nat. Med.* **4**: 1032-1037.

Smith, T. F., Srinivasan, A., Schochetman, G., Marcus, M. and Myers, G. 1988. The phylogenetic history of immunodeficiency viruses. *Nature* **333**: 573-575.

Soto-Ramirez, L. E., Renjifo, B., McLane, M. F., Marlink, R., O'Hara, C., Sutthent, R., Wasi, C., Vithayasai, P., Vithayasai, V., Apichartpiyakul, C., Auewarakul, P., Pena Cruz, V., Chui, D. S., Osathanondh, R., Mayer, K., Lee, T. H. and Essex, M. 1996. HIV-1 Langerhans' cell tropism associated with heterosexual transmission of HIV. *Science* **271**: 1291-1293.

Srinivasan, A., York, D., Ranganathan, P., Ferguson, R., Butler, D., Jr., Feorino, P., Kalyanaraman, V., Jaffe, H., Curran, J. and Anand, R. 1987. Transfusion-associated AIDS: donor-recipient human immunodeficiency virus exhibits genetic heterogeneity. *Blood* **69**: 1766-1770.

Swofford, D.L. 1991. PAUP: Phylogenetic Analysis Using Parsimony, version 3.1.1. Illinois Natural History Survey, Champaign, Ill.

Triques, K., Bourgeois, A., Saragosti, S., Vidal, N., Mpoudi-Ngole, E., Nzilambi, N., Apetrei, C., Ekwalanga, M., Delaporte, E. and Peeters, M. 1999. High diversity of HIV-1 subtype F strains in Central Africa. *Virology* **259**: 99-109.

Triques, K., Bourgeois, A., Vidal, N., Mpoudi-Ngole, E., Mulanga-Kabeya, C., Nzilambi, N., Torimiro, N., Saman, E., Delaporte, E. and Peeters, M. 2000. Near-full-length genome sequencing of divergent African HIV type 1 subtype F viruses leads to identification of a new HIV-1 subtype designated K. *AIDS Res. Hum. Retroviruses* **16**: 139-151.

Tscherning, C., Alaeus, A., Fredriksson, R., Bjorndal, A., Deng, H., Littman, D. R., Fenyo, E. M., Albert, J. 1998. Differences in chemokine coreceptor usage between genetic subtypes of HIV-1. *Virology* **241**: 181-188.

Ubolyam, S., Ruxrungtham, Sirivichayakul, S., Okuda, K. and Phanuphak, P. 1994. Evidence of three HIV-1 subtypes in subgroups of individuals in Thailand. *Lancet* **344**: 485-486.

van Harmelen, J., Wood, R., Lambrick, M., Rybicki, E. P., Williamson, A. L., Williamson, C. 1997. An association between HIV-1 subtypes and mode of transmission in Cape Town, South Africa. *AIDS* **11**: 81-87.

Vartanian, J. P., Meyerhans, A., Asjo, B. and Wain-Hobson, S. 1991. Selection, recombination, and G->A hypermutation of human immunodeficiency virus type 1 genomes. *J. Virol.* **65**: 1779-1788.

Wain-Hobson, S., Sonigo, P., Guyader, M., Gazit, A. and Henry, M. 1995. Erratic G->A hypermutation within a complete caprine arthritis- encephalitis virus (CAEV) provirus. *Virology* **209**: 297-303.

Zagury, J. F., Franchini, G., Reitz, M., Collalti, E., Starcich, B., Hall, L., Fargnoli, K., Jagodzinski, L., Guo, H. G., Laure, F., Arya, S. K., Josephs, S., Zagury, D., Wong-Staal, F. and Gallo, R. C. 1988. Genetic variability between isolates of human immunodeficiency virus (HIV) type 2 is comparable to the variability among HIV type 1. *Proc. Natl Acad. Sci. USA* **85**: 5941-5945.

Zhang, L., Huang, Y., He, T., Cao, Y. and Ho, D. D. 1996. HIV-1 subtype and second-receptor use. *Nature* **383**: 768.

Zhu, T., Wang, N., Carr, A., Wolinsky, S. and Ho, D. D. 1995. Evidence for coinfection by multiple strains of human immunodeficiency virus type 1 subtype B in an acute seroconvertor. *J. Virol.* **69**: 1324-1327.

Zhu, T., Korber, B. T., Nahmias, A. J., Hooper, E., Sharp, P. M. and Ho, D. D. 1998. An African HIV-1 sequence from 1959 and implications for the origin of the epidemic. *Nature* **391**: 594-597.

HIV SEQUENCE SIGNATURES AND SIMILARITIES

Bette Korber

Theoretical Biology and Biophysics Group,
Los Alamos National Laboratory Los Alamos, NM 87545 USA

1. INTRODUCTION

A fundamental use of sequence similarities is to screen GenBank and other sequence databases in search of homologues and motifs that retain similar features to the protein or gene under study (Myers and Farmer, 1996). A further use for broad sequence database searches, particularly for HIV studies, is to reveal contaminations with HIV laboratory strains among sequences that would otherwise be assumed to be valid. While search tools like BLAST (*http://www.ncbi.nlm.nih.gov/BLAST* (Altschul *et al.*, 1990; 1997) were not designed with this specific application in mind, such tools can be used extremely efficiently to add a critical level of quality control to HIV sequencing studies (Kuiken and Korber, 1998).

There are also many biologically interesting applications of signature analysis for HIV (which could, by analogy, be applied to other rapidly evolving infectious agents) that depend upon comparisons of one alignment of HIV sequences with defined characteristics to another with contrasting characteristics. One can consider whether particular patterns in sequence variation are associated with transmission patterns and risk factors on an epidemic scale. Or one can focus down on a specific transmission event. Signature pattern analysis, used in conjunction with epidemiological evidence and phylogenetic analysis, was applied to the transmission study of the case of the Florida dentist and six of his patients. The patients were found to be infected by a virus that was highly related to the virus infecting the dentist (Ou *et al.*, 1992; Korber and Myers, 1992; Ciesielski *et al.,* 1992). People have also looked for viral signatures related to tissue tropism, particularly in viral reservoirs that are distinct and have biological barriers that inhibit free traffic of virus, for example the peripheral blood in contrast to the central nervous system

(Korber *et al.*, 1994; Kuiken *et al.*, 1995; van 't Wout *et al.*, 1998) or reproductive tract (Delwart *et al.*, 1998). Rather than focusing exclusively on amino acid changes to distinguish these sets, some studies have incorporated methods that define positions that are conserved in one set relative to another (Korber *et al.*, 1994; Kuiken *et al.*, 1995). Using this strategy, we defined a brain-conserved viral sequence signature that shared features of "macrophage-tropic" strains (Korber *et al.*, 1994). (Macrophage-tropic strains were found in brain and blood, and T-cell tropic forms were found only in the blood.) This result has been misinterpreted in the literature. For example, seven of eight signature amino acids from brain-derived viruses that we described in 1994 are evident in the alignments of CNS-derived sequences presented by van 't Wout and her colleagues (1998). Yet it is stated in the text that there was no evidence of the CNS-associated viral signature pattern we had previously defined (Korber *et al.*, 1994). (Perhaps this could be viewed as a question of whether the glass is 7/8 full, or 1/8 empty.) Signature patterns can be applied to studies of biological phenotypes. Probably the most well known example of an HIV signature pattern that strongly correlates with a biological characteristic is the pattern associated with the syncytium-inducing (SI) phenotype of the virus. The V3 loop (a stretch of protein sequence in the third variable domain of the envelope glycoprotein gp120, bounded by a Cys-Cys disulfide bond) has a signature of the positively charged amino acids in two particular positions that are associated with the SI phenotype (Fouchier *et al.*, 1992; De Wolf *et al.*, 1994). The NSI and SI phenotypes reflect viral HIV chemokine receptor usage (Choe *et al.*, 1996; Deng *et al.*, 1996; CCR5 and CXCR4, respectively), critical for viral entry (for reviews, see Doms and Moore, 19970, and Fenyo *et al.*, 1997), which in turn dictates the susceptibility of particular cell-types to infection. The chemokine receptor usage, CCR5 versus CXCR4, as would be expected, has similar amino acid signatures as the NSI versus SI phenotype (Speck *et al.*, 1997; Xiao *et al.*, 1998). But as is typical for biological phenomena, a simple answer is not the complete answer, and correlations of substitutions in certain positions in the V3 loop can not fully explain the complexity of viral co-receptor usage. For example the V1 and V2 loops have been shown experimentally to also be involved, and CCR5 and CXCR4 are only two of several chemokine receptors that the virus can use for entry (Hoffman *et al.*, 1998; Ross and Cullen, 1998; Doms and Moore, 1997)

The following sections describe some results in the literature pertaining to the applications described above, computational methods, where to find and how to use different tools, and examples. The examples shown here to illustrate the methods will all be based on the HIV protease protein and its coding region within the *pol* gene. Protease was selected to illustrate concepts in this review in part because it is a small protein, 99 amino acids long, and as such it is simple to consider. The main reason for selecting protease, however, is that it is of great interest as a therapeutic target, and studies of drug resistance may be particularly amenable to signature analysis. Drug resistance mutational patterns are complex, requiring multiple substitutions. For a class of drugs there can be many mutational routes for viral acquisition of resistance, (Hammond *et al.*, 1997; Condra, 1998; the Los Alamos National Laboratory drug resistance database: *http://hiv-web.lanl.gov/*; the Stanford University drug resistance database: *http://hivdb.stanford.edu/hiv/*), some of which tend to arise sequentially (Molla *et al.*, 1996; Craig *et al.*, 1998). Some of these mutations directly inhibit drug binding to protease, while others are compensatory mutations for those that affect the active site of the enzyme (Schock *et al.,* 1996). Despite this complexity, there are obvious ways to divide sequences into defined sets for study, related to therapy, such as the following three cases: (a) sequences amplified directly from blood samples, both before and after implementation of therapy; (b) sequences from drug-resistant versus non-resistant strains as defined by viral culture; or (c) viral sequences obtained pre-therapy and stratified according to outcome. In the third case, sequences could be grouped together from individuals that eventually responded well to therapy, and compared to sequences from those that rapidly acquired drug resistance.

2. PROTEIN AND NUCLEIC ACID DATABASE SEARCHES

There are several reasons one might wish to search the existing nucleotide and protein databases using HIV sequences as a query. The first of these is essentially the same reason most of these searches are conducted for genes in any organism: to look for homologues, sequences with shared evolutionary history, or to look for shared amino acid motifs which may be associated with a functional domain. There are several ways to search for protein homologies, and once such a search is launched, through any one of many web-based search tools which can serve as a portal, a tremendous information network can be accessed [see the summary and links at the National Biotechnology Information Facility (NBIF) for a very useful summary of software and internet resources *http://www.nbif.org*]. The Blocks database is an excellent starting point for protein searches. Blocks searches scan sets of defined protein families with a Query sequence (Henikoff and Henikoff, 1994; *http://blocks.fhcrc.org*). Blocks are aligned, ungapped protein segments comprised of highly conserved regions, automatically extracted from groups of proteins that are documented in the Prosite database. The Prosite database contains organized sets of protein families and domains of significant sites, patterns and profiles that can allow one to quickly find associations between a query sequence and known proteins with conserved amino acid motifs and similar function (Bairoch *et al.,* 1997; *http://www.expasy.ch*). Using an HIV-1 protease consensus sequence to search the Blocks database, the block with the greatest protein similarity, not surprisingly, includes other viral proteases. One can move down through lower scoring matches and uncover aspartyl proteases from *Homo sapiens* and other organisms. Aspartyl proteases comprise the functional family to which HIV-1 protease belongs; this has been known essentially since the first HIV-1 sequences were obtained. Aspartyl proteases can be found in vertebrates, fungi, plants, retroviruses and some plant viruses. Many retroviruses encode in the *pol* gene an aspartyl protease, homodimers of a chain length of about 95 to 125 amino acids, essential to viral protein processing (Rawlings and Barrett, 1995; *http://www.expasy.ch/*). A more direct route to information about this family of proteins is through searching Prosites with key words viral and protease; the Prosite database can help reveal the essential features of aspartyl proteases (*http://www.expasy.ch/*). An intriguing observation was recently made based on similarity searching with protease amino acids. Protease inhibitors can have serious side effects in people taking them for extended periods, including lipodystrophy, hyperlipidaemia and insulin resistance (Carr *et al.,* 1998a). Following the logic that most protease inhibitors bind near the active site of protease, and that these inhibitors are also probably binding; to a human lipid regulatory protein, a search was initiated for a human protein with the protease motif. Carr and colleagues (1998b) used a 12 amino acid sequence spanning the active site or protease (KEALLDTGADDT) to query the mammalian protein and genome databases. Two interesting similarities were found, in proteins that were not aspartyl proteases, but that had similarities to the active site of protease and are involved in lipid metabolism. One similarity was with the cytoplasmic retinoic-acid binding protein (CRABP-1) and the other was lipoprotein-receptor-related protein (LRP). It has not yet been determined if protease inhibitors binding to these proteins are the cause of the side effects of the antiviral therapy, but the homology search provides a reasonable hypothesis to ultimately test experimentally. Other general searches of specific subsets of sequences in the databases (mammalian, human, a particular human chromosome, ...) can be conducted through the Blast service at the National Center for Genome Resources (Altschul *et al.,* 1997: *http://seqsim.ncgr.org/newBlast.html*); the Tentative Human Consensus (THC) Sequences from the Institute for Genomic Research (TIGR) can also be searched using BLASTn: *http://www.tigr.org/tdb/*. As mentioned earlier, for HIV there is an important pragmatic reason for database searches: to look for potential matches to lab strains, and so to identify sequences which arise from a contamination event that occurred in the course of the experimental procedures ultimately required for viral sequence generation. A review of methods that can be used to identify contamination, examples, and a web interface with interactive tools (Kuiken and Korber, 1998) can be found at *http://hiv-web.lanl.gov/*. Polymerase chain reaction (PCR) amplification is the experimental step most susceptible to contamination problems, but viral culture contamination and sample mix-up can also introduce problems (Kuiken and Korber, 1998; Frenkel *et al.,* 1998). If only a fragment of *pol* sequence is available, like the 300 bases of protease, it can be difficult to distinguish viral sequences from different individuals; the region is highly conserved and the 300 bases may not yield enough information to easily and clearly spot potential contaminations (Kuiken and Korber, 1998).

General database BLAST searches (Altschul *et al.,* 1997; *http://www.ncbi.nlm.nih.gov/BLAST*) using an HIV-1 protease sequence as a query are overwhelmed with the signal from closely related HIV proteases; there are ~450 HIV database protease sequences (*http://hiv-web.lanl.gov/*) as of this writing, and all are, naturally, highly similar to a HIV-1 protease query. Such lists can be of great use for identification of lab strain contaminations, as they are ranked according to the relative extent of the similarity. Scores also include information about how likely a sequence relationship is to have occurred by chance alone. The score of greatest interest for contamination checks is the percent identity, or frequency of perfect matches. (A BLAST search of all HIV sequences with output sorted according to this field, as opposed to a conventional BLAST output, can be performed at the HIV sequence database: *http://hiv-web.lanl.gov/*.) If the identity score between the query sequence newly generated in a study and a lab strain identified through a BLAST search is high (a rule of thumb for most regions would be >98% similar), it is time to consider carefully the possibility that the lab strain might be contaminating the sample (Frenkel *et al.,* 1998; Kuiken and Korber, 1998). The BLASTn output seen in Figure 1 is an example of such a case. The possibility of contamination would need to be considered, particularly if the sequence was recurrent in sequence sets

```
                                                                   Score     E
Sequences producing significant alignments:                        (bits)  Value

gb|K03455|HIVHXB2CG  Human immunodeficiency virus type 1 (HXB2),...   565  e-159
gb|U31398|HIV1U31398  Human immunodeficiency virus type 1, clone...   565  e-159
gb|K02012|HIVH3BH5  Human immunodeficiency virus type 1, isolate...   557  e-157
emb|X01762|REHTLV3  Human T-cell leukaemia type III (HTLV-III) p...   557  e-157

Human immunodeficiency virus type 1 (HXB2), complete genome;
            HIV1/HTLV-III/LAV reference genome
            >gi|2801501|gb|AF033819|AF033819 HIV-1 complete genome
            Length = 9719

 Score =  565 bits (285), Expect = e-159
 Identities = 294/297 (98%), Positives = 294/297 (98%)

Query: 1    cctcaggtcactctttggcaacgacccctcgtcacaataaagataggggggcaactaaag 60
            ||||||||||||||||||||||||||||||||||||||||||||||||||||||||||||
Sbjct: 2253 cctcaggtcactctttggcaacgacccctcgtcacaataaagataggggggcaactaaag 2312


Query: 61   gaagctctattagatacaggagcagaggatacagtattagaagaaatgagtttgccagga 120
            |||||||||||||||||||||||||| |||||||||||||||||||||||||||||||||
Sbjct: 2313 gaagctctattagatacaggagcagatgatacagtattagaagaaatgagtttgccagga 2372


Query: 121  agatggaaaccaaaaatgatagggggaattggaggttttatcaaagtaagacagtatgat 180
            ||||||||||||||||||||||||||||||||||||||||||||||||||||||||||||
Sbjct: 2373 agatggaaaccaaaaatgatagggggaattggaggttttatcaaagtaagacagtatgat 2432


Query: 181  aagatactcatagaaatctgtggacataaagctataggtacagtattagtaggacctaaa 240
             ||||||||||||||||||||||||||||||||||||||||||||||||||||||||| |
Sbjct: 2433 cagatactcatagaaatctgtggacataaagctataggtacagtattagtaggacctaca 2492


Query: 241  cctgtcaacataattggaagaaatctgttgactcagattggttgcactttaaatttt 297
            |||||||||||||||||||||||||||||||||||||||||||||||||||||||||
Sbjct: 2493 cctgtcaacataattggaagaaatctgttgactcagattggttgcactttaaatttt 2549
```

Figure 1 Results of an HIV-1 protease BLASTn search. A BLASTn output of a search of GenBank using a protease sequence that is very similar, but not identical, to HXB2 as the query. The three highest matching scores are shown, as well as the BLASTn alignment of HXB2, the highest match.

from several different individuals, or was very unusual compared to the other sequences from the same sample. The Expect (E) value in Figure 1, is e-159, or 10^{-159}. This indicates the probability of finding this degree of similarity by chance alone in a search of the database; of course it is vanishingly small, as the sequences are nearly identical across 297 bases. The Expect value will be similarly small for any comparison of HIV sequences to each other, and this score isn't particularly indicative of problems. While BLAST searches of GenBank are overwhelmed with the most highly related HIV sequences, BLAST searches using HIV-1 protease as a query of human sequences (excluding viral sequences) in GenBank reveal other more remote similarities. Few of these, however, are statistically significant; in other words the degree of similarity might be uncovered by chance alone, with the exception of a few proteases in a protein search. The way to draw these limited but potentially interesting similarities out of a BLAST search is through using a restricted search (humans only, for example) and setting a high E value (say 1000), so similarities that might happen by chance will be permitted. These kinds of similarities can be of interest for scanning for potential for molecular mimicry (e.g., Trujillo *et al.,* 1993), however such results must be interpreted cautiously (Myers and Farmer, 1996).

3. EPIDEMIOLOGICAL TRACKING

There are multiple examples in the literature of cases where genetically distinctive forms of the virus circulate in different risk groups within the same geographic region. Often these forms are distinguishable through phylogenetic analysis (Lukashov *et al.*, 1996; Leigh Brown *et al.*, 1997; Clemente-Estable *et al.*, 1998; Lukashov *et al.*, 1998) and sometimes infections in different risk groups are associated with different HIV subtypes (Mastro *et al.*, 1997; Kunanusont *et al.*, 1995; Liitsola *et al.*, 1998). Subtypes are genetically clearly distinctive forms of the virus that have such clear phylogenetic associations they have been given their own nomenclature, those identified to date are given letter designations, subtypes A to K (The Joint U. N. Programme on HIV AIDS, 1997; Triques *et al.*,2000; Robertson *et al.*, 2000).

For example, in South Africa, C subtype dominates in the heterosexual epidemic, and B subtype in HIV infected homosexuals (van Harmelen *et al.*,1997). In Thailand the B subtype has dominated the epidemic among IV drug users (IVDUs), while cases resulting from sexual transmission tended to be a mosaic genome of subtypes A and E (circulating recombinant form AE) (Kunanusont *et al.*,1995; Ou *et al.*,1993); this distinction, however, appears to be blurring over time, with AE increasingly observed in young drug users (Subbarao *et al.*,1998; Kitayaporn *et al.*,1998).

An intriguing signature pattern associated with IVDU transmission has been defined in HIV-1 infected European IVDUs (Kuiken and Goudsmit, 1994; Kuiken *et al.*,1996), and this pattern has been used to show that HIV in heterosexuals is associated less frequently with IVDUs, and more often with homosexuals (Lukashov *et al.*,1998). This signature was originally described as two silent nucleotide substitutions (i.e., those that do not result in amino acid change), although additional risk-associated signature nucleotides were later described (Kuiken *et al.*,1996; Lukashov *et al.*,1996). The virus containing the nucleotide associated with IV drug use threads its way through the drug user epidemic in many European countries, suggesting that the separate epidemics moving through the different at-risk groups in Europe are connected across geographic boundaries. Furthermore the IVDUs in Europe are linked to those in the USA (Lukashov *et al.*,1996). The fact that the polymorphism is silent at the protein level strongly suggests that this pattern depends upon founder effects, and has little if anything to do with functional distinctions. These signature patterns were defined with the aid of the program VESPA (Korber and Myers, 1992; *http://hiv-web.lanl.gov/*), which highlights distinctive positions in two alignments by summarizing the frequencies of nucleotides or amino acids in each position. Principal coordinate analysis was also used in these studies, a procedure used to find meaningful patterns in sequence data with no a priori knowledge about the sequences. The procedure summarizes the variation in the sequences in a limited number of dimensions. A dimension is a combination of positions in a sequence that behave similarly, and can sometimes be linked to biological properties of the viruses under study, like the risk factor for transmission in the studies by Lukashov and Kuiken described above (Kuiken *et al.*,1996; Lukashov *et al.*,1996). The method was developed by J.C. Gower in 1966 (Gower, 1992), and software to implement the approach (PCOORD) was developed by D. Higgins (1992). A web interface for this software is available at the Los Alamos HIV sequence database web site: *http://hiv-web.lanl.gov/*.

4. TRANSMISSION STUDIES

The case of the Florida dentist, who apparently infected six of his patients during invasive dental procedures (Ou *et al.*,1992) is an example of a carefully considered signature pattern analysis, used in conjunction with phylogenetic analysis and classical epidemiology, to test whether his patients were infected with his virus (Myers, 1994). In this study, phylogenetic analysis confirmed the association of the dentist's viral sequences with those of his patients relative to HIV positive individuals from the surrounding area who provided a local control set for comparison (Ou *et al.*,1992; Crandall, 1995). Careful epidemiological investigation evaluated the patients' risks of exposure and determined that there were no other likely sources of infection, for the six patients who had virus that was very similar to the dentist (Ciesielski *et al.*,1992; 1994).

Signature patterns were used in this case to develop a statistic that could be used to directly compare the local control population with the dental patients (Ou *et al.*,1992; Korber and Myers, 1992). A bootstrap value on a phylogenetic tree (Felsenstein, 1985), or a likelihood ratio test applied to branching pattern in maximum likelihood trees (Holmes *et al.*,1993), can establish the degree of certainty about a phylogenetic association, or branch point, on a tree. But for a molecular epidemiological study where transmission linkage needs to be established, it is important to take into consideration the number of individuals sampled, as well as the relative sequence relatedness, from the study subjects and from the local control population. This way one can test the possibility that the phylogenetic clusters observed may be merely a reflection of an unexpected geographic or risk-group association. In the dentist's case, first a viral signature was established for the dentist — it was defined as the particular amino acids that were conserved among the dentist's viral sequences, but rarely found in the background sequence set available in the HIV database. An important feature of the strategy was that the dentist's viral signature was defined relative to the Los Alamos database, independent of the local controls, so that a comparison between the local controls and the dentist's patients was not biased. Then the dentist's viral signature was compared in turn to the local control population's sequences, and to the HIV infected dental patients' sequences. The number of amino acids shared with the signature was determined for each study sequence. A non-parametric rank-order statistic was used to test the null hypothesis that the infected dental patients were indistinguishable from the local control population. The evidence linking the dentist to his patients was overwhelming by every measure. The signature analysis rejected the null hypothesis with a very high degree of significance, because the six infected patients with no other clear risks shared more of the signature amino acids than any of 35 local controls (Ou *et al.*,1992; Korber and Myers, 1992).

An alternative to signature analysis for a rank order statistical comparison of local controls to a putative transmission case would be a rank comparison of genetic distances, which has the advantage of using sequence data from all positions. Signatures provide a compelling way to use the most overtly distinctive features of an index case sequence, however, and can add different kinds of information for consideration. Signatures can incorporate information concerning how rare a particular signature amino acid is in a background set (Korber and Myers, 1992) as well as include insertions and deletions that are generally excluded from phylogenetic analysis. For example, for many years people have used the notorious and very distinctive "QR" insertion in the V3 loop as a flag to spot IIIB or LAI derived contaminations in sequence sets. Signatures are frequently noted in transmission studies, though generally they are used less formally than in the Florida dentist case; they are a striking form of sequence similarity that is hard to ignore (Belec *et al.*, 1998; Ahmad *et al.*, 1995; Ou *et al.*, 1993).

5. EXAMPLES BASED ON A DRUG RESISTANCE STUDY

5.1 Data and Previous Conclusions: an Indinavir Study

This section will provide illustrations of how signature analysis can be applied to a study of HIV sequences in the context of developing drug resistance. Examples are provided to illustrate the concepts, using a previously published sequence data set (Condra *et al.*, 1996) (sequence accession numbers U71606-U72026).

In the original study by Condra and colleagues (1996), 21 patients were given indinavir monotherapy, and 17 of these developed drug resistance during the study period. 421 sequences were generated from samples taken from the study subjects through the course of the study. An elegant analysis strategy was initially employed. For each of the 30 positions that varied in more than 4 patients, the ln CIC_{95} (how much drug was necessary to give 95% inhibition of viral spread in culture) was plotted as a function of the mutational frequency of each site for each patient. The mutational frequency was defined as the proportion of clones that differed from the B subtype consensus. Significantly non-zero slopes indicated a correlation between the substitutions at particular sites and resistance. A weighted mixed-effects regression was used allowing the intercept to differ for each patient, and weighting by the number of time points tested for each patient. Bonferroni's correction for multiple tests was used. This led the authors to identify variability in multiple positions in protease that were correlated with the acquisition of indinavir resistance: L10, K20, L24, M46, I54, L63, A71, V82, I84, and L90. They further noted that single substitutions have no effect on indinavir susceptibility, and that V82 to A, F, or T had the strongest association with resistance.

Additional studies were conducted on this sequence data set to look for co-varying substitution patterns (Leigh Brown *et al.*, 1999), using strategy originally applied to the V3 loop (Korber *et al.*, 1993). Five pairs of positions had amino acids that co-varied, positions 71 and 82, 54 and 82, 10 and 82, 10 and 54, and 63 and 64.

5.2 Synonymous and Nonsynonymous Substitution Analysis

In this section the data from indinavir resistance data set described above (Condra *et al.*, 1996) will be used to re-explore emerging amino acid mutations in the context of drug therapy, relative to the background variability of protease. As the data was already studied carefully to look for variability after the initiation of therapy, other points will be considered in this section. Of the 21 original study subjects from the indinavir study, 13 had sequences that were available from both pre- and post-therapy baseline time points, and it is this subset of 13 subjects that will be used to illustrate signature pattern analysis and synonymous and nonsynonymous substitution analysis here. The sequences from these subjects were compared internally, pre- to post-initiation of therapy mutational patterns from within the same patient. Figure 2 examines only the frequency of nonsynonymous substitutions for all pairwise comparisons of sequences from samples before and after initiation of therapy, and Figure 3 examines both synonymous and nonsynonymous patterns of substitutions. The two figures provide different ways of viewing substitution patterns on a codon-by-codon basis. The method used for estimating the synonymous and nonsynonymous substitutions as defined by Nei and Gojobori (1986) was used for this study, implemented through the SNAP program, available at: *http://hiv-web.lanl.gov/*. This method calculates the average synonymous (d_s) and nonsynonymous substitution (d_n) rate and the ratio of the two, spanning a stretch of aligned sequence (Nei and Gojobori, 1986). [Although the assumptions about substitution patterns using this model are relatively simplistic and can give systematic biases (Muse, 1996; Ina, 1995), if the overall variation is not too great, they can provide reasonable estimates.] SNAP calculates the d_s and d_n rates, and incorporates a statistic developed by Ota and Nei (1994) for determining the variance of these measures. But SNAP also allows one to calculate the mutational behavior of each codon in terms of synonymous and nonsynonymous substitutions, providing a codon-by-codon examination of the evolutionary patterns across a given gene.

The fraction of nonsynonymous substitutions at each codon before and after the start of therapy is shown in Figure 2, compared to the nonsynonymous substitution patterns when protease is not under the selective pressure of drug therapy. Four positions stand out as particularly diagnostic of emerging drug resistance in Figure 2: M46, I54, V82, and I84. Two others appear less useful as predictors of resistance, as they are also variable in the absence of therapy, K20 and L63. This does not mean they are not important, however; for example, mutations in positions V82, I84, M46, and L63 have been shown to influence drug resistance by complementary mechanisms: M46 and L63 substitutions act to increase the catalytic efficiency of protease, and V82 and I84 are located in the binding cleft of the enzyme and block the binding of protease inhibitors (Schock *et al.*,1996).

In Figure 3, the cumulative pattern of both synonymous and nonsynonymous substitutions is shown. The codon positions 1-99 are represented along the abscissa. The fraction of substitutions of particular class (synonymous or nonsynonymous) at each codon is calculated, and the cumulative sum of these is plotted from left to right across the protein. Regions of steep slope are highly variable, and flat regions are highly conserved. The global variability of protease and the pre-indinavir set from Condra and his colleagues track well, with the global variability, of course, displaying more extensive variation. There are a few distinctive regions in the control sets, such as a region that appears variable in the global set and conserved in the B subtype set studied here, between codons 64 and 72. There is a region that is conserved even at the synonymous level spanning positions 24 to 48, which may be indicative of RNA structural constraints. A very striking feature of Figure 3 is the steep slopes in the Condra study sequence comparison, pre- and post-indinavir, at the four positions M46, I54, V82, I84, that are embedded in two perfectly conserved regions in both background data sets. Other points worth considering and shown in Figure 3 are the amino acid residues that are contact

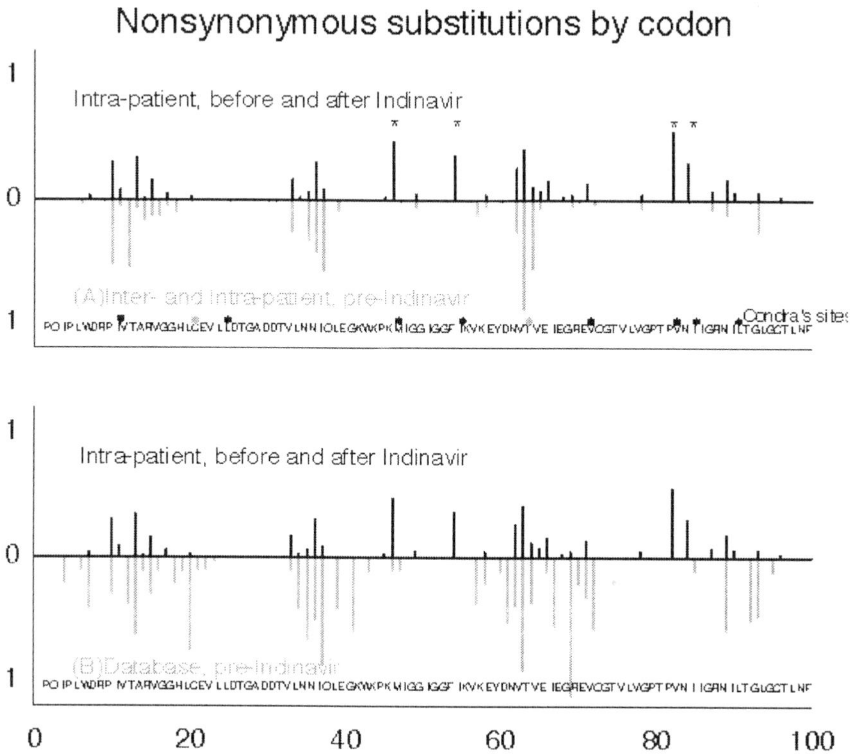

Figure 2 Average nonsynonymous substitutions observed for each codon, in 13 patients, before and after indinavir monotherapy. The top, dark lines show the data for the pre-and post- therapy comparisons in both panels. The upper panel (A) compares this data to all pre-indinavir time points from the same study, both inter- and intra-patient. The lower panel (B) compares this data to the overall variation found in protease of all RT sequences in the 1997 database reference set, including international isolates, these are essentially derived from untreated individuals. The positions that were found by Condra, *et al.*, (1996) to correlate with resistance are indicated on the top protease sequence by bullets. The four positions M46, I54, V82, I84 are almost invariant in the background data sets, hence appear to be the cleanest markers for emerging drug resistance (these positions are marked with as asterisk). Two of the sites, K20 and L63, were highly variable even in the background sets. These sites are indicated by a lighter colored dot above the reference sequence.

points for indinavir (*http://www.ncbi.nlm.nih.gov/*), and an overlay of CTL epitopes which can be found for any HIV protein at the Los Alamos HIV Immunology database (Korber *et al.*,1998; *http://hiv-web.lanl.gov*). Peptide stimulation of CTL from 5 of 6 HIV$^+$ A*0201 individuals had CTL reactivity for the A2 peptide shown (Konya *et al.,* 1997). This epitope is centered in a very conserved region that also happens to be a focus of resistance mutations; this illustrates the potential for drug resistance acquisition to also drive immunological escape.

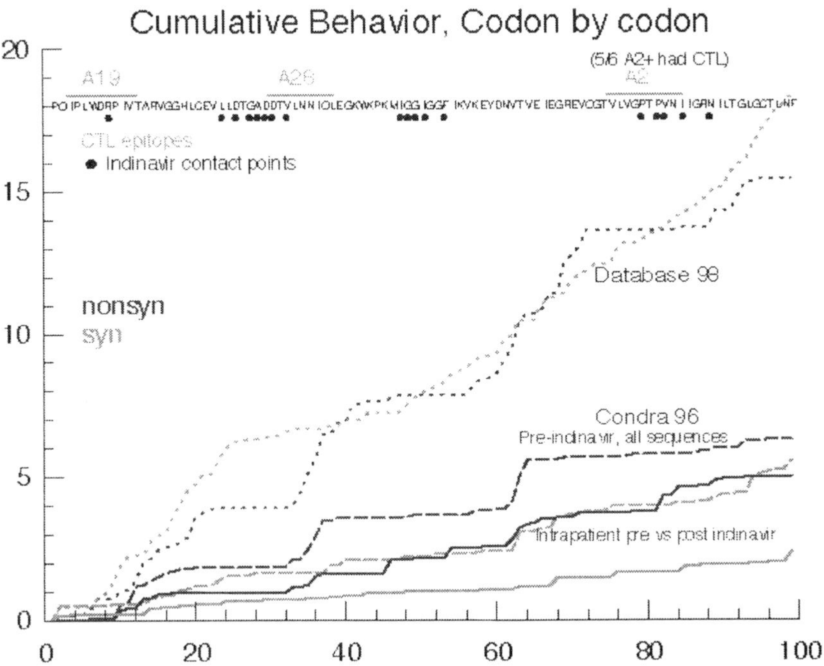

Figure 3 The cumulative synonymous and nonsynonymous substitutions spanning with protease gene. For each codon, the numbers of synonymous or nonsynonymous substitutions are calculated for all pairwise comparisons, and then this number is divided by the number of comparisons. Moving across protease from codon 1 to codon 99, these numbers are added to the previous values. Steep parts of the plot indicate regions of rapid change, level regions, slow change. The dark lines track nonsynonymous changes, the light lines, synonymous changes. If synonymous changes were completely neutral, the slope should be even throughout, yet there is a relatively level region between codons 20-46. The slope of nonsynonymous change varies radically; large stretches of the protein are invariant, even in the global set. Two control sets are included. The dotted lines represent the analysis of Los Alamos database set of protease sequences from 1997. The dashed lines represent the indinavir treatment group sequences, within and between patient comparisons, pre-therapy. The much more rapid rise for the database set is because of the much greater overall global variability — proteases were included from all clades. The final set that was studied was intra-patient, before and after treatment with indinavir, indicated by the solid lines. The slopes are the most gradual for this set, as all comparisons were within a patient. What is of interest here are the marked regions of variation in extremely conserved stretches in the control sets. The contact residues for indinavir and Protease (within 5 Å) are indicated by bullets under the sequence, and CTL epitopes in protease are also shown.

5.3 Signature Analysis

This section will be illustrated using the same test indinavir data set to explore how signatures might be used to address the question of pre-disposition to rapid development of resistance to indinavir (Condra *et al.*, 1996). The pre-therapy time point sequences were used to look for sites that might predispose an individual to rapid development of resistance. The definition of resistance was based on Condra and coworkers; < 400 CIC_{95} was the criteria for susceptible virus. In this example, "slow" was defined by week 24 virus < 400 CIC_{95}, and "rapid" was defined by week

24 virus $\geq$ 400 CIC_{95}. There were only 5 patients that rapidly developed indinavir resistant virus, and 7 that slowly developed resistance with pre-treatment sequence data available from this study. A consensus sequence from the pre-indinavir time point was used to represent each patient. This sample is too small to effectively address the question we are asking, and no statistically significant differences were observed, but the data sets still serve as an illustration of the methods. Signature analysis such as this could be conducted on larger sets to see whether there are particular patterns found in non-treated patients that predispose an individual to resistance.

Two types of signatures were estimated from the available sequences. The first was a simple signature generated with the program (VESPA, Korber and Myers, 1992: *http://hiv-web.lanl.gov/*). VESPA provides tallies of the amino acid frequencies at different positions, and when used to compare two different sets of data, it will highlight positions where the most common amino acid in the alignment is different for the two sets. A Fisher's exact test, or a chi-squared test, can be employed to check the p-value for differences observed in amino acid distribution. Interpretation of p-values should be made bearing in mind multiple test issues (see the next section).

The other type of signature was defined using the program ENTROPY (Korber, *et al.,*1994), (code available on request). ENTROPY compares the variability of amino acids at every position, and was first used to identify a relatively conserved set of positions with a significantly lower entropy in brain- than in blood-derived viral sequences. These sites were a brain "signature pattern", or a non-contiguous set of amino acids in the V3 region conserved in viral sequences derived from brain tissue. These positions were also conserved among macrophage-tropic strains. The entropy $H(i)$ (Blahut, 1987) is defined in terms of the probabilities, $P(s_i)$, of the different symbols, s, that can appear at a given position i (e.g. in this case we are considering the symbols s = A,S,L ... for the twenty amino acids: Ala, Ser, Leu ...). $H(i)$ is defined as:

$$H(i) = - \sum_{s=A,S,L\ldots} P(s_i) \log P(s_i).$$

Entropy takes into account both the variety and frequency of observed amino acids in each position. The program incorporates a Monte Carlo re-sampling strategy to test statistical significance (see the next section), and allows the user to either use all amino acids, or to group amino acids according to chemical classes.

6. LIMITATIONS AND STATISTICAL CONSIDERATIONS

Distinctive signatures, such as the association between positive charge in certain positions in the V3 loop and the SI phenotype and CXCR4 co-receptor usage, can be so clear and overt that one does not need to look further than a simple sequence alignment to spot them (Fouchier *et al.,*1992). It could be argued that unless a pattern is that clear, it is probably going to be too complex for useful biological interpretation or as a predictive tool. But even clear patterns become virtually impossible to find by eye in large alignments. Sequencing techniques have improved enormously in recent years; dozens of sequences thousands of bases long are now within the scope of a single study, and new technologies will steadily advance these numbers. If the problem of identifying distinctive characteristics in large sets of long sequences and determining their statistical significance is to be addressed at all, it will depend on systematic computational analysis. Furthermore, the statistical power when using large datasets will be enhanced, and more subtle distinctions may give a high level of significance, and ultimately prove to be important.

Scanning a long sequence for signatures or distinctive regions of conservation or variability is a situation where statistical issues related to multiple tests must be taken into consideration. Each variable position in an alignment will be tested and compared. To illustrate the problem, imagine lining up a row of 1000 coins, and tossing each coin 10 times. If one of the coins gives 10 heads, it would be in accord with what one might expect by chance alone, as this will happen (on average) one time in 2^{10}, or once every 1,024 coins. Thus 1/1000 coins giving 10/10 heads would be a poor argument for the existence of magic coins. On the other hand, if you had reason to suspect that there might be a magic heads-only coin in the set, that coin would be the best candidate for further study.

There are a number of strategies one can take to compensate for multiple tests. For example, Bonferroni's correction is often used, which essentially requires that the stringency of the p-value used for statistical significance be divided by the number of tests being done. So, in a stretch of protein like protease, about 100 amino acids long, a p-value threshold of 0.0005 would be used for the individual sites in order to achieve an effective p-value of .05 for the full test. This is a conservative standard, and may be unduly so in sequence signature studies. In particular, many sites are essentially invariant, and arguably should be excluded from consideration. As an alternative approach, one can use a Monte Carlo random-with-replacement strategy to generate 100s or 1000s of pseudo-sequence sets, constructed from the real data, for comparison (essentially a bootstrap, Efron and Tibshirani, 1993). To do this, apply the signature test to the original data and to each of the random data sets, and count how many random data sets exhibit a signature pattern in any position that is as distinctive as the pattern observed in the real data. This count, divided by the total number of random sets, provides the level of significance for the signature test (Korber *et al.*, 1994).

Another issue for this kind of analysis is the influence of the underlying phylogenetic structure and genetic lineages of the viruses under consideration. If the phylogenetic relationships are poorly defined for the sequence set in question, (e.g., the sequences are essentially randomly distributed approximating a star phylogeny, and the genetic distance between the viruses under consideration roughly comparable), then signature analyses as described here are reasonable. Signatures, however, can be simple manifestations of phylogenetic relationships. This situation can be interesting under some circumstances, such as the signature associated with IVDU HIV infections in Europe, however it could also overwhelm attempts to define sequence patterns that relate to biological functions. Another approach to this problem would be to define signature patterns by incorporating a phylogenetic tree, and taking into account reconstructed ancestral sequences. In general, there are two basic strategies one might take to study a phenomenon like the relationship between mutational patterns and drug resistance. The first, an alternative to signature patterns, would be to cluster sequences with no a priori knowledge of phenotype (e.g., drug resistance) incorporated into the analysis, and then go back and look for clusters which are associated with phenotype. Using this strategy, the clustering pattern itself might be the basis for predicting drug resistance, or one might try to decipher the components that are critical for determining the cluster. This may also reveal things about the merits of a particular clustering approach. The second alternative is signature analysis as described in this chapter. It is a conceptually simple approach, and there are ample ways available to assist in the analysis and to estimate measures of statistical confidence. The validity of sequence signatures for issues like predicting appropriate drug regimens from pre-treatment viral sequences, in an effort to tailor the medications to optimize the chances for success, is yet to be determined. This strategy could benefit by the development of methods that incorporate phylogenetic information. It holds promise as a practical approach for interpreting the large numbers of sequences currently being generated through drug trials.

ACKNOWLEDGMENTS

With thanks to James Theiler, Carla Kuiken, Brian Foley, Robert Funkhouser, Satish Pillai, and Vijaja Doddi for help with this manuscript and sound advice. BK is supported by an NIH-DOE Interagency Agreement Y01-AI4058-03, and by the Pediatric AIDS Foundation as an Elizabeth Glaser Scientist.

REFERENCES

Ahmad, N., Baroudy, B. M., Baker, R. C., and Chappey, C. 1995. Genetic analysis of human immunodeficiency V3 region isolates from mothers and infants transmission. *J. Virol.* **69**: 1001-1012.

Altschul, S. F., Gish, W., Miller, W., and Lipman, D. J. 1990. Basic local alignment search tool. *J. Mol. Biol.* **215**: 403-410.

Altschul, S. F., Madden, T. L., Schaffer, A. A., Zhang, J., Zhang, Z., Miller, W., Lipman, D. J. 1997. Gapped BLAST and PSI-BLAST: a new generation of protein database search programs. *Nucl. Acids Res.* **25**: 3389-3402.

Bairoch, A., Bucher, P. and Hofmann, K. 1997 The PROSITE database, its status in 1997. *Nucl. Acids Res.* **25**: 217-221.

Belec, L., Mohamed, A. S., Müller-Trutwin, M. C., Gilquin, J., Gutmann, L., Safar, M., Barre-Sinoussi, F. and Kazatchkine, M. D. 1998. Genetically related human immunodeficiency virus type 1 in three adults of a family with no identified risk factor for intrafamilial transmission. *J. Virol.* **72**: 5831-5839.

Blahut, R. E. 1987. *Information Theory and Statistics.* Addison-Wesley, Reading, MA.

Carr, A., Samaras, K., Burton, S., Law, M., Freund, J., Chisholm, D. J. and Cooper, D. A. 1998a. A syndrome of peripheral lipodystrophy, hyperlipidaemia and insulin resistance in patients receiving HIV protease inhibitors. *AIDS* **12**: F51-F58.

Carr, A., Samaras, K., Chisholm, D. J. and Cooper, D. A. 1998b. Pathogenesis of HIV-1-protease inhibitor-associated peripheral lipodystrophy, hyperlipidaemia, and insulin resistance. *Lancet* **351**: 1881-1883.

Choe, H., Farzan, M., Sun, Y., Sullivan, N., Rollins, B., Ponath, P. D., Wu, L., Mackay, C. R., LaRosa, G., Newman, W., Gerard, N., Gerard, C. and Sodroski, J. 1996. The P-chemokine receptors CCR3 and CCR5 facilitate infection by primary HIV-1 isolates. *Cell* **85**: 1135-1148.

Ciesielski, C., Marianos, D., Ou, C.-Y., Dumbaugh, R., Witte, J., Berkelman, R., Gooch, B., Myers, G., Luo, C.-C., Schochetman, G., Howell, J., Lasch, A., Bell, K., Economou, N., Scott, B., Furman, L., Curran, J. and Jaffe, H. 1992. Transmission of human immunodeficiency virus in a dental practice. *Ann. Intern. Med.* **116**: 798-805.

Ciesielski, C. A., Marianos, D. W., Schochetman, G., Witte, J. J. and Jaffe, H. W. 1994. The 1990 Florida dental investigation. The press and the science. *Ann. Intern. Med.* **121**: 886-888.

Clemente-Estable, M., Merzouki, A., Arella, M. and Sadowski, I. J. 1998. Distinct clustering of HIV type 1 sequences derived from injection versus noninjection drug users in Vancouver, Canada. *AIDS Res Hum Retroviruses* **14**: 917-919.

Condra, J. H. 1998. Resistance to HIV protease inhibitors. *Haemophilia* **4**: 610-615.

Condra, J. H., Holder, D. J., Schleif, W. A., Blahy, O. M., Danovich, R. M., Gabryelski, L. J., Graham, D. J., Laird, D., Quintero, J. C., Rhodes, A., Robbins, H. L., Roth, E., Shivaprakash, M., Yang, T., Chodakewitz, J. A., Deutsch, P. J. Leavitt, R. Y., Massari, F. E., Mellors, J. W., Squires, K. E., Steigbigel, R. T., Teppler, H. and Emini, E. A. 1996. Genetic correlates of in vivo viral resistance to indinavir, a human immunodeficiency virus type 1 protease inhibitor. *J. Virol.* **70**: 8270-8276.

Craig, C., Race, E., Sheldon, J., Whittaker, L., Gilbert, S., Moffatt, A., Rose, J., Dissanayeke, S., Chirn, G. W., Duncan, I. B. and Cammack, N. 1998. HIV protease genotype and viral sensitivity to HIV protease inhibitors following saquinavir therapy. *AIDS* **12**: 1611-1618.

Crandall, K. 1995. Intraspecific phylogenetics: support for dental transmission of human immunodeficiency virus. *J. Virol.* **69**: 2351-2356.

De Wolf, F., Hogervorst, E., Goudsmit, J., Fenyo, E. M., Rubsamen-Waigmann, H., Holmes, H., Galvao-Castro, B., Karita, C. W. E. and Sempala, S. D. 1994. Syncytium-inducing and non-syncytium-inducing capacity of human immunodeficiency virus type 1 subtypes other than B: phenotypic and genotypic characteristics. WHO Network for HIV Isolation and Characterization. *AIDS Res. Hum. Retroviruses* **10**: 1387-1400.

Delwart, E. L., Mullins, J. I., Gupta, P., Learn, G. H., Holodniy, M., Katzenstein, D., Walker, B. D. and Singh, M. K. 1998. Human immunodeficiency virus type 1 populations in blood and semen. *J. Virol.* **72**: 617-623.

Deng, H., Liu, R., Ellmeier, W., Choe, S., Unutmaz, D., Burkhart, M., Di Marzio, P., Marmon, S., Sutton, R. E., Hill, C. M., Davis, C. B., Peiper, S. C., Schall, T. J., Littman, D. R. and Landau, N. R. 1996. Identification of a major co-receptor for primary isolates of HIV-1. *Nature* **381**: 661-666.

Doms, R. W. and Moore, J. P. 1997. HIV-1 coreceptor use: A molecular window into viral tropism. In *Human Retroviruses and AIDS 1997* (Korber, B., Hahn, B., Foley, B., Mellors, J. W., Leitner, T., Myers, G., McCutchan, F. and Kuiken, C. L., eds). Los Alamos National Laboratory, Los Alamos, NM.

Efron, B. and Tibshirani, R. J. 1993. *An Introduction to the Bootstrap.* Chapman and Hall, New York, NY.

Felsenstein, J. 1985. Confidence limits on phylogenies: an approach using the bootstrap. *Evolution* **39**: 783-791.

Fenyo, E. M., Schuitemaker, H., Asjo, B. and McKeating, J. 1997. The history of HIV-1 biological phenotypes, past, present and future. In *Human Retroviruses and AIDS 1997* (Korber, B., Hahn, B., Foley, B., Mellors, J. W., Leitner, T., Myers, G., McCutchan, F. and Kuiken, C. L., eds). Los Alamos National Laboratory, Los Alamos, NM.

Fouchier, R. A. M., Groenink, M., Kootstra, N. A., Tersmette, M., Huisman, H. G., Miedema, F. and Schuitemaker, H. 1992. Phenotype-associated sequence variation in the third variable region of the human immunodeficiency virus type 1 gp120 molecule. *J. Virol.* **66**: 3183-3187

Frenkel, L., Mullins, J., Learn, G., Manns-Arcuino, L., Herring, B., Kalish, M., Steketee, R., Thea, D., Nichols, J., Liu, S., Harmache, A., He, X., Muthui, D., Madan, A., Hood, L., Haase, A., Zupancic, M., Staskus, K., Wolinsky, S., Krogstad, P., Zhao, J., Chen, I., Koup, R., Ho, D., Korber, B., Apple, R. J., Coombs, R. W., Pahwa, S. and Roberts, N. J., Jr. 1998. Genetic evaluation of suspected cases of transient HIV-1 infection of infants. *Science* **280:** 1073-1077.

Gower, J. C. 1992. Some distance properties of latent root and vector methods used in multivariate analysis. *Biometrika* **53**: 325-328.

Hammond, J., Larder, B., Schinazi, R. and Mellors, J. 1997. Mutations in retroviral genes associated with drug resistance. In *Human Retroviruses and AIDS 1997* (Korber, B., Hahn, B., Foley, B., Mellors, J. W., Leitner, T., Myers, G., McCutchan, F. and Kuiken, C. L., eds). Los Alamos National Laboratory, Los Alamos, NM.

Henikoff, S. and Henikoff, J. G. 1994. Protein family classification based on searching a database of blocks. *Genomics* **19**: 97-107.

Higgins, D. G. 1992. Sequence ordinations: a multivariate analysis approach to analysing large sequence data sets. *Comput. Appl. Biosci.* **8**: 15-22.

Hoffman, T. L., Stephens, E. B., Narayan, O. and Doms, R. W. 1998. HIV type I envelope determinants for use of the CCR2b, CCR3, STRL33, and APJ coreceptors. *Proc. Natl. Acad. Sci. USA* **95**: 11360-11365.

Holmes, E. C., Zhang, L. Q., Simmonds, P., and Leigh Brown, A. J. 1993. Molecular investigation of human immunodeficiency virus (HIV) infection in a patient of an HIV-infected surgeon. *J. Inf. Dis.* **167**: 1411-1414.

Ina, Y. 1995. New methods for estimating the numbers of synonymous and nonsynonymous substitutions. *J. Mol. Evol.* **40**:190-226.

Joint U. N. Programme on HIV AIDS, The. 1997 Implications of HIV variability for transmission: scientific and policy issues. *AIDS* **11**: UNAIDS1-S15.

Kitayaporn, D., Vanichseni, S., Mastro, T. D., Raktham, S., Vaniyapongs, T., Des Jarlais, D. C., Wasi, C., Young, N. L., Sujarita, S., Heyward, W. L. and Esparza, J. 1998. Infection with HIV-1 subtypes B and E in injecting drug users screened for enrollment into a prospective cohort in Bangkok, Thailand. *J. Acquir. Immun. Defic. Syndr. Hum. Retrovirol.* **19**:289-295.

Konya, J., Stuber, G., Bjorndal, A., Fenyo, E. M. and Dillner, J. 1997. Primary induction of human cytotoxic lymphocytes against a synthetic peptide of the human immunodeficiency virus type 1 protease. *J. Gen. Virol.* **78**:2217-2224.

Korber, B., Brander, C., Walker, B., Koup, R., Moore, J., Haynes, B. and Myers, G. 1998. *Molecular Immunology Database 1998*. Los Alamos National Laboratory, Los Alamos, NM.

Korber, B. and Myers, G. 1992. Signature patterns analysis: a method for assessing viral sequence relatedness. *AIDS Res. Hum. Retroviruses* **8**: 1549-1558.

Korber, B. T., Farber, R. M., Wolpert, D. H. and Lapedes, A. S. 1993. Covariation of mutations in the V3 loop of human immunodeficiency virus type 1 envelope protein: an information theoretic analysis. *Proc. Natl. Acad. Sci. USA* **90**: 7176-7180.

Korber, B. T., Kunstman, K. J., Patterson, B. K., Furtado, M., McEvilly, M. M., Levy, R. and Wolinsky, S. T. 1994. HIV-1 sequence differences between blood and simultaneously obtained brain biopsy samples: conserved elements in the V3 region of the envelope protein of brain-derived sequences. *J. Virol.* **68**:7467-7481.

Kuiken, C. L., Cornelissen, M. T., Zorgdrager, F., Hartman, S., Gibbs, A. J. and Goudsmit, J. 1996. Consistent risk group-associated differences in human immunodeficiency virus type 1 vpr, vpu and V3 sequences despite independent evolution. *J. Gen. Virol.* **77**: 783-792.

Kuiken, C. L. and Goudsmit, J. 1994. Silent mutation pattern in V3 sequences distinguishes virus according to risk group in Europe. *AIDS Res. Hum. Retroviruses* **10**:319-320.

Kuiken, C. L., Goudsmit, J., Weiller, G. F., Armstrong, J. S., Hartman, S., Portegies, P., Dekker, J. and Cornelissen, M. 1995. Differences in human immunodeficiency virus type 1 V3 sequences from patients with and without AIDS dementia complex. *J. Gen. Virol.* **76**: 175-180.

Kuiken, C. L., and Korber, B. T. 1998. Sequence quality control. In *Human Retroviruses and AIDS 1998* (Korber, B., Hahn, B., Foley, B., Mellors, J. W., Sodroski, J., McCutchan, F. and Kuiken C. L. eds). Los Alamos National Laboratory, Los Alamos, NM.

Kunanusont, C., Foy, H. M., Kreiss, J. K., Rerks-Ngarm, S., Phanuphak, P., Raktham, S., Pau, C. P. and Young, N. L. 1995. HIV-1 subtypes and male-to-female transmission in Thailand. *Lancet* **345**:1078-1083.

Leigh Brown, A. J., Korber, B. and Condra, J. H. 1999. Association between amino acids in the evolution of HIV type 1 protease sequences under indinavir therapy. *AIDS Res. Hum. Retroviruses* **15**: 247-253.

Leigh Brown, A. J., Lobidel, D., Wade, C. M., Rebus, S., Phillips, A. N., Brettle, R. P., France, A. J., Leen, C. S., McMenamin, J., McMillan, A., Maw, R. D., Mulcahy, F., Robertson, J. R., Sankar, K. N., Scott, G., Wyld, R. and Peutherer , J. F. 1997. The molecular epidemiology of human immunodeficiency virus type 1 in six cities in Britain and Ireland. *Virology* **235**: 166-177.

Liitsola, K., Tashkinova, I., Laukkanen, T., Korovina, G., Smolskaja, T., Momot, O., Mashkilleyson, N., Chaplinskas, S., Brummer-Korvenkontio, H., Vanhatalo, J., Leinikki, P. and Salminen, M. O. 1998. HIV-1 genetic subtype A/B recombinant strain causing an explosive epidemic in injecting drug users in Kaliningrad. *AIDS* **12**: 1907-1919.

Lukashov, V. V., Kuiken, C. L., Vlahov, D., Coutinho, R. A. and Goudsmit, J. 1996. Evidence for HIV type 1 strains of U.S. intravenous drug users as founders of AIDS epidemic among intravenous drug users in northern Europe. *AIDS Res. Hum. Retroviruses* **12**: 1179-1183.

Lukashov, V. V., Op de Coul, E. L., Coutinho, R. A. and Goudsmit, J. 1998. HIV-1 strains specific for Dutch injecting drug users in heterosexually infected individuals in The Netherlands. *AIDS* **12**: 635-641.

Mastro, T. D., Kunanusont, C., Dondero, T. J. and Wasi, C. 1997. Why do HIV-1 subtypes segregate among persons with different risk behaviors in South Africa and Thailand? *AIDS* **11**: 113-116.

Molla, A., Korneyeva, M., Gao, Q., Vasavanonda, S., Schipper, P. J., Mo, H. M., Markowitz, M., Chernyavskiy, T., Niu, P., Lyons, N., Hsu, A., Granneman, G. R., Ho, D. D., Boucher, C. A., Leonard, J. M., Norbeck, D. W. and Kempf, D. J. 1996. Ordered accumulation of mutations in HIV protease confers resistance to ritonavir. *Nat. Med.* **2**:760-766.

Muse, S. V. 1996. Estimating synonymous and nonsynonymous substitution rates. *Mol. Biol. Evol.* **13**: 105-114.

Myers, G. 1994. Molecular investigation of HIV transmission. *Ann. Intern. Med.* **121**: 889-890.

Myers, G. and Farmer, A.. 1996. HIV alignments, database searches, and structure predictions. In *Human Retroviruses and AIDS 1996* (Myers, G., Korber, B. T., Foley, B. T., Jeang, K. T. and Mellors, J. W. eds). Los Alamos National Laboratory, Los Alamos, NM.

Nei, M., and Gojobori, T. 1986. Simple methods for estimating the numbers of synonymous and nonsynonymous nucleotide substitutions. *Mol. Biol. Evol.* **3**: 418-426.

Ota, T. and Nei, M. 1994. Variance and covariances of the numbers of synonymous and nonsynonymous substitutions per site. *Mol. Biol. Evol.* **11**: 613-619.

Ou, C.-Y., Ciesielski, C.A., Myers, G., Bandea, C.I., Luo, C.-C., Korber, B.T.M., Mullins, J.I., Schochetman, G., Berkelman, R., Economou, A.N., Witte, J.J., Furman, L.J., Satten, G.A., MacInnes, K.A., Curran, J.W., Jaffe, H.W., Moore, J., Villamarzo, Y., Schable, C., Shpaer, E.G., Liberti, T., Lieb, S., Scott, R., Howell, J., Dunbaugh, R., Lasch, A., Kroesen, B., Ryan, L., Bell, K., Munn, V., Marianos, D. and Gooch, B. 1992. Molecular epidemiology of HIV transmission in a dental practice. *Science* **256**: 1165-1171.

Ou, C. Y., Takebe, Y., Weniger, B. G., Luo, C. C., Kalish, M. L., Auwanit, W., Yamazaki, S., Gayle, H. D., Young, N. L. and Schochetman, G. 1993. Independent introduction of two major HIV-1 genotypes into distinct high-risk populations in Thailand. *Lancet* **341**:1171-1174.

Rawlings, N. D. and Barrett, A. J. 1995. Families of aspartic peptidases, and those of unknown catalytic mechanism. *Methods Enzymol.* **248**: 105-120.

Robertson, D. L., Anderson, J. P., Bradac, J. A., Carr, J. K., Foley, B., Funkhouser, R. K., Gao, F., Hahn, B. H., Kalish, M. L., Kuiken, C., Learn, G. H., Leitner, T., McCutchan, F., Osmanov, S., Peeters, M., Pieniazek, D., Salminen, M., Sharp, P. M., Wolinsky, S. and Korber, B.. 2000. HIV-1 nomenclature proposal. *Science* **288:** 55.

Ross, T. M. and Cullen, B. R. 1998. The ability of HIV type 1 to use CCR-3 as a coreceptor is controlled by envelope V1/V2 sequences acting in conjunction with a CCR-5 tropic V3 loop. *Proc. Natl. Acad. Sci. USA* **95**:7682-7686.

Schock, H. B., Garsky, V. M. and Kuo, L. C. 1996. Mutational anatomy of an HIV-1 protease variant conferring cross-resistance to protease inhibitors in clinical trials. Compensatory modulations of binding and activity. *J. Biol. Chem.* **271**: 31957-31963.

Speck, R. F., Wehrly, K., Platt, E. J., Atchison, R. E., Charo, I. F., Kabat, D., Chesebro, B. and Goldsmith, M. A. 1997. Selective employment of chemokine receptors as human immunodeficiency virus type 1 coreceptors determined by individual amino acids within the envelope V3 loop. *J. Virol.* **71**: 7136-7139.

Subbarao, S., Limpakamjanarat, K., Mastro, T. D., Bhumisawasdi, J., Warachit, P., Jayavasu, C., Young, N. L., Luo, C. C., Shaffer, N., Kalish, M. L. and Schochetman, G. 1998. HIV type 1 in Thailand, 1994-1995: persistence of two subtypes with low genetic diversity. *AIDS Res. Hum. Retroviruses* **14**: 319-327.

Triques, K., Bourgeois, A., Vidal, N., Mpoudi-Ngole, Mulanga-Kabeya, E. C., Nzilambi, N., Torimiro, N., Saman, E., Delaporte, E. and Peeters, M. 2000. Near-full-length genome sequencing of divergent African HIV type 1 subtype F viruses leads to the identification of a new HIV type 1 subtype designated K. *AIDS Res. Hum. Retroviruses* **18**: 139-151.

Trujillo, J. R., McLane, M. F., Lee, T.-H. and Essex, M. 1993. Molecular mimicry between the human immunodeficiency virus type 1 gp120 V3 loop and human brain proteins. *J. Virol.* **67**:7711-7715.

van Harmelen, J., Wood, R., Lambrick, M., Rybicki, E. P., Williamson, A. L. and Williamson, C. 1997. An association between HIV-1 subtypes and mode of transmission in Cape Town, South Africa. *AIDS* **11**:81-87.

van 't Wout, A. B., Ran, L. J., Kuiken, C. L., Kootstra, N. A., Pals, S. T. and Schuitemaker, H. 1998. Analysis of the temporal relationship between human immunodeficiency virus type 1 quasispecies in sequential blood samples and various organs obtained at autopsy. *J. Virol.* **72**:488-496.

Xiao, L., Owen, S. M., Goldman, I., Lal, A. A., deJong, J. J., Goudsmit, J. and Lal, R. B. 1998. Adaptation to promiscuous usage of CC and CXC-chemokine coreceptors in vivo correlates with HIV-I disease progression. *Virology* **240**:83-92.

GRAPHICAL METHODS FOR EXPLORING SEQUENCE RELATIONSHIPS

Georg F. Weiller

Bioinformatics Group, Research School of Biological Sciences, Australian National University
Canberra, ACT 2601 Australia

1. INTRODUCTION

Large bodies of information-rich data, such as molecular sequences, often contain unexpected characteristics. The investigation of this kind of data usually benefits from a data-driven exploratory analysis phase whereby a variety of plausible conjectures can be investigated and alternatives compared, modified and refined until a consistent picture emerges. The ability to test new ideas quickly is important as this allows a large number of alternatives to be explored. The initial questions are typically quite broad, and designed to lead towards the more specific properties of the data. We may want to scan a large volume of data and hope to see trends, symmetries or anomalies. The graphical representation of the various properties of the data is especially helpful here, as it allows us to comprehend information quickly, intuitively fill in missing data points, and see the general trends as well as local variations.

This chapter discusses the advantages of graphical representation for the exploratory analysis of HIV sequences using distance plots and phylogenetic profiles together with interactive computer graphics programs (Weiller, 1998a). Interactive graphs not only display aspects of the data, but also become part of the program's user interface. The investigator can interrogate or modify the graph directly and so indirectly specify some of the parameters required to produce the graph, or obtain further information on the underlying data. While the graphical representation supports an intuitive understanding of a large amount of data, the interaction with the graph helps to quickly identify specific aspects, explore the parameter space, and compare alternative hypotheses.

2. EXPLORING SEQUENCE RELATIONSHIPS

Many different comparison metrics can be used to estimate sequence relationships, for example a particular scoring scheme for character changes as in parsimony programs, likelihood functions, or specific distance metrics. Amongst the latter is the well known Jukes and Cantor formula (Jukes and Cantor, 1969) which uses the overall frequency of observed sequence differences to estimate the number of mutations that have occurred per site, thereby considering the possibility that multiple changes have occurred at the same position. Subsequently, more elaborate methods have been developed that are able to distinguish between transitional and transversional changes or other classes of substitutions (Gojobori *et al.*, 1990). However all methods have shortcomings that may limit their validity in specific situations. The biggest challenges are presented by the fact that sequences are often biased toward specific nucleotides, contain mixtures of conserved and variable sites, have varying functional, structural and evolutionary constraints, and may even constitute mosaics of recombined regions with different historical backgrounds.

Once established, the sequence relationships are usually represented as dendrograms. Many different methods of tree construction have been developed (see Weiller *et al.*, 1995 and Swofford *et al.*, 1996 for an overview), however a tree diagram can only represent one comparison metric. Trees constructed with different methods often differ, raising the possibility that several different, sometimes contradictory signals have been detected. When such discrepancies have been found they are often attributed to limitations of the methods used. However, different sequence regions, codon positions and sites may indeed exhibit different sequence relationships, and establishing which single method is adequate may not always be possible. Several different trees could be compared in some cases, however this would be quite complicated particularly when different tree topologies are involved. Multivariate analysis methods such as Multidimensional Scaling or Principal Coordinate Analysis (PCA) are often better suited to detect and separate different signals in the data (for an introduction to these methods see Everitt and Dunn, 1992) and several computer programs are available to perform such analyses (Higgins, 1992; Rohlf, 1993). The Distance plot method, which is described in more detail below, provides an alternative and more transparent technique, and an example where both distance plots and PCA were utilized to distinguish HIV 1 sequences obtained from hemophiliacs, homosexuals and intravenous drug users can be found in Kuiken and colleagues (1994). These methods are specifically aimed at exploring the contribution of different sequence substitutions and may be helpful in detecting evolutionary trends in a set of sequence data. Yet, similar to other types of phylogenetic analysis methods, they treat sequences as coherent entities that represent some sort of taxonomic unit. However, this is not always justified as varying evolutionary constraints can shape some sequence regions differently and genetic recombination or gene conversion events can combine sequence fragments from different origins. Studies that infer phylogenetic histories from sequences should therefore always ensure that the phylogenetic signal of all sequences is consistent and homogenous. Conversely, detecting regions of different phylogenetic relationships forms the basis of most recombination studies.

Several methods for detecting genetic recombination in sequences have been developed in recent years. Stephens' (1985) and Sawyer's (1989) methods examine the distribution of polymorphic sites in a set of aligned sequences to detect likely recombination events, while Hein (1993) developed a heuristic extension of the parsimony principle to infer phylogenies from recombinant sequences. Several other methods assume prior knowledge of correct phylogenetic relationships that are not distorted through recombination events. Fitch and Goodman's (1991) phylogenetic scanning method, for example, calculates the support for a set of assumed reference trees at different intervals and displays the results graphically. Salminen and coworkers (1995) have developed the bootscanning method, a graphical method that uses a sliding window to compare candidate sequences with a set of non-mosaic reference sequences. Siepel and colleagues (1995) have taken a similar approach. The last two methods have been extensively used in the analysis of HIV sequences whereby consensus sequences, profiles or representatives of the known HIV-1 subtypes have been used as reference sequences.

Recently developed graphical methods make fewer assumptions and handle large numbers of sequences and are therefore better suited for sequence exploration. Among these are the methods devised by Jakobsen and Easteal (1996), Jakobsen, Wilson and Easteal (1997), McGuire, Wright and Prentice (1997) and Weiller's phylogenetic profile method (1998b). While the first two methods exploit compatibility matrices and partition matrices to plot the consistency of the phylogenetic signal between all individual columns of a multiple sequence alignment, the latter two employ sliding windows to match phylogenetic signals obtained in two adjacent sequence regions. Examining either two single columns or two regions of the alignment at a time, all these methods measure the agreement of the phylogenetic signals obtained. The phylogenetic profile method differs in that it analyses every sequence individually, thereby partially insulating the recombination signal observed in one sequence from recombinations or noise obtained in the comparison of other sequences. All methods then produce a graphical representation that indicates the regions where genetic recombination might have occurred. While all these methods are relatively good in pinpointing sites of genetic recombination only the phylogenetic profile method also indicates which sequences are the recombinants. The latter information is especially helpful during the data exploration phase, as sequences that have consistent phylogenies can be quickly identified. Contrasting these sequences with their putative recombinant counterparts greatly improves the detection of recombination events.

It is important to note that none of these graphical methods attempt to evaluate the statistical significance of the recombination signals detected. This is left to a later analysis, and several statistical tests have been devised to check the support for a previously identified putative recombination. All these tests require carefully assembled data sets containing several sequences that are known to be non-recombinant. The maximum chi-square test designed by Maynard Smith (1992) for instance assumes only a single recombination site in the alignment although several recombination sites can be examined in successive stages. The Kishino-Hasegawa test (1989) can be used to test for significant differences between two maximum likelihood trees and is easily applied, as it now is included in the

PHYLIP package (Felsenstein, 1993). The likelihood ratio test of Huelsenbeck and Bull (1996) may also be used to test conflicting phylogenetic signals.

David Robertson maintains a comprehensive and perpetually updated list of recombination analysis computer programs (*http://grinch.zoo.ox.ac.uk/RAP_links.html*).

3. ANALYZING DISTANCE PLOTS WITH DIPLOMO

The DIPLOMO program and its application to HIV sequence analysis has been described previously (Weiller and Gibbs, 1995) and only a brief account of the underlying distance plot principle is given here. Distance plots display the distance data represented by two distance matrices in the form of a simple scatter plot (Figure 1). Such graphs show not only how well the two comparison measures correlate but also their distribution and specific groupings or outliers. Distance plots allow sequence differences to be dissected into their component parts. For example, nucleotide differences can be separated into transitions and transversions, or they may

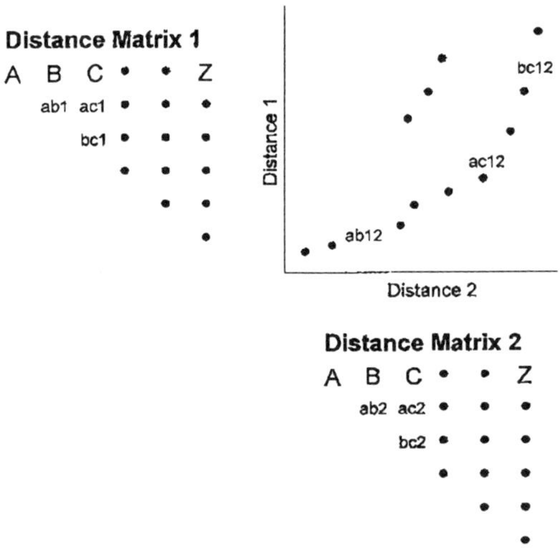

Figure 1 The distance-plot principle. The distance matrices 1 and 2 contain different pairwise distances estimates of the taxa A to Z. Each comparison of two taxa is plotted at y and x coordinates representing the corresponding distance value of matrix 1 and 2 respHectively. (from Weiller and Gibbs, 1995).

be further separated. Depending on the comparison measures used insight into evolutionary processes and trends are obtained. For example comparisons of synonymous and non-synonymous sequence changes provide clues of the selection pressure that individual taxa experience. Comparisons of amino acid differences and nucleotide differences or between different codon positions may provide similar information.

Outliers or clusters of comparisons that exhibit distinct trends are of particular interest. To be identified, individual data points must be associated with the taxa they represent. The DIPLOMO computer program (DIstance PLOt MOnitor) has been designed to generate and analyze distance plots. For instance, specific data points or groups of data points can be selected using a mouse and the corresponding data points are immediately identified in a separate monitor window. Comparisons comprising specific groups of sequences can be represented using a variety of colors and symbols. These labels are retained when alternative distance matrices are plotted, making it easy to quickly compare many different distance measures.

An analysis aimed at exploring mutational tendencies of different HIV and SIV strains as well as of the different genes in these strains is described below. To simplify comparisons between different subgroups and genes, only strains for which the complete genome is available, and only subgroups that are represented by more than one sequence were considered. Multiple sequence alignments were constructed for *gag*, *pol*, *vif*, and *env* as follows. Initial sequence alignments for the HIV-1, HIV-2/SIV and SIVagm groups were obtained from the HIV database (Korber *et al.*, 1997). For each gene, the three alignments were combined using the profile alignment feature of CLUSTALW (Thompson *et al.*, 1994). As it was intended to analyze separate codon positions and synonymous substitutions, all regions encoding overlapping genes were removed from the alignments. Also removed were some very variable regions that were poorly aligned or contained mostly alignment gaps, and triples that contained either stop codons or undetermined nucleotides. The remaining alignments consisted of 975 columns for *gag*, 2670 for *pol*, 411 for *vif* and 2202 for *env*. Using these alignments a number of distance matrices using different distance measures were constructed using the MEGA (Kumar *et al.*, 1994) and DisCalc (Weiller, in prep.) programs. These were combined in a single file suitable for analysis with DIPLOMO. A number of distance plots comparing different distance measures and genes were then compared using the interactive graphics options provided by the program. Some of the plots that do not require color representations are given below. The distance plot on the left in Figure 2 compares the percentage of nucleotide differences in the 3rd codon position (horizontal axis) with the 1st plus 2nd codon positions (vertical axis). It becomes obvious that the data set contains discrete groups that have different ratios of these mutations. As most substitutions in the 3rd codon position are synonymous the distance plot was contrasted with a plot comparing the proportion of synonymous substitutions per synonymous site (horizontal) and non-synonymous substitutions per non-synonymous site (vertical) given on the right of the figure. Note that both plots are very similar, indicating that both distance measures yield very similar results. This was confirmed by plots that compare 3rd codon position and synonymous changes, or 1st and 2nd codon position and non-synonymous changes, both of which resulted in straight diagonals (not shown).

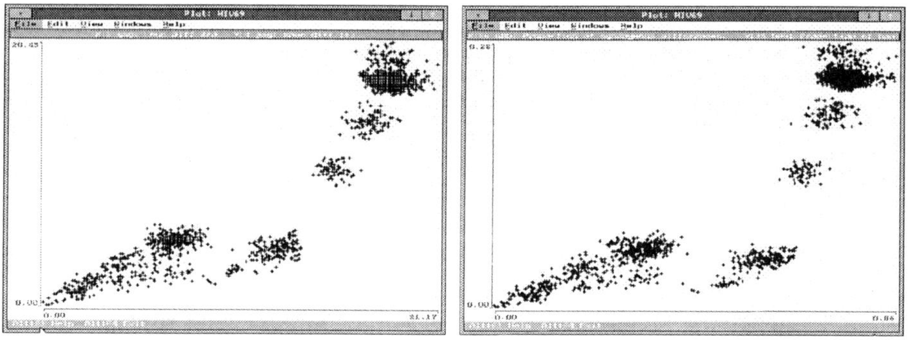

Figure 2 DIPLOMO distance plot of HIV and SIV sequences. The plot on the left compares the percentage of nucleotide differences in the 3rd codon position (horizontal axis) with the 1st plus 2nd codon positions (vertical axis). The plot on the right compares the proportion of synonymous mutations (horizontal) and non-synonymous mutations (vertical). The data set comprises the following 62 sequences: *HIV-1 group M*: B-MN, B-AD8, B-RF, B-BCSG3, B-JRFL, B-JRCSF, B-OYI, B-CAM1, B-LAI, B-HXB2R, B-YU2, B-YU10, B-SF2, B-D31, B-MANC, B-HAN, F-93BR020, H-90CR056, C-ETH2220, AG-92NG083, AG-92NG003, AG-IBNG, A-U455, A-92UG037, A-Q23, AE-93TH253, AE-CM240, AE-90CF402, AD-MAL, D-NDK, D-ELI, D-94UG114; *HIV-1 group O*: O-MVP5180, O-ANT70; *HIV-2 subtype A*: A-BEN, A-GH1, A-D194, A-UC2, A-ROD, A-NIHZ, A-ISY, A-ST, A-CAM2, A-MDS, A-KR; *HIV-2 subtype B*: B-UC1, B-D205, B-EHOA; *SIVmac*: SD-MM251, SD-MM32H, SD-MM1A11AA, SD-MM142, SD-MM239; *SIVsm*: SD-SMMH4, SD-SMMH9; *SIVpbj*: SD-SMMPBJ14.441, SD-SMMPBJB14.15, SD-SMMPBJ6.6; *SIVagm*: VER-TYO, VER-155, VER-3, VER963.

Figure 3 illustrates a step in the identification of the individual data points with the highlighted data-points identified as comparisons between the HIV-1 group M (clade B) and group O sequences. The region of the distance plot representing the comparisons of more closely related sequences is also given in Figure 4 (top left). In the top right plot intra-group comparisons between HIV-1 sequences are represented by the symbol *1* (group M) and *O* (group O, only 2 sequences). Note that all HIV-1 sequences have a high ratio of non-synonymous to synonymous mutations, which is indicative of the evolutionary instability and the rapid change of these viruses. The bottom left panel identifies the analogous HIV-2 intra-subtype comparisons with comparisons between subgroup A sequences labeled *A* and comparisons between subgroup B sequences labeled *B*. Comparisons between both subgroups are shown as +. Note the HIV-2 comparisons are quite separated from HIV-1 comparisons, indicating that HIV-2 is somewhat more stable than HIV-1 as it does not change its amino-acid sequence so freely. The bottom right panel gives the more complex case of the SIV sequences. The comparisons between the SIVagm sequences, all isolated from the vervet species (*Ceropithecus aethiops pygerythrus*) of African green monkeys, are given as *G*. These viruses appear to be evolutionarily much more stable as they have a particularly low ratio of non-synonymous to synonymous mutations. By contrast, comparisons between SIVmac sequences (*M*) isolated for macaque monkeys have an extremely high ratio of non-synonymous to synonymous mutations. SIVsm sequences derived from sooty mangabeys fall into quite different

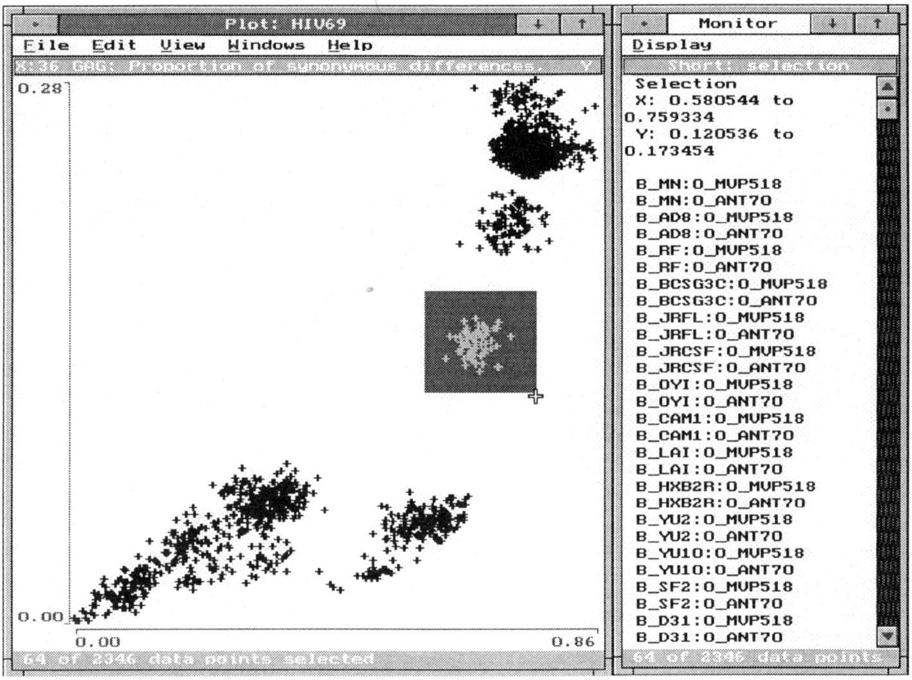

Figure 3 DIPLOMO screen showing the plot- and monitor-windows. The distance plot is equivalent to the right plot in Figure 2. The acronyms of the sequence pairs associated with each datapoint highlighted in the plot-window are displayed in the monitor-window on the right. These are all HIV-1 group M (clade B)/group O sequences pairs. The cluster of data points directly above the highlighted group represents the comparisons between the sequences of the HIV-2/SIV group and SIVagm sequences, and the cluster above these compares the HIV-1 with HIV-2 or SIV sequences, which form the most distant sequence pairs in the data set.

groups. The comparisons between the PBJ isolates (SMMPBJ14.41, SMMPBJ14.15 and SMMPBJ6.6) are shown as *P*. These viruses are derived from SMMH9 and can infect macaques where they are highly pathogenic (Courgnaud *et al.*, 1992). Note that these viruses exhibit the highest ratio of non-synonymous to synonymous mutations in the data set. Comparisons of the PBJ sequences to their parental SMMH9 sequence are shown as *9* while those to SMMH4 comparisons are displayed as *4*. The latter resemble the SMMH9 vs. SMMH4 comparison labeled *S* and are positioned similar to the HIV-2 intra subtype comparison (*A* , *B*) discussed above.

Various observations can be made of the data presented so far. The most remarkable one is that the positioning of the sequences in the distance plots by and large corresponds with the pathogenicity of the particular virus. While PBJ viruses (P) can kill infected macaques within 8 days (Dewhurst *et al.*, 1990) the SIVmac viruses (M) kill macaques within months of infection. HIV-1 (1) kills infected humans in about 10 years while HIV-2 (A and B) is usually less virulent. Most benign are SIVagm (G) which appear not to cause disease in their natural host. Shpaer and Mullins (1993) have previously observed a correlation between the pathogenicity of

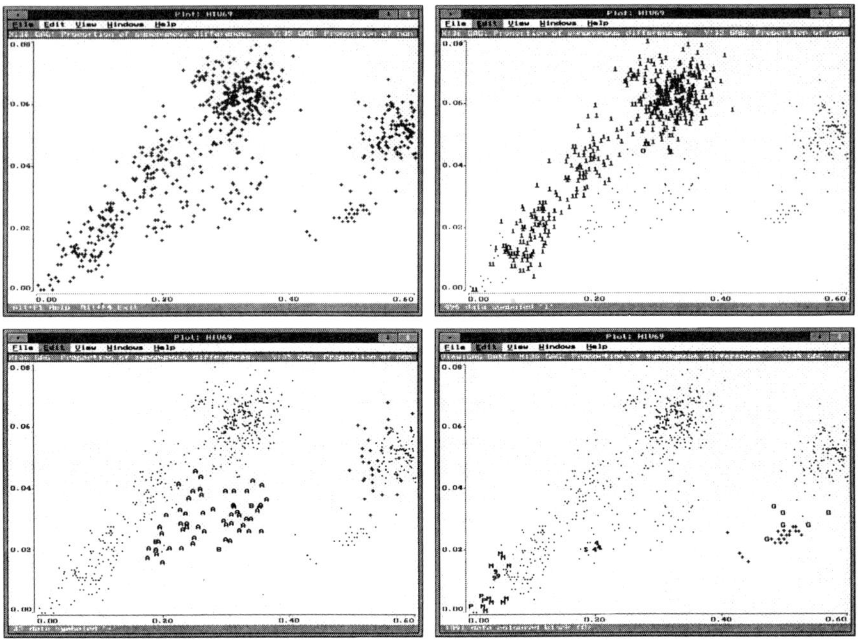

Figure 4 Identification of data-points. The distance plots are equivalent to the right plot in Figure 3, however only the region up to 0.08 synonymous mutations per synonymous site and 0.6 non-synonymous mutations per non-synonymous site is given. The sequence comparisons have been labeled as follows: *1*: HIV-1-M/HIV-1-M; *O*: HIV-1-O/HIV-1-O; *A*: HIV-2-A/HIV-2-A; *B*: HIV-2-B/HIV-2-B; *G*: SIVagm/SIVagm; *P*: SIBpbj/SIVpbj; *9*: SIVpbj/SIV-SMMH9; *4*: SIVpbj/SIV-SMMH4; *S*: SIV-SMMH4/SIV-SMMH9 (see text).

lentiviruses and the silent replacement ratio. The positioning of viruses in the plot also reflects the approximate time the virus has had to adjust to their host. While the PBJ viruses were isolated shortly after being experimentally passed from sooty mangabeys to macaques (Courgnaud *et al.*, 1992), SIVmac has arisen from an earlier host-switch involving the same species (Novembre *et al.*, 1992). Many researchers agree that HIV-1 has entered the human population less than 100 years ago while evidence suggests that SIVagm has co-evolved with the different species of African green monkeys for at least several thousand years (Allan *et al.*, 1991).

The distance plots in Figure 5 contrast the evolutionary trends of different genes for the same viruses. While the plot in the top left represents and combines the information presented for *gag* in Figure 4, others present analogous plots for *env* (top right), *vif* (middle left), and *pol* (middle right). Note that the relative positions of HIV-1-intra group and HIV-2/SIV intra subtype comparison (*1*, *O*, *A*, *B*, *M*, *P*, *S*, *4*, *9*) remain relatively steady, while the position of the SIVagm (G) and the HIV-2/SIV inter subtype comparisons (+) vary to a larger degree. The largest overall difference can be found between the plots for *gag* and *pol* and the two genes were compared directly. The bottom left plot shows that the proportion of synonymous changes in *pol* (horizontal) and *gag* (vertical) correlate well. The analogous comparison of non-synonymous mutations (bottom right) clearly differs, as the

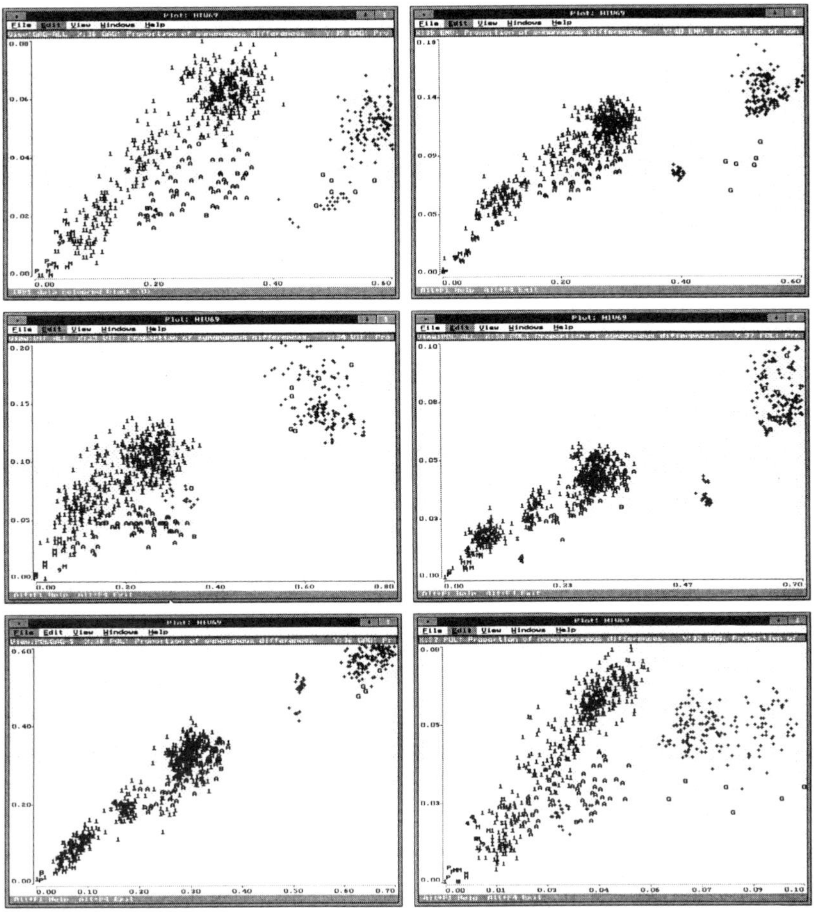

Figure 5 Distance plots of different genes. All taxa and labels are as in Figure 4. The top 4 plots compare the proportion of synonymous mutations (horizontal) and non-synonymous mutations (vertical) of *gag* (top left), *env* (top right), *vif* (middle left), and *pol* (middle right). Bottom left: plot giving the proportion of synonymous changes in *pol* (horizontal) and *gag* (vertical). Bottom right: plot giving the proportion of non-synonymous changes in *pol* (horizontal) and *gag* (vertical).

SIVagm sequences (G) and HIV-2/SIV inter subtype comparisons (+) assume a quite different position, revealing that the *gag* gene of these viruses is more conserved than *pol*. The converse is true for most of the other viruses in the data set. These observations agree with the findings made by Seibert and coworkers (1995).

4. DISENTANGLING THE MOSAIC STRUCTURE OF HIV SEQUENCES

The influence of recombination in the evolution of HIV is well established and many case have been published (see Robertson *et al.*, 1997). Most of these studies were restricted to detecting inter-subtype recombinations as they relied on methods that require non-recombinant reference sequences. Finding such sequences may not always be possible and some reference sequences may constitute mosaics themselves. Methods that do not require such prototypes should therefore be able to find recombination sites not detected so far and in particular help identify intra-subtype recombination as well as recombinations involving subtypes not yet characterized. An example where the phylogenetic profile method has been used to explore the reticulate nature of HIV-1 is given below.

5. ANALYZING PHYLOGENETIC PROFILES WITH PHYLPRO

The methodical details of the phylogenetic profile method have been presented elsewhere (Weiller, 1998b). Briefly, the method introduces the "phylogenetic correlation" measure that quantifies the coherence of the sequence interrelationships in two different regions of a multiple alignment. Positions in which sequence relationships in the upstream region clearly differ from their downstream counterpart exhibit low phylogenetic correlations and are likely recombination sites. For each individual sequence in the alignment, the phylogenetic correlations are computed at every position using a sliding window technique. The plot of the phylogenetic correlations against the sequence positions is termed a "phylogenetic profile," and the profiles of all individual sequences are typically superimposed in a single diagram. Such profiles support the identification of individual recombinant sequences as well as recombination hotspots.

To determine the phylogenetic correlation of a given test-sequence at a given test-location, the method defines two sequence windows located immediately before and after the test-location, and determines the genetic distances (e.g. p-distances) between the test-sequence and all other sequences within the window limits, resulting in two vectors of distance data. If the test-sequence relates to the other sequences in a similar way in both windows then the two distance vectors will correlate well. The linear (Pearson) correlation coefficient was used to determine the phylogenetic correlations for all phylogenetic profiles presented here, thus phylogenetic correlations are limited to values between +1.0 and -1.0, with high positive values representing a high phylogenetic correlation and values around 0 representing uncorrelated relationships that indicate sites of genetic recombination. Values near -1.0 correspond to inverse correlations and were rarely observed.

Several parameters such as the distance metric used influence the creation of phylogenetic profiles. The width of the windows used for determining the sequence relationships is probably the most important parameter. While too narrow windows increase the influence of stochastic variations and therefore the noise level in the resulting plots, excessively large windows might encompass several recombination sites and consequently limit the resolution of the method. Suitable window sizes depend on the sequence data and must be determined empirically. The size of the analysis windows in PhylPro can be defined not only by the length of a sequence, but also by the number of sequence differences encountered. In this case the length of the segment in which the test-sequence is compared varies for each background sequence. This technique, which ensures that the window size suffices in all pairwise comparisons, has been employed for all profiles given here.

As the method does not rely on a predefined reference sequence, every sequence in the data set is used to provide the background information against which the test sequence is measured. A recombination in one or more of the background sequences can also lower the phylogenetic correlation obtained for the test sequence, and further considerations are required to identify which sequence carries the recombination junction. Fortunately, in most analyses it is reasonable to assume that at a particular position of the alignment only one or a minority of sequences in the data set represent the recombinant variant. In these situations the true recombinant can easily be distinguished as its phylogenetic correlation is clearly lower than that of the other sequences. Furthermore, removal of the recombinant sequence(s) will increase the phylogenetic correlation of the background sequences.

Obviously, the individual recombination sites are best identified when one recombinant sequence is compared with a background of sequences that are themselves not recombinant but represent a balanced set of taxa encompassing and discriminating the spectrum of sequences that have recombined. However as a considerable proportion of HIV sequences may be recombinant finding genuine non-recombinant reference sequences is problematic. A phylogenetic profile of a data set can help to identify sequences with largely coherent phylogeny suited to provide a background against which likely recombinants can be tested.

In the following example phylogenetic profiles of the *gag* locus in HIV-1 have been produced using the PhylPro program. For this analysis the *gag* sequence alignment has been obtained from the HIV database. The alignment has been imported into PhylPro, and an initial data set containing all group M sequences has been defined (126 sequences). After visual examination of the alignment in the sequence window, 17 sequences that exhibited extended regions of missing data have been excluded to simplify the analysis. A variety of phylogenetic profiles for the remaining 109 sequences have been created using different distance metrics, analysis windows, and correlation measures. While the profiles obtained differed in minor details they all showed that the phylogenetic signal in many of these sequences is quite inconsistent, indicating that recombination has played a substantial role in their evolution. The simple proportion of different nucleotides (p-distance) and a window that contained 20 nucleotide differences for each pairwise comparison appeared to be well suited for the data-set, and all profiles presented here were created with these parameters. The upper part of Figure 6 gives a phylogenetic profile of the 109 *gag* sequences. The profile plot shows that a large number of sequences have very poor phylogenetic correlation suggesting that the data set may contain many recombination sites. As individual recombination junctions are best examined when one or a few recombinants are compared against a background of sequences with consistent phylogeny, the next steps were aimed at identifying candidate background sequences. This was done by removing 26 sequences that contained sites of particularly low phylogenetic correlation (boxed in

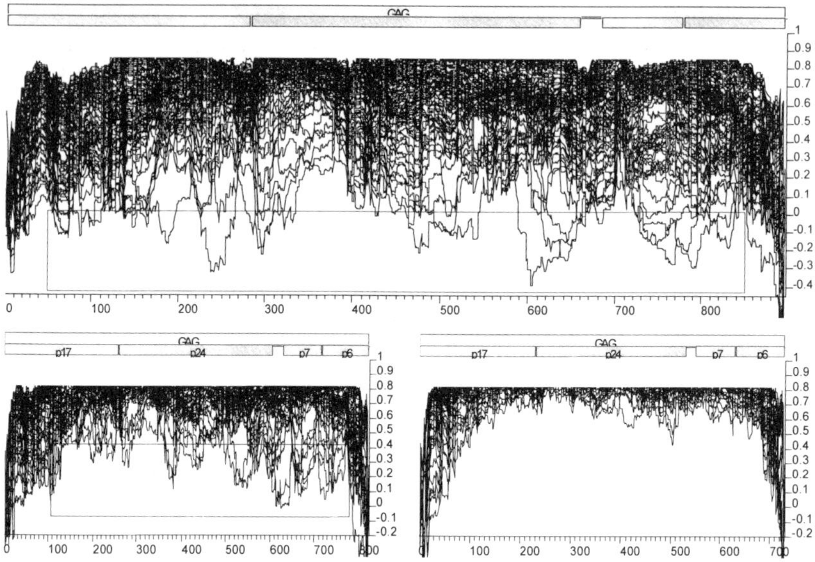

Figure 6 Phylogenetic profiles of the *gag* gene of HIV-1. Phylogenetic profiles including a graphical representation of the sequence features printed above the plot were created with the PhylPro program using the parameters as given in the text. The vertical axis gives the phylogenetic correlation from –0.4 to 1 on the top, and –0.2 to 1 in the bottom plots. The horizontal axis gives the positions of the variable columns in the alignment from which the phylogenetic correlations were calculated. Top: profile containing all 109 *gag* sequence profiles. Sequences with profiles extending into the boxed area are excluded from the profile on the bottom left. Bottom left: profiles of the remaining 83 sequences. Sequences with profiles extending into the boxed area are excluded from the profile on the bottom right. Bottom right: Profiles of the remaining 69 background sequences. The sequences are as follows with "*" indicating the 26 sequences removed initially and "**" the 14 sequences removed in the second step: A_U455**, A92UG037, A_VI59, A_VI310, A_VI57, A_K112, A_K88, A_K29, A_K7, A_K98, A_K89, A_VI132, A_VI415, A_CI4, A_LBV23**, A_CI20, A_CI59, A_LBV2310, A_CI51, A_IC144, A_DJ258, A_UG266**, A_Q23, B_C18MBC, B_HXB2R, B_LAI-LW123, B_AD8, B_WR27, B_SF2, B_BZ167, B_PH153, B_PH136, B_TB132, B_BZ190, B_LAI, B_MN, B_JH31, B_JRCSF, B_JRFL, B_OYI, B_NY5, B_NL43, B_CDC41, B_HAN, B_CAM1, B_RF, B_D31, B_UG280, B_YU2, B_YU10, B_BCSG3C, B_P896, B_GAG46, B_MANC, B_GAG314, B_WEAU160, B_3-13, B_4-7, B_3202A12, B_3202A21, C_ETH2220*, C_92BR025*, C_UG268**, C_SM145**, C_ZAM20**, C_DJ259**, C_VI313**, D_84ZR085, D_94UG114.1**, D_ELI, D_Z2Z6, D_NDK, D_VI205**, D_G109, D_K3*, D_UG274**, D_UG270*, D_SE365**, D_VI203*, F_93BR020.1*, F_VI174*, F_VI69*, F_BZ162*, F_VI325*, G_LBV217**, H_VI525*, H_VI557*, H_90CR056*, AC_RW009.6*, AC_ZAM184*, AD_K124*, AD_CI32*, AD_G141*, ADI_MAL*, AE_CM240, AE_93TH253, AE_TN2431, AE_TN2431, AE_TN245, AE_CM238, AG_VI354*, AG_LBV105*, AG_IBNG, AG_VI191*, AG_92NG083.2*, AG_92NG003.1**, AGI_Z321*, BF_BZ200*, BF_93BR029.4*.

Figure 6, top) using PhylPro's interactive graphics features. Next a new profile of the remaining 83 sequence was created (Figure 6 bottom left) and 14 more sequences with regions of poor phylogenetic correlation (boxed) were excluded. The profile of the remaining 69 sequences is given in Figure 6 (bottom right). Several attempts to exclude more sequences have not resulted in a significant improvement of the overall phylogenetic consistency of the data set. Although it appears very likely that many of the 69 sequences also contain recombinations their phylogenetic correlations are clearly better than those of the sequences removed in the preceding steps, and they are therefore suitable as a background against which the previously excluded sequences can be tested.

The previously removed sequences have then been reintroduced into the data set individually. Figure 7 gives many of these phylogenetic profiles, each plot containing the 69 background sequences plus the reintroduced sequence. In each of

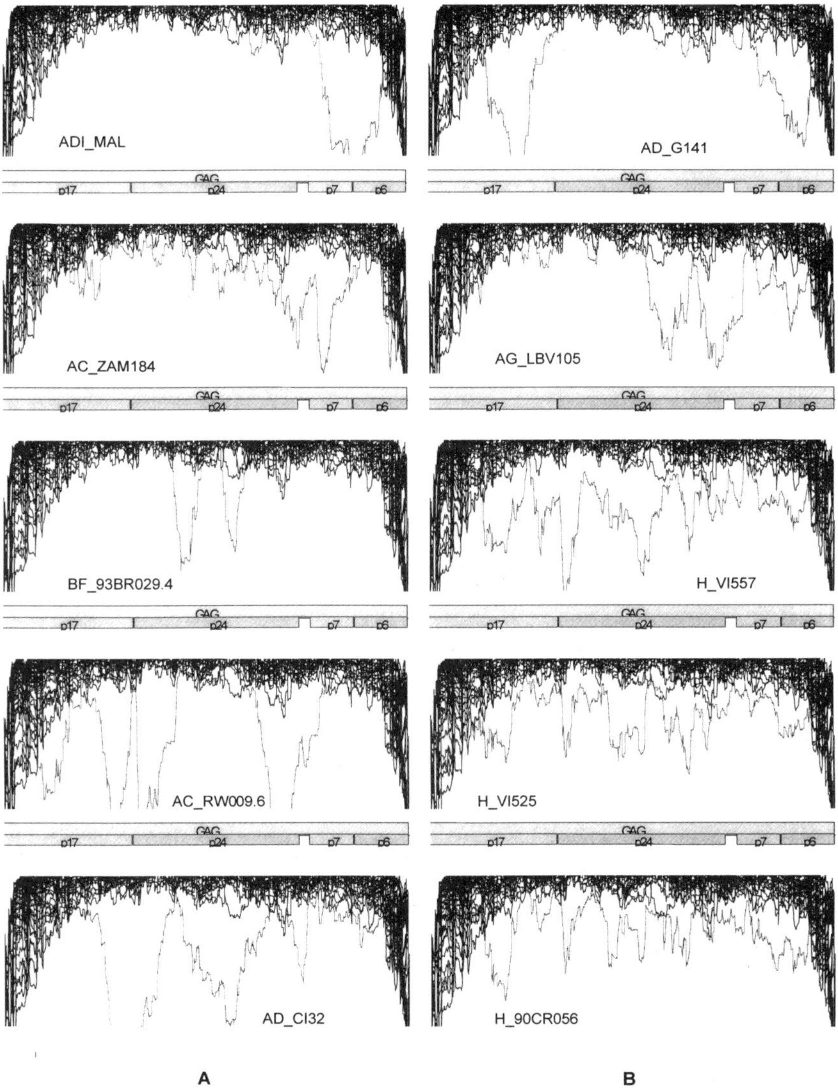

Figure 7 Phylogenetic profiles of the *gag* gene of HIV-1. Each plot contains the 69 background sequences plus one reintroduced sequence. The axes correspond to the bottom right plot of Figure 6. The acronym of the reintroduced sequence is given in each plot. (See text.)

the plots the profile of the reintroduced sequence very clearly suggests the site where the recombination(s) took place. To my knowledge, only 4 of these putative recombination sites have previously been reported (ADI_MAL, AC_ZAM184, BF_93BR029.4, AC_RW009.6, Figure 7 column A) and the profiles are in agreement with the previous findings. However many of the additional recombination signals shown here for the first time are similarly strong. Although further confirmation is required, it is possible that the requirement for non-recombinant reference sequences has severely limited previous studies. Note that

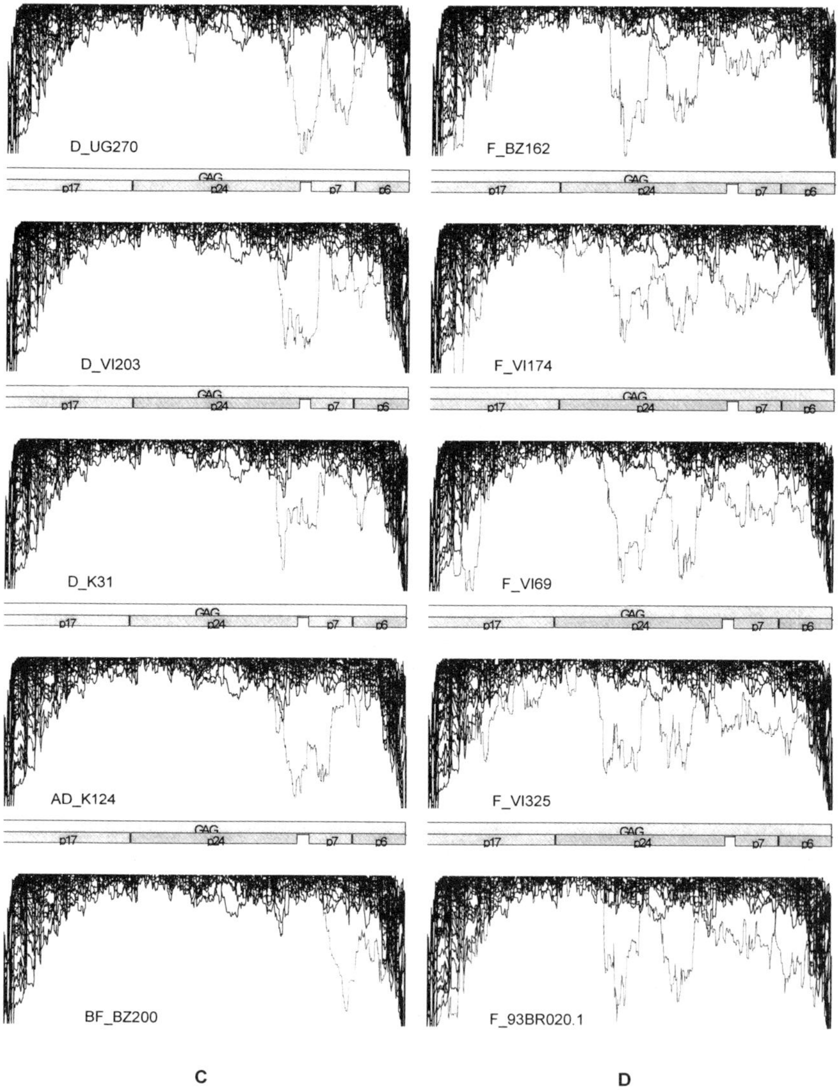

Figure 7, cont'd. Phylogenetic profiles of the *gag* gene of HIV-1

the data set of the 69 background sequences used here only contain sequences assigned to the subtypes A, B, and D. A preliminary analysis suggests that all subtype H sequences are potentially recombinant as they exhibit a large number of potential recombination sites (Figure 7, lower 3 plots in column B). The comparisons of the profiles of the three subtype H sequences shows that their profiles are very similar, featuring the same potential recombination sites. Using

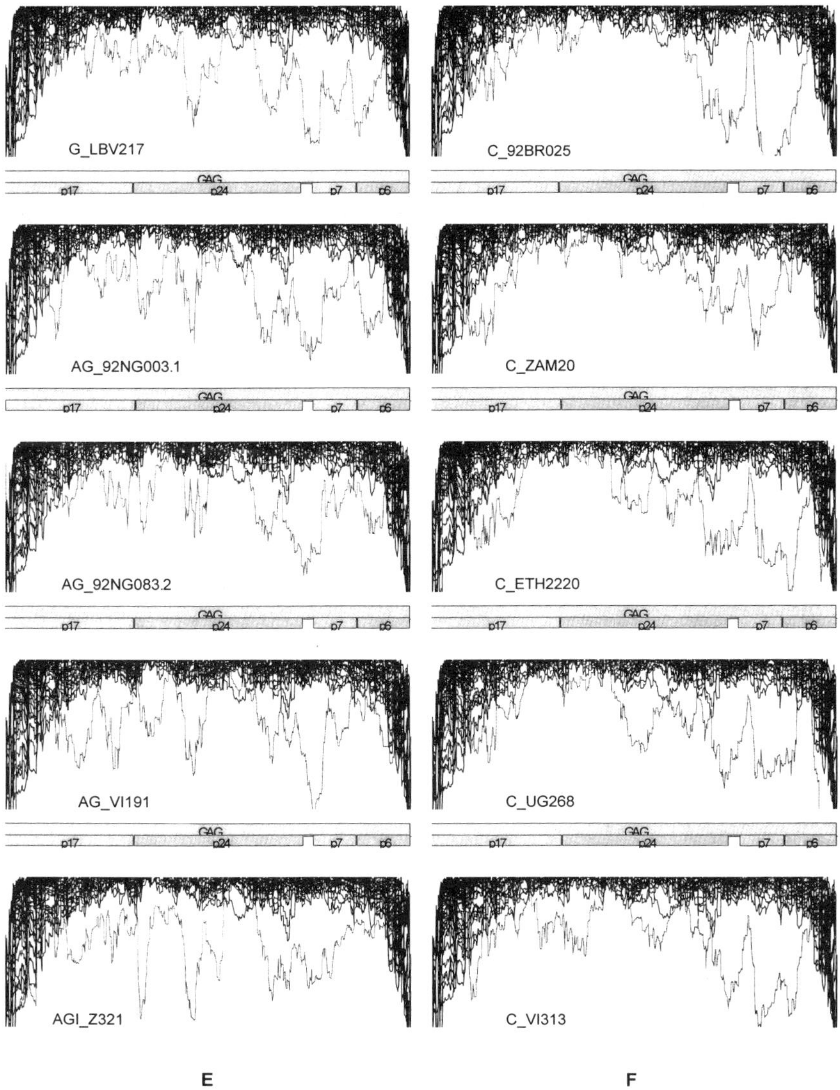

Figure 7, cont'd. Phylogenetic profiles of the *gag* gene of HIV-1

PhylPro's "relationships"-feature it was observed that all the potential sites combine subtype A and subtype B characteristics, and a similar analysis involving the *pol* gene has given an analogous result (data not shown). Leitner and colleagues (1997) have also observed the close association of subtype H and A and the possibility that subtype H was founded by an A/B hybrid should be further investigated. A similar situation might apply to some sequences currently assigned to the subtypes D (column C), F (column D), G (column E) and C (column F). The profiles of many viruses indicated within each of these columns are quite similar suggesting that they are descendants of common recombinant founder viruses.

In PhylPro, possible recombination sites can be selected in the graph and the underlying sequence alignment displayed. Most of the recombination sites indicated in Figure 7 were found to map to regions containing many gaps in the original alignment. This is encouraging, as alignment gaps are indicative of sequence recombination. Note however that gaps do not interfere with profile creation. PhylPro's ability to map the profiles to known sequence features revealed a further phenomenon: most putative recombinations are located close to the boundaries of the genes contained in the *gag* locus (p17, p24, p7, p6). In addition two hotspots of recombination are suggested inside p24 as can be seen in BF_93BR029.4 in column A, and also in subtype F in column D and subtype G in column E (1 site), and some others of Figure 7. Gibbs, Waterhouse and Weiller (1997) have previously observed that recombinations in viruses occur frequently not only between distinct genes but also between distinct protein domains. To examine whether the two putative recombination hot-spots in *gag* are located on domain boundaries, the sequence positions were mapped to the structure of the amino-terminal core domain of p24 as determined by Gitti and colleagues (1996). It was observed that one of the presumed hotspots maps to a region upstream of the cyclophilin-binding site (CryA). A more detailed examination mapped the crossover site to the beginning of helix 4 in some of the sequences, and to the end of helix 4 in others. The second hotspot could be mapped to the end of the core domain of p24 thus adding further support to the observation of Gibbs, Waterhouse and Weiller (1997). As more structural information becomes available, it will become possible to determine whether most HIV recombinations occur on domain boundaries.

The purpose of this text has been to demonstrate the potential of graphical methods in data exploration and it must be noted that the presumed recombination sites and evolutionary suggestions given above represent initial stages of an analysis that need further confirmation.

ACKNOWLEDGEMENTS

I thank A. Van den Borre for help with data analysis and production of figures.

REFERENCES

Allan, J. S., Short, M., Taylor, M. E., Su, S., Hirsch, V. M., Johnson, P. R., Shaw, G. M. and Hahn, B. H. 1991. Species-specific diversity among simian immunodeficiency viruses from African green monkeys. *J. Virol.* **65:** 2816-2828.

Courgnaud, V., Laure, F., Fultz, P. N., Montagnier, L., Brechot, C. and Sonigo, P. 1992. Genetic differences accounting for evolution and pathogenicity of simian immunodeficiency virus from a sooty mangabey monkey after cross- species transmission to a pig-tailed macaque. *J. Virol.*: **66**: 414-419.

Dewhurst, S., Embretson, J. E., Anderson, D. C., Mullins, J. I. and Fultz, P.N. 1990. Sequence analysis and acute pathogenicity of molecularly cloned SIVSMM-PBj14. *Nature* **345**: 636-640.

Everitt, B. S. and Dunn, G. 1992. *Applied Multivariate Analysis*. Oxford University Press, New York.

Felsenstein, J. 1993. *PHYLIP: Phylogeny Inference Package*. University of Washington, Seattle.

Fitch, D. H. A. and Goodman, M. 1991. Phylogenetic scanning: a computer-assisted algorithm for mapping gene conversions and other recombinational events. *CABIOS* **7**: 207-215.

Gibbs, M. J., Waterhouse, P. M. and Weiller, G. F. 1997. Analysis of natural viral recombination may assist the design of new virus resistance traits, In *Commercialisation of Transgenic Crops: Risk, Benefit and Trade Considerations*. (McLean, G.D., Evans, G., Waterhouse, P.M. and Gibbs, M.J., eds.) Australian Government Publishing Service, Canberra, ACT, Australia.

Gitti, R. K., Lee, B. M., Walker, J., Summers, M. F., Yoo, S. and Sundquist, W. I. 1996. Structure of the amino-terminal core domain of the HIV-1 capsid protein. *Science* **273**:231-235.

Gojobori, T., Moriyama, E. N. and Kimura, M. 1990. Statistical methods for estimating sequence divergence, In *Methods in Enzymology* (Doolittle, R.F., ed.) Academic Press, Orlando, FL.

Hein, J. 1993. A heuristic method to reconstruct the history of sequences subject to recombination. *J. Mol. Evol.* **36**: 396-405.

Higgins, D. G. 1992. Sequence ordinations: a multivariate analysis approach to analysing large sequence data sets. *Comput. Appl. Biosci.* **8**: 15-22.

Huelsenbeck, J. P. and Bull, J. J. 1996. A likelihood ratio test to detect conflicting phylogenetic signal. *Syst. Biol* **45:** 92-98.

Jakobsen, I. B. and Easteal, S. 1996. A program for calculating and displaying compatibility matrices as an aid in determining reticulate evolution in molecular sequences. *Comput. Appl. Biosci.* **12**: 291-295.

Jakobsen, I. B., Wilson, S. R. and Easteal, S. 1997. The partition matrix: exploring variable phylogenetic signals along nucleotide sequence alignments. *Mol. Biol. Evol.* **14**: 474-484.

Jukes, T. H. and Cantor, C. R. 1969. Evolution of protein molecules, In *Mammalian Protein Metabolism* (Munro, H.N., ed.), Academic Press, New York.

Kishino, H. and Hasegawa, M. 1989. Evaluation of the maximum likelihood estimate of the evolutionary tree topologies from DNA sequence data, and the branching order in hominoidea. *J. Mol. Evol.* **29**:170-179.

Korber, B., Foley, B., Leitner, T., McCutchan, F., Hahn, B., Mellors, J. W., Myers, G. and Kuiken, C. 1997. *Human retroviruses and AIDS 1997. A Compilation and Analysis of Nucleic Acid and Amino Acid Sequences*. Los Alamos National Laboratory, Los Alamos, NM.

Kuiken, C. L., Nieselt-Struwe, K., Weiller, G. F. and Goudsmit, J. 1994. Quasispecies behavior of human immunodeficiency virus type 1: sample analysis of sequence data. *Meth. Mol. Genet.* **4**: 100-119.

Kumar, S., Tamura, K. and Nei, M. 1994. MEGA: Molecular Evolutionary Genetics Analysis software for microcomputers. *Comput. Appl. Biosci.* **10**: 189-191.

Leitner, T., Korber, B., Robertson, D., Gao, F. and Hahn, B. 1997. Updated proposal of reference sequences of HIV-1 genetic subtypes, In *Human retroviruses and AIDS 1997. A Compilation and Analysis of Nucleic Acid and Amino Acid Sequences*. (Korber, B., Foley, B., Leitner, T., McCutchan, F., Hahn, B., Mellors, J. W., Myers, G. and Kuiken, C., eds.) Los Alamos National Laboratory, Los Alamos, NM.

Maynard Smith, J. 1992. Analyzing the mosaic structure of genes. *J. Mol. Evol.* **34**: 126-129.

McGuire, G., Wright, F. and Prentice, M. J. 1997. A graphical method for detecting recombination in phylogenetic data sets. *Mol. Biol. Evol.* **14**: 1125-1131.

Novembre, F. J., Hirsch, V. M., McClure, H. M., Fultz, P. N. and Johnson, P.R. 1992. SIV from stump-tailed macaques: molecular characterization of a highly transmissible primate lentivirus. *Virology* **186**: 783-787.

Robertson, D. L., Gao, F., Hahn, B. H. and Sharp, P. M. 1997. Intersubtype recombinant HIV-1 sequences, In *Human retroviruses and AIDS 1997. A Compilation and Analysis of Nucleic Acid and Amino Acid Sequences.* (Korber, B., Foley, B., Leitner, T., McCutchan, F., Hahn, B., Mellors, J. W., Myers, G. and Kuiken, C., eds.) Los Alamos National Laboratory, Los Alamos, NM.

Rohlf, F. J. 1993. *NTSYS-pc. Numerical Taxonomy and Multivariate Analysis System.* Exeter Software, Setauket, New York.

Salminen, M. O., Carr, J. K., Burke, D. S. and McCutchan, F. E. 1995. Identification of breakpoints in intergenotypic recombinants of HIV type 1 by bootscanning. *AIDS Res. Hum. Retrovir.* **11**: 1423-1425.

Sawyer, S. 1989. Statistical tests for detecting gene conversion. *Mol. Biol. Evol.* **6**: 526-538.

Seibert, S. A., Howell, C. Y., Hughes, M. K. and Hughes, A. L. 1995. Natural selection on the gag, pol, and env genes of human immunodeficiency virus 1 (HIV-1). *Mol. Biol. Evol.* **12**: 803-813.

Shpaer, E. G. and Mullins, J. I. 1993. Rates of amino acid change in the envelope protein correlate with pathogenicity of primate lentiviruses. *J. Mol. Evol.* **37**: 57-65.

Siepel, A. C., Halpern, A. L., Macken, C. and Korber, B. T. 1995. A computer program designed to screen rapidly for HIV type 1 intersubtype recombinant sequences. *AIDS Res. Hum. Retrovir.* **11**: 1413-1416.

Stephens, J. C. 1985. Statistical methods of DNA sequence analysis: detection of intragenic recombination or gene conversion. *Mol. Biol. Evol.* **2**: 539-556.

Swofford, D. L., Olsen, G. J., Waddell, P. J. and Hillis, D. M. 1996. Phylogenetic inference, In *Molecular Systematics* (Hillis, D. M., Moritz, C. and Mable, B. K., eds.) Sinauer Associates, Inc., Sunderland, MA.

Thompson, J. D., Higgins, D. G. and Gibson, T. J. 1994. CLUSTAL W: improving the sensitivity of progressive multiple sequence alignment through sequence weighting, position-specific gap penalties and weight matrix choice. *Nucl. Acids Res.*. **22**: 4673-4680.

Weiller, G. F. 1998a. Interactive computer graphics for analysing molecular sequences, In: *Advances in Computational Life Sciences* (Michalewicz, M.T., ed.) CSIRO Publishing, Melbourne.

Weiller, G.F. 1998b. Phylogenetic profiles: a graphical method for detecting recombinations in homologous sequences. *Mol. Biol. Evol.* **15**: 326-335.

Weiller, G.F. and Gibbs, A. 1995. DIPLOMO: the tool for a new type of evolutionary analysis. *CABIOS* **11**:535-540.

Weiller, G.F., McClure, M.A. and Gibbs, A.J. 1995. Molecular phylogenetic analysis. In: *Molecular basis of virus evolution.* (Gibbs, A.J., Calisher, C.H. and Garcia-Arenal, F., eds.), Cambridge University Press, Cambridge, pp. 553-585.

QUANTIFYING HETEROGENEITY IN THE HIV GENOME

Hildete Prisco Pinheiro* and
Françoise Seillier-Moiseiwitsch[§]

*Department of Statistics, State University of Campinas,
13081-970 Campinas, São Paulo, Brazil
[§]Department of Biostatistics, University of North Carolina at Chapel Hill,
Chapel Hill, NC 27599 USA

1. INTRODUCTION

Comparing the variability in sets of HIV sequences is key to understanding selection pressures and other dynamical processes these sequences are subject to. Here are sample questions the methodology described in this paper is geared to tackle. Is the viral population more homogeneous in plasma than in semen? Are therapeutic agents having an impact on viral diversity? In any one individual does the viral diversity remain constant over time?

We review a number of diversity measures commonly used in population genetics for both within-population and between-population comparisons (Section 2). We also consider different distance metrics for pairwise sequence comparisons (Section 3). We outline their characteristics and assess their appropriateness for HIV sequence data. In particular, we evaluate the relevance of each measure for specific types of hypotheses. We then describe two approaches that allow the inclusion of important covariate information in the analysis. One approach deals with correlating distance matrices and covariates (Section 4). The other approach involves covariates through an analysis of variance (Section 5). We end the paper with a data analysis to illustrate this novel CATANOVA methodology (Section 6).

2. GENETIC DIVERSITY MEASURES

Many metrics have been constructed to measure distances. Examples of such metrics are the Hamming, Nei and Mahalanobis distances (Jorde, 1980; Lalouel, 1980; Chakraborty and Rao, 1991; Seillier-Moiseiwitsch *et al.*, 1994 and references therein). They fall into two classes: they measure genetic variability either within or between populations.

In the present context, a locus is simply a position along the genome and the alleles the amino acids or nucleotides at that position. A population is a set of sequences with common characteristics, as stipulated by the objective of the study. For instance, if the goal of the study is to characterize the global variation in the envelope gene (*env*), the population of interest is made up of sequences collected in different countries, with the relevant subgroups consisting of sequences from the same clade. To illustrate further, in a longitudinal study of diversity, each individual defines a population, and sets of sequences sampled at the same time form the subgroups to be compared. The metrics described below are appropriate when considering a single position or a small number of positions so that the total number of possible alleles is not too large (with several positions, an allele is made up of a series of labels, one for each position). In particular, they are suitable for studying jointly the positions along protease that harbor resistance mutations to a specific protease inhibitor.

2.1 Within-Population Measures

2.1.1 Gene Diversity

The Simpson Index of ecological diversity (Simpson, 1949), also known as gene diversity (Nei, 1972; Lewontin, 1972) is a measure of genetic variation at a specific locus. If $p_1, p_2, \ldots, p_k$, represent the true relative frequencies of k alleles at a locus in a population, the gene diversity at that locus is

$$h = 1 - \sum_{i=1}^{k} p_i^2 \tag{1}$$

This is in fact the probability that two genes randomly chosen from the population are dissimilar in their allelic type. Its sample estimate $\hat{h}$ is obtained by replacing p_i by the sample relative frequency $\hat{p}_i$. Though the bias in this estimator is small for even moderate sample sizes ($n \approx 100$), it is recommended to use

$$\frac{n}{n-1}(1 - \sum_{i=1}^{k} \hat{p}_i^2) \tag{2}$$

which is unbiased (Nei and Roychoudhury, 1974; Nei, 1978).

2.1.2 Shannon-Information Index

The Shannon-Information Index

$$h_S = -\sum_{i=1}^{k} p_i \log_e p_i \tag{3}$$

finds its origin in the concept of entropy functions in information science and theoretical physics. It has also been used in the context of evolutionary and ecological studies (Lewontin, 1972; Rao, 1982a,b; Magurran, 1988). Again, its sample counterpart is biased (Hutcheson, 1970; Bowman *et al.*, 1971). Adjusting for this bias is not straightforward, as it is a function of the p_i's. However,

$$\hat{h}_S = -\sum_{i=1}^{k} \hat{p}_i \log_e \hat{p}_i + \frac{k-1}{n} \tag{4}$$

corrects most of the bias: it is now of order n^{-2} and thus can be ignored when $n >$ 100 (Peet, 1974).

It is important to point out that for evolutionary studies this measure has some disadvantages. First, its biological meaning is not clear. Second, even though its minimum value is zero (for a constant locus), it can take very large values. When each of the k alleles is equally frequent in the population, h_s attains its maximum value $\log_e k$. Its sampling distribution depends on the underlying relative frequencies. For independent loci, because of the additive property of the logarithms of products, the h_s's measured over each locus can be averaged to evaluate the average diversity per locus within a population.

2.1.3 *Unified Measures of Diversity*

In an effort to understand their mathematical properties, measures of genetic variation have been classified under different mathematical frameworks. All measures considered here belong to a general class with sound mathematical foundations.

Rao (1982c) formulated a characterization theorem which states that for a single locus with k different alleles occurring with relative frequencies $p_1, p_2, \ldots p_k$ in a population, if a measure of diversity $h(\mathbf{p}) = h(p_1, p_2, \ldots p_k)$ satisfies the following postulates:

(a) $h(\mathbf{p})$ is symmetric with respect to the components of $\mathbf{p}$ and attains its maximum when all k categories are equally frequent
(b) $h(\mathbf{p})$ admits partial derivatives up to the second order of the k - 1 independent components of $\mathbf{p}$ and the matrix of second partial derivatives, $(h''_{ij}(\mathbf{p}) = (h''_{ij}(\mathbf{p}))$ for $i, j = 1, 2,\ldots, k\text{-}1$ with $h''_{ij}(\mathbf{p}) = \partial^2 h(\mathbf{p})/\partial p_i \partial p_j$, is continuous and not null at $\mathbf{p} = \mathbf{e} \equiv (k^{-1}, k^{-1},\ldots, k^{-1})$
(c) $h\{(\mathbf{p}+\mathbf{e})/2\} = {}^1/_2\{h(\mathbf{p}) + h(\mathbf{e})\} = c\{h(\mathbf{e}) - h(\mathbf{p})\}$ where c is a constant

then $h(\mathbf{p})$ must be of the form

$$h(\mathbf{p}) = a\left[1 - \sum_{i=1}^{k} p_i^2\right] + b \tag{5}$$

where $a > 0$ and b are constants. This theorem therefore essentially characterizes the gene diversity index (1) and its relatives.

The Shannon-Information Index (3) and the four indices (Rao, 1982a; Rao and Boudreau, 1984),

α-order entropy of Havrda and Charavát

$$h_\alpha(\mathbf{p}) = [1 - \sum_{i=1}^{k} p_i^\alpha]/[2^{\alpha-1} - 1] \text{ for } \alpha > 0 \text{ and } \alpha \neq 1 \tag{6}$$

paired Shannon entropy

$$h_p(\mathbf{p}) = -\sum_{i=1}^{k} p_i \log_e p_i - \sum_{i=1}^{k} (1 - p_i) \log_e (1 - p_i), \tag{7}$$

α-degree entropy of Renyi

$$h_R(\mathbf{p}) = (1-\alpha)^{-1} \log_e (\sum_{i=1}^{k} p_i^\alpha) \text{ for } 0 < \alpha < 1, \tag{8}$$

γ-entropy function

$$h_\gamma(\mathbf{p}) = [1 - (\sum_{i=1}^{k} p_p^{1/\gamma})^\gamma]/[1 - 2^{\gamma-1}] \text{ for } \gamma > 0, \gamma \neq 1, \tag{9}$$

satisfy the following two conditions:

C_1: $h(\mathbf{p}) = 0$ if and only if all components of $\mathbf{p}$ are zero except one (i. e, $p_i = 1$ for one i and the remaining p_i's are all zero)

C_2: $h\{\lambda\mathbf{p} + (1 - \lambda)\mathbf{q}\} \geq \lambda h(\mathbf{p}) + (1 - \lambda)h(\mathbf{q})$, with equality if and only if $\mathbf{p} = \mathbf{q}$ (concavity property).

The index

$$h_H(\mathbf{p}) = \left[\sum_{i=1}^{k} p_i^\alpha\right]^{1/(1-\alpha)} \tag{10}$$

is the basis of another approach to unify diversity measures (Hill, 1973). For various values of α, it reduces to well-known indices.

In comparing DNA or amino-acid sequences, the above measures of diversity involve one specific locus at a time. They would therefore not be suitable to study relatively large genomic segments. If we assume independence among positions, the Shannon-Information Index and its paired version can be measured over all loci, because of the additive property of the logarithms of products. Generally in DNA sequences, however, dependencies among neighboring nucleotide positions are well-established (Tavaré and Giddings, 1989). Linked positions have been identified in HIV amino-acid sequences (Korber *et al.*, 1993; Bickel *et al.*, 1996; Karnoub *et al.*, 1999).

2.2 Between-Population Measures

2.2.1 Mahalanobis Distance

For a population α, let the allele relative frequencies at a k-allelic locus be represented by the k-dimensional vector $\mathbf{p}_\alpha = (p_{\alpha 1}, p_{\alpha 2}, \ldots, p_{\alpha k})$. For two populations α and β, the Mahalanobis distance is

$$D^2 = \sum_{i=1}^{k-1} \sum_{j=1}^{k-1} (p_{\alpha i} - p_{\beta i}) \mathbf{V} (p_{\alpha j} - p_{\beta j}) \tag{11}$$

where $\mathbf{V}$ is the $(k - 1) \times (k - 1)$ symmetric matrix with entries based on the pooled allele frequencies $^1/_2(p_{\alpha i} + p_{\beta i})$ $(i = 1, 2, \ldots, k)$ (Mahalanobis, 1936). $\mathbf{V}$ is in fact the inverse of the multinomial variance-covariance submatrix based on the independent entries of the probability vector $^1/_2(\mathbf{p}_\alpha + \mathbf{p}_\beta)$: the (i, j) element is

$$v_{ij} = \begin{cases} 2/(p_{\alpha k} + p_{\beta k}) & \text{for } i \neq j, i, j = 1, 2, \ldots, k-1 \\ 2/(p_{\alpha i} + p_{\beta i}) + 2/(p_{\alpha k} + p_{\beta k}) & \text{for } i = j. \end{cases} \tag{12}$$

Substituting (12) in (11) yields

$$D^2 = \sum_{i=1}^{k} \frac{(p_{\alpha i} - p_{\beta i})^2}{2(p_{\alpha i} + p_{\beta i})}, \tag{13}$$

which is equivalent to Sanghvi's distance (1953). When $\{p_{\alpha i} - p_{\beta i}, i = 1, \ldots, k\}$ are small, (11) is close to Bhattacharyya's distance θ^2 (Bhattacharyya, 1946) defined by

$$\cos\theta = \sum_{i=1}^{k} \sqrt{p_{\alpha i} p_{\beta i}} \tag{14}$$

These distances are intended for classifying populations rather than studying evolution. Under a stated evolutionary model, they do not follow a specific pattern: for instance, they do not necessarily increase with divergence time (Chakraborty and Rao, 1991).

2.2.2 *Nei's Distance*

When investigating distances among multiple populations, the matrix $\mathbf{V}$ in (11) involves averages over all population relative frequencies. Hence, the pairwise distances (13) are functions of relative frequencies in populations other than the two being compared. This prompted Nei (1972) to consider

$$D_m = \frac{1}{2} \sum_{i=1}^{k} (p_{\alpha i} - p_{\beta i})^2 \tag{15}$$

as the minimum distance between populations α and β. Written as

$$D_m = (\sqrt{J_{\alpha\alpha}} - \sqrt{J_{\beta\beta}})^2 + 2\sqrt{J_{\alpha\alpha} J_{\beta\beta}} (1 - \cos\theta_{\alpha\beta}) \tag{16}$$

where $J_{\alpha\alpha}$ is the probability of two genes being identical when both are sampled from population α, i.e. $J_{\alpha\alpha} = \sum_{i=1}^{k} p_{\alpha i}^2, J_{\beta\beta} = \sum_{i=1}^{k} p_{\beta i}^2$, and $\theta_{\alpha\beta}$ is the angle between the allele relative-frequency vectors of the two populations (Rao, 1982c), it is clear that it depends on both the difference in diversity between the two populations and the angle between their relative-frequency vectors. The standard genetic distance (Nei, 1972)

$$D_N = \log_e \frac{J_{\alpha\beta}}{\sqrt{J_{\alpha\alpha} J_{\beta\beta}}} \tag{17}$$

where $J_{\alpha\beta}$ is the probability of two genes being identical when one is sampled from population α and the other from population β, i.e. $J_{\alpha\beta} = \sum_{i=1}^{k} p_{\alpha i} p_{\beta i}$, only depends on the angle $\theta_{\alpha\beta}$ since

$$D_N = -\log_e \cos \theta_{\alpha\beta} \tag{18}$$

(Rao, 1982c). Extension to several loci is straightforward: for D_m, it is computed for each locus and then these locus-specific statistics are averaged, while for D_N the J's are now average values over the loci considered. D_N is useful in evolutionary studies as it can be predicted under various models (in terms of evolutionary time and effective population sizes). In a cluster analysis of populations, because D_N is not a proper metric, Rao (1982c, 1984) advocates the use of $\theta_{\alpha\beta}$.

D_m and D_N are estimated via the method of moments by

$$\hat{D}_m = (1-\hat{J}_{\alpha\beta}) - \frac{1}{2}[(1-\hat{J}_{\alpha\alpha}) + (1-\hat{J}_{\beta\beta})] \text{ and } \hat{D}_N = \log_e \frac{\hat{J}_{\alpha\beta}}{\sqrt{\hat{J}_{\alpha\alpha}\hat{J}_{\beta\beta}}} \tag{19}$$

replacing $\mathbf{p}_\alpha$ and $\mathbf{p}_\beta$ by their sample estimates $\hat{\mathbf{p}}_\alpha$ and $\hat{\mathbf{p}}_\beta$ in the J's to give the $\hat{J}$'s (Nei and Roychoudhury, 1974; Nei, 1978). The biases in the resulting estimators are respectively

$$\frac{1}{4}\left[\frac{1-J_{\alpha\alpha}}{n_\alpha} + \frac{1-J_{\beta\beta}}{n_\beta}\right] \text{ and } \frac{1}{4}\left[\frac{1-J_{\alpha\alpha}}{n_\alpha J_{\alpha\alpha}} + \frac{1-J_{\beta\beta}}{n_\beta J_{\beta\beta}}\right] \tag{20}$$

To reduce this bias, Nei (1978) replaces $\hat{J}_{\alpha\alpha}$ and $\hat{J}_{\beta\beta}$ by $J^*_{\alpha\alpha}$ and $\hat{J}^*_{\beta\beta}$ where

$$J^*_{\alpha\alpha} = \frac{2n_\alpha \sum_i \hat{p}^2_{\alpha i} - 1}{2n_\alpha - 1} \text{ and } J^*_{\beta\beta} = \frac{2n_\beta \sum_i \hat{p}^2_{\beta i} - 1}{2n_\beta - 1} \tag{21}$$

The biases are now of order n^{-2}.

2.3 Testing Hypotheses

For within-population studies, it is possible to construct large-sample tests (utilizing the moments of the multinomial distribution) for generalized diversity measures (Nayak, 1983) or any regular function of $\sum_i p_i^\alpha$ (Chakraborty and Rao, 1991). With these tests, one can compare diversity at two loci or in two populations. For instance, let $\hat{D}_1$ and $\hat{D}_2$ be estimates of diversity, respectively, among n_1 NSI and n_2 SI sequences at a specific protease site involved in drug resistance. If $\hat{\sigma}_1^2$ and $\hat{\sigma}_2^2$ denote the variance estimates for D_1 and D_2 respectively, then the following test

$$\frac{\hat{D}_1 - \hat{D}_2}{\left(\hat{\sigma}_1^2 + \hat{\sigma}_2^2\right)^{1/2}} \sim N(0,1) \tag{22}$$

can be utilized to compare diversity between two populations with different biological characteristics, provided each individual contributes a single sequence to this study. These tests can also help in assessing whether the diversity at a particular locus fits what would be expected under a specific model of evolution. For example, when sampling n random sequences from a population that has reached steady state, the expectation and variance of the diversity measure are computed under the assumption of neutrality (Kimura and Crow, 1964): e.g., for (1)

$$E(h) = \frac{n-1}{n}\frac{\theta}{\theta+1} \tag{23}$$

and

$$V(h) = \frac{n-1}{2n^3(1+\theta)}\left[\frac{5n(n-1)+2}{1+\theta} - \frac{8(n-1)(n-2)}{2+\theta} + \frac{3(n-2)(n-3)}{3+\theta} - \frac{n-1}{n^3}\left(2+\frac{n-1}{1+\theta}\right)\right] \tag{24}$$

where $\theta = 4N_e\mu$ with N_e the effective population size and μ the mutation rate per locus per generation (Chakraborty and Fuerst, 1979; Chakraborty and Griffiths, 1982). Then

$$\frac{\hat{h} - \hat{E}(h)}{\hat{V}(h)^{1/2}} \sim N(0,1) \tag{25}$$

can be used as a test of neutrality (Nei *et al.*, 1976; Fuerst *et al.*, 1977).

For between-population comparisons, a test based on Sanghvi's distance

$$2n_\alpha n_\beta \sum_{i=1}^{k} \frac{(p_{\alpha i} - p_{\beta i})^2}{n_\alpha p_{\alpha i} + n_\beta p_{\beta i}} \sim X_{k-1}^2 \tag{26}$$

is suitable to test the hypothesis that $D(\alpha, \beta) = 0$ when the sample size is large (Nei, 1987). To ensure that the asymptotic approximation is valid, it is necessary to combine the rare alleles into a single group. To test contrasts of distances or compare multiple populations, Nei's minimum distance is the only measure that can easily take into account several loci (Chakraborty, 1985; Nei, 1987). For instance, one can test $D(\alpha, \beta) = D(\gamma, \delta)$ using

$$\frac{1}{R}\sum_{r=1}^{R} \frac{\hat{D}_m(\alpha,\beta;r) - \hat{D}_m(\gamma,\delta;r)}{\sqrt{Var(\hat{D}_m(\alpha,\beta;r) - \hat{D}_m(\gamma,\delta;r)}} \sim t_{R-1} \tag{27}$$

where r indexes the loci (Chakraborty, 1985).

3 DNA SEQUENCE DISTANCES

For greatly divergent sequences, for instance sequences from different individuals or from different subtypes, multiple substitutions may have occurred at highly polymorphic sites. The observed number of nucleotide differences is thus likely to underestimate the actual number of substitutions undergone since divergence.

Hence, for comparative studies of DNA sequences (e.g., reconstructing phylogenetic relationships and evaluating the rate of evolution), statistical methods for estimating the number of nucleotide substitutions rely on models of molecular evolution. When sequences originate from the same individual, it is unlikely that the same sites have been repeatedly hit by mutations. The Hamming distance (see Section 3.3) then provides a good estimate of the actual distance between sequences.

3.1 Model-based Distances

Consider two homologous sequences, which diverged some time t ago. Denote by $I(t)$ the probability that two nucleotide bases at corresponding sites at time t are identical. Assume that the substitution rate is the same for all pairs of nucleotides and constant over time (Jukes and Cantor, 1969). Call this rate α. Then, for a given site, the probability that two homologous nucleotides remain identical at time $t + 1$ when they are identical at time t is $[(1-3\alpha)^2 + 3\alpha^2]I(t)$. This probability statement involves two mutually exclusive events: both bases change into two other identical bases with probability $3\alpha^2$, and both nucleotides remain unchanged with probability $(1-3\alpha)^2$. Furthermore, the probability that two nucleotide sites become identical at time $t+1$ when they are different from each other at time t is equal to $[2\alpha(1-3\alpha) + 2\alpha^2][1 - I(t)]$. This again consists of two mutually exclusive events: a change occurs at one of the sites and the other site remains unchanged with probability $2\alpha(1-3\alpha)$, and both bases change into two other identical bases simultaneously with probability $2\alpha^2$. Hence,

$$I(t+1) = ((1-3\alpha)^2 + 3\alpha)\, I(t) + (2\alpha\,(1-3\alpha) + 2\alpha^2)\,(1 - I(t)) \tag{28}$$

With the initial condition $I(0)= 1$,

$$I(t) = {}^1/_4[1 + 3(1 - 8\alpha + 16\alpha^2)^t] \tag{29}$$

Since α is usually small, the terms in α^2 are negligible and

$$I(t) \approx {}^1/_4[1 + 3(1-8\alpha)^t] \approx 1 - {}^3/_4\,(1 - e^{-8\alpha t}) \tag{30}$$

(Nei, 1975; Gojobori *et al.*, 1990). Let K be the evolutionary distance, i. e. the mean number of nucleotide substitutions accumulated per site at time t.

$$K = 2 \times 3\alpha t = -{}^3/_4 \ln(1 - {}^4/_3\, F_D) \tag{31}$$

where $F_D = 1 - I(t)$. The standard error of K is

$$\sigma_k = \frac{\sqrt{\frac{1}{n} F_D (1 - F_D)}}{1 - {}^4\!/_3\, F_D} \tag{32}$$

where n is the total number of sites compared (Kimura and Ohta, 1972).

When one compares several sequences from each of two groups (generically referred to as species), one must consider the possibility that any two sequences may be descended from different sequences in the ancestral population. Let S be a measure of within-species similarity. It depends on the population size and the mutation rate. With the ancestral population in equilibrium (with respect to drift and mutation), S is expected to remain constant over time at some value $\hat{S}$. Taking into account the within-species variation, one can modify the Jukes-Cantor distance to

$$K_W = \frac{3}{4}\ln\left(\frac{4\hat{S}-1}{4I-1}\right), \tag{33}$$

which measures the distance between the extant and ancestral populations. S is estimated from the variation observed in each of the two extant species (Cockerham, 1984; Weir and Basten, 1990). Suppose n_i sequences are sampled from population i. Define $r_{ijj'}$ as the proportion of homologous bases that are the same in sequences j and j'. The sample similarity within population i is

$$\tilde{S}_i \equiv \frac{1}{n_i(n_i-1)}\sum_{i=1}^{n_i}\sum_{j\neq j'} r_{ijj'} \quad \text{and} \quad \tilde{S} = \frac{1}{2}(\tilde{S}_1 + \tilde{S}_2) \tag{33}$$

If sjj' is the number of like bases between sequence j in population 1 and sequence j' in population 2, the between-population similarity is estimated by

$$\tilde{I} = \frac{1}{n_1 n_2}\sum_{j=1}^{n_1}\sum_{j'=1}^{n_2} s_{jj'}\,. \tag{35}$$

In the Jukes-Cantor model, all mutations occur at the same rate, which is highly unlikely in practice. More realistic models have been proposed (see Gojobori *et al.*, (1990) for a review). A summary of the expressions for K and the instantaneous nucleotide substitution rates under these models appear in Tables 1 and 2.

Table 1: Expected number of nucleotide substitutions under different models of evolution.

Number of Parameters	Formula for estimating K
1	$-\frac{3}{4}\ln(1-F_D)$
2	$-\frac{1}{2}\ln\left\{(1-2P-Q)\sqrt{1-2Q}\right\}$
3	$-\frac{1}{4}\ln\left\{1-2P-2\overline{Q})(1-2P-2R)(1-2\overline{Q}-2R)\right\}$
5	$-\frac{1}{4}\ln\left\{\frac{(S_{13}-\overline{Q}_1)(S_{24}-\overline{Q}_2)-[(P-R)/2]^2}{w\ (1-w)}\left[1-\frac{P+R}{2w(1-w)}\right]^{8w(1-w)-1}\right\}$
6	$-pq\ln\left(\frac{B_1}{pq}\right)-\frac{2q_A q_T}{p}\ln\left[\frac{p}{3q_A q_T}\left(F_{12}-B_1+\frac{3E_{12}}{B_1}\right)\right]$ $-\frac{2q_C q_G}{q}\ln\left[\frac{q}{3q_C q_G}\left(F_{34}-B_1+\frac{3E_{34}}{B_1}\right)\right]$

In the two-parameter model (Kimura, 1980), α and β are the rates of transition and transversion. Let $P + Q = F_D$, where

$$P \equiv P(t) \equiv {}^1/_4 - {}^1/_2 e^{-4(\alpha+\beta)t} + {}^1/_4 e^{-8\beta t} \text{ and } Q \equiv Q(t) \equiv {}^1/_2 - {}^1/_2 e^{-8\beta t}. \tag{36}$$

P and Q represent the proportions of nucleotide sites with, respectively, transition and transversion differences between the two sequences compared. Under this model, $k = \alpha + 2\beta$ is the number of nucleotide substitutions per site per year, and K

$= 2kt$ is the total number of nucleotide substitutions per site between two sequences which diverged from their common ancestor t years ago.

In the three-parameter model (Kimura, 1981), $P(t)$ denotes the proportion of sites showing **TC** or **AG** nucleotide pairs at time t in the two sequences considered, $\overline{Q}(t)$ the proportion of sites with **TA** or **CG**, and $R(t)$ the proportion of sites with **TG** or **CA**:

$$P \equiv P(t) \equiv [1 - e^{-4(\alpha+\beta)t} - e^{-4(\alpha+\gamma)t} + e^{-4(\beta+\gamma)t}]/4$$
$$\overline{Q} \equiv \overline{Q}(t) \equiv [1 - e^{-4(\alpha+\beta)t} + e^{-4(\alpha+\gamma)t} - e^{-4(\beta+\gamma)t}]/4 \quad (37)$$
$$R \equiv R(t) \equiv [1 + e^{-4(\alpha+\beta)t} - e^{-4(\alpha+\gamma)t} - e^{-4(\beta+\gamma)t}]/4$$

Now, the total number of nucleotide substitutions per site is $K = 2(\alpha + \beta + \gamma)t$.

Table 2 Models of evolution with different numbers of free parameters.

	Substituted Nucleotide			
Original Nucleotide	**A**	**T**	**C**	**G**
One-parameter model				
A	–	α	α	α
T	α	–	α	α
C	α	α	–	α
G	α	α	α	–
Two-parameter model				
A	–	β	β	α
T	β	–	α	β
C	β	α	–	β
G	α	β	β	–
Three parameter model				
A	–	β	γ	α
T	β	–	α	γ
C	γ	α	–	β
G	α	γ	β	–
Five-parameter model				
A	–	γ	δ	α
T	γ	–	α	δ
C	ε	β	–	γ
G	β	ε	γ	–
Six-parameter model				
A	–	α_1	α	α
T	β_1	–	α	α
C	β	β	–	α_2
G	α	α	β_2	–

In the *five-parameter model* (Takahata and Kimura, 1981), w represents the fraction of **A** + **T** in the two sequences. S_{13}, S_{24}, $\overline{Q}_1$, Q_1, P and Q are the fractions of sites having, respectively, **AA** or **TT**, **CC** or **GG**, **AT**, **GC**, **CT** or **AG**, and **GT** or **AC** nucleotide pairs.

The *six-parameter model* (Gojobori *et al.*, 1982) is based on Kimura's three parameter model (Kimura, 1981). There, q_A, q_T, q_C, and q_G stand, respectively, for the contents of **A**, **T**, **C** and **G** in the sequences under study:

$$
\begin{aligned}
&p = q_A + q_T\,,\ q = q_C + q_G\,,\ B_1 = pq - (x_{AC} + x_{AG} + x_{TC} + x_{TG})\\
&E_{12} = (q_A q - x_{AC} - x_{AG})(q_T q - x_{TC} - x_{TG})\\
&E_{34} = (q_C p - x_{AC} - x_{TC})(q_G p - x_{AG} - x_{TG}) \qquad (38)\\
&F_{12} = x_{AA} + x_{TT} - x_{AT} - p^2 + 3q_A q_T, \text{ and}\\
&F_{34} = x_{CC} + x_{GG} - x_{CG} - q^2 + 3q_C q_G
\end{aligned}
$$

where x_{ii} represents the fraction of sites having the same base pair i, and $2\,x_{ij}$ $(i \neq j)$ the fraction of sites having different base pairs i and j (i, j = **A**, **C**, **T**, **G**).

A substitution process is said to be time-reversible if the probability of starting at nucleotide i and changing to nucleotide j in a time interval is the same as the probability of starting at j and changing to i in the same period, i.e., time reversibility requires that $p_{ij(t)}\pi_i = p_{ji(t)}\pi_j$ for all i and j and all t; π_i is the equilibrium probability of nucleotide i (Li, 1997). Note that time reversibility holds for the one-, two- and three-parameter models described in Table 2, since their transition matrices are symmetric and we are assuming equal base frequencies (i.e., π_i=1/4 for all i). For the five- and six-parameter models the time-reversibility property does not hold: the matrices are no longer symmetric. In phylogenetic reconstruction, if the process is time reversible, any node or point of the tree can be taken as the ancestral node. This is due to the so-called pulley principle (Felsenstein, 1981). However, when the process is not time reversible, we must select a sequence to root the tree.

All models described so far assume that the base frequencies $\{\pi_A, \pi_C, \pi_T, \pi_G\}$ are equal. Extensions of the above models with this constraint removed have been proposed. For example, Felsenstein's model (1981)

$$
\begin{pmatrix}
- & \alpha\pi_T & \alpha\pi_C & \alpha\pi_G\\
\alpha\pi_A & - & \alpha\pi_C & \alpha\pi_G\\
\alpha\pi_A & \alpha\pi_T & - & \alpha\pi_G\\
\alpha\pi_A & \alpha\pi_T & \alpha\pi_C & -
\end{pmatrix}
$$

corresponds to the one-parameter model, and the model of Hasegawa *et al.* (1985)

$$
\begin{pmatrix}
- & \beta\pi_T & \beta\pi_C & \alpha\pi_G\\
\beta\pi_A & - & \alpha\pi_C & \beta\pi_G\\
\beta\pi_A & \alpha\pi_T & - & \beta\pi_G\\
\alpha\pi_A & \beta\pi_T & \beta\pi_C & -
\end{pmatrix}
$$

to the two-parameter model. These models are special cases of the general reversible Markov-chain model (Tavaré, 1986)

$$\begin{pmatrix} - & \theta_1\pi_T & \theta_2\pi_2 & \theta_3\pi_G \\ \theta_1\pi_A & - & \theta_4\pi_C & \theta_5\pi_G \\ \theta_2\pi_A & \theta_4\pi_T & - & \theta_6\pi_G \\ \theta_3\pi_A & \theta_5\pi_T & \theta_6\pi_C & - \end{pmatrix}$$

3.2 Log Determinant Distance

In order to compute distances between sequences with different nucleotide or amino-acid compositions, Lockhart and coworkers (1994) introduced the Log Determinant Distance. It relies on the so-called divergence matrix F_{xy}. For sequences x and y, its (i, j) element is the proportion of sites in which x is of category (i. e, nucleotide or amino acid) i while y has j. The sum of all matrix elements is 1. The LogDet distance between x and y is defined as

$$d_{xy} \equiv \ln(\det F_{xy}) \tag{39}$$

where det stands for the determinant of a matrix. To assign a distance of zero between a sequence and itself, this quantity is modified: for nucleotide sequences, it becomes

$$d'_{xy} \equiv -\frac{1}{4}\ln\left(\frac{\det F_{xy}}{\sqrt{(\det F_{xx})(\det F_{yy})}}\right) \tag{40}$$

Note that when the four base frequencies are equal, det F_{xx} = det F_{yy} = $(1/4)^4$, and the expected value of the LogDet Distance is the mean number of substitutions per site.

3.3 Hamming Distance

The Hamming distance is largely used for descriptive purposes (see Seillier-Moiseiwitsch *et al.*, 1994, for a review). Let $\mathbf{X}_i = (X_{i1}, X_{i2},\ldots, X_{iK})'$ be a vector representing sequence i of length K. X_{ik} is thus the label of the nucleotide or amino acid appearing at position k. Consider $\mathbf{X}_i$ and $\mathbf{X}_{i'}$ The Hamming distance $H_{ii'}$ is

$$\begin{aligned} H_{ii'} &= \frac{1}{K}\sum_{k=1}^{K}\delta(X_{ik} \neq X_{i'k}) \\ &= \frac{1}{K} \times \text{ number of positions where } X_i \text{and } X_{i'} \text{ differ} \end{aligned} \tag{41}$$

where δ denotes the indicator function (i.e., $\delta(A) = 1$ if event A is true and 0 otherwise). While this distance should only be treated as a descriptive statistic in many situations, it provides a reasonable estimate of the actual distance when the sequences are closely related (i.e., they are separated by few rounds of replication so that a specific position having undergone both a forward and a backward mutation is highly unlikely).

4 RELATING DISTANCE MATRICES TO COVARIATES

4.1 Mantel's Correlation Test

A test comparing two distance matrices was introduced by Mantel (1967) to assess space and time clustering of diseases. Matrix comparison is commonly used to detect a correspondence between two types of measured distances. Such distances can be genetic, morphological or geographical.

In order to understand the procedure, consider an example from Manly (1985). Suppose that a population is sampled at four colonies and morphological proportions are determined. These proportions are displayed in a 4×4 distance matrix where the entry in row i and column j is the morphological distance between colony i and colony j. Distances between colonies with similar proportions are small while distances between colonies with very different proportions are large. The morphological distance matrix is

$$\mathbf{M} = \begin{bmatrix} 0.0 & 1.0 & 1.4 & 0.9 \\ 1.0 & 0.0 & 1.1 & 1.6 \\ 1.4 & 1.1 & 0.0 & 0.7 \\ 0.9 & 1.6 & 0.7 & 0.0 \end{bmatrix}$$

As usual, diagonal elements are zero because they represent distances from colonies to themselves, and the matrix is symmetric as the distance from colony i to colony j must be the same as the distance from colony j to colony i. In the context of HIV sequences, the rows and columns of **M** relate to single sequences or groups of sequences, and the elements of **M** are distances computed with any of the metrics described in Section 3.

Now, suppose that an environmental variable is measured on each colony, which results in a matrix of environmental distances between the colonies, as for instance

$$\mathbf{E} = \begin{bmatrix} 0.0 & 0.5 & 0.8 & 0.6 \\ 0.5 & 0.0 & 0.5 & 0.9 \\ 0.8 & 0.5 & 0.0 & 0.4 \\ 0.6 & 0.9 & 0.4 & 0.0 \end{bmatrix}$$

For DNA strings, this matrix could record, for instance, whether they were sampled from the same compartment (with "0" if they are and "1" if they are not).

Mantel's test was constructed to investigate whether the elements of **M** and **E** are correlated. Let the morphological distance from colony i to colony j be m_{ij}, with $m_{ij} = m_{ji}$ and $m_{ii} = 0$. This pattern also applies to **E**. The test statistic is

$$Z = \sum_i \sum_j m_{ij} e_{ij}. \tag{42}$$

Its distribution is obtained by taking the colonies in a random order for one of the matrices, i.e., the matrix **M** remains as it is and a random permutation is chosen for the colonies in **E** (call this matrix $\mathbf{E}_R$). Z is then calculated from **M** and $\mathbf{E}_R$. By repeating this procedure using all different random orderings for $\mathbf{E}_R$, we get the randomized distribution of Z. The idea is that, if environmental and morphological distances are not correlated, then **E** is just like one of the randomly ordered matrices $\mathbf{E}_R$ and the observed Z is a typical randomized Z value. If the distances have a positive (or negative) correlation, the observed Z will tend to be greater (or smaller) than the randomized values.

When there are only a few colonies, it is possible to calculate all randomized Z values. As the number of colonies increases, it becomes impractical to enumerate all the randomized Z values. Then, one can carry out the Mantel test in one of two ways. One generates a very large number of randomized $\mathbf{E}_R$ matrices and the resulting empirical distribution of Z values is an estimate of the true randomized distribution. Alternatively, the mean $E(Z)$ and variance $V(Z)$ of the randomized Z are calculated, and the distribution of

$$Z^* = \frac{Z - E(Z)}{V(Z)^{1/2}} \tag{43}$$

is approximated by the standard normal distribution. In the above example, $Z^* = 0.92836$, and Pr(random $Z^* \geq$ observed Z^*) = 0.0789.

Consider the general situation where **M** and **E** are of order $L \times L$. To calculate $E(Z)$ and $V(Z)$, the following quantities are needed:

$$A_M = \sum_{i=1}^{L}\sum_{j=1}^{L} m_{ij} \qquad B_M = \sum_{i=1}^{L}\sum_{j=1}^{L} m_{ij}^2 \qquad D_M = \sum_{i=1}^{L}\left(\sum_{j=1}^{L} m_{ij}\right)^2$$

$$G_M = A_M^2 \qquad H_M = D_M - B_M \qquad K_M = G_M + 2B_M - 4D_M$$

$$A_E = \sum_{i=1}^{L}\sum_{j=1}^{L} e_{ij} \qquad B_E = \sum_{i=1}^{L}\sum_{j=1}^{L} e_{ij}^2 \qquad D_E = \sum_{i=1}^{L}\left(\sum_{j=1}^{L} e_{ij}\right)$$

$$G_E = A_E^2 \qquad H_E = D_E - B_E \qquad K_E = G_E + 2B_E - 4D_E.$$

Mantel (1967) showed that for the randomized Z

$$E(Z) = \frac{A_M A_E}{L(L-1)} \tag{44}$$

and

$$V(Z) = \frac{2B_M B_E + \dfrac{4H_M H_E}{L-2} + \dfrac{K_M K_E}{(L-2)(L-3)} - \dfrac{G_M G_E}{L(L-1)}}{L(L-1)} \tag{45}$$

The link between Z and the ordinary Pearson correlation coefficient between the elements of **M** and **E** is as follows

$$r = \frac{Z - E(Z)}{\left[\left(B_M - \dfrac{G_M^2}{L(L-1)}\right)\left(B_E - \dfrac{G_E^2}{L(L-1)}\right)\right]^{1/2}} \tag{46}$$

4.2 Analysis of Molecular Variance

To study molecular variation, Excoffier and colleagues (1992) used DNA haplotype divergence measures and performed an analysis of variance on the matrix of squared distances among all pairs of haplotypes. This methodology can be applied to HIV sequences when, for example, one wants to compare rates of mutations between and within HIV subtypes.

4.2.1 Distance Metric

Compare each sequence to the consensus sequence and look for differences. The outcome is binary: let

$$y_s = \begin{cases} 0 \text{ if there is no mutation at site } s \\ 1 \text{ if there is a mutation at site } s. \end{cases} \tag{47}$$

For a sequence with S sites, the whole sequence is considered an S-dimensional boolean vector $\mathbf{y}' = (y_1\, y_2 \ldots y_S)$. The difference between two sequences j and k is $(\mathbf{y}_j - \mathbf{y}_k)$:

$$(\mathbf{y}'_j - \mathbf{y}'_k)' = [(y_{1j} - y_{1k})\,(y_{2j} - y_{2k}) \ldots (y_{Sj} - y_{Sk})]. \tag{48}$$

Define the Euclidean distance metric δ^2_{jk} between sequences j and k as

$$\delta^2_{jk} = (\mathbf{y}_j - \mathbf{y}_k)'\, \mathbf{W}\, (\mathbf{y}_j - \mathbf{y}_k), \tag{49}$$

where $\mathbf{W}$ is a matrix of differential weights for the various sites. $\mathbf{W}$ can take several forms. If all sites are assumed independent and equally informative, $\mathbf{W} = \mathbf{I}$, the identity matrix, and the distance metric is equal to the number of nucleotide differences. When $\mathbf{W}$ is diagonal, i.e., $W = \text{diag}\{w_s^2\}$ (weighting sites differentially but treating them as independent), (49) can be rewritten as

$$\delta^2_{jk} = \sum_{s=1}^{S} w_s^2 (y_{sj} - y_{sk})^2. \tag{50}$$

4.2.2 Partitioning the Distance Matrix

Consider N sequences from I populations (e.g., subtypes, individuals) and the associated distance matrix $\mathbf{D}^2$ partitioned into a series of submatrices corresponding to particular subdivisions:

$$\mathbf{D}^2 = \begin{bmatrix} \begin{bmatrix} \mathbf{D}^2_{11} & \mathbf{D}^2_{12} \\ \mathbf{D}^2_{21} & \mathbf{D}^2_{22} \end{bmatrix} & \cdots & \mathbf{D}^2_{1I} \\ & \cdots & \mathbf{D}^2_{2I} \\ \vdots \quad\quad \vdots & \cdots & \vdots \\ \mathbf{D}^2_{I1} \quad \mathbf{D}^2_{I2} & \cdots & \mathbf{D}^2_{II} \end{bmatrix}$$

The elements of the block-diagonal submatrices $\mathbf{D}^2_{ii}$ are the pairwise squared distances δ^2_{jk} between sequences from the same (i^{th}) population, and those of the off-diagonal matrix blocks $\mathbf{D}^2_{ii'}$ are the pairwise squared distances between sequences from populations i and i'. Sequences may also be grouped at higher levels, according to geographical areas or routes of transmission.

The conventional sum of squares (SS_{Total}) may be written, barring a constant ($2N$), as the sum of squared differences between all pairs of N items

(Hoeffding, 1948). In the multidimensional case, SS_{Total} becomes a sum of squared deviations (SSD) from the centroid of the space. Thus,

$$SSD_{Total} = \sum_{j=1}^{N} (\mathbf{y}_j - \bar{\mathbf{y}})' \mathbf{W} (\mathbf{y}_j - \bar{\mathbf{y}}) = \frac{1}{N} \sum_{j=1}^{N-1} \sum_{k>j}^{N} (\mathbf{y}_j - \mathbf{y}_k)' \mathbf{W} (\mathbf{y}_j - \mathbf{y}_k), \tag{51}$$

or

$$SSD_{Total} = \frac{1}{2N} \sum_{j=1}^{N} \sum_{k=1}^{N} (\mathbf{y}_j - \mathbf{y}_k)' \mathbf{W} (\mathbf{y}_j - \mathbf{y}_k) = \frac{1}{2N} \sum_{j=1}^{N} \sum_{k=1}^{N} \delta_{jk}^2 . \tag{52}$$

If sequences are arranged into populations and populations are nested within groups (defined a priori), a linear model can be utilized

$$\mathbf{y}_{jig} = \mu + \mathbf{a}_g + \mathbf{b}_{ig} + \mathbf{e}_{jig} \tag{53}$$

where $\mathbf{y}_{jig}$ refers to individual j ($j = 1,\ldots, N_{ig}$) in population i ($i = 1,\ldots, I_g$) in group g ($g = 1,\ldots, G$). μ is the unknown expected value of $\mathbf{y}_{jig}$ $\mathbf{a}$ the group effect, $\mathbf{b}$ the population effect and $\mathbf{e}$ the individual-within-population effect. All these effects are assumed additive, random, uncorrelated, and the corresponding variance components are σ_a^2, σ_b^2 and σ_e^2, respectively.

As in the classical analysis of variance, for any choice of hierarchical partition of N sequences into strata,

$$SSD_{Total} = SSD_{AmongStrata} + SSD_{WithinStrata} \tag{54}$$

In the present context, this decomposition is referred to as an Analysis of Molecular Variance (AMOVA). For illustration, we partition the total sum of squared deviations SSD_{Total} into components of variation within populations (SSD_{WP}), among populations within groups ($SSD_{AP/WG}$) and among groups (SSD_{AG}):

$$SSD_{WD} = \sum_{g=1}^{G} \sum_{i=1}^{Ig} \frac{\sum_{j=1}^{N_{ig}} \sum_{k=1}^{N_{ig}} \delta_{jk}^2}{2N_{ig}} \tag{55a}$$

$$SSD_{AP/WG} = \sum_{g=1}^{G} \left(\frac{\sum_{i=1}^{I_g} \sum_{j=1}^{N_{ig}} \sum_{i'=1}^{I_g} \sum_{k=1}^{N_{i'g}} \delta_{jk}^2}{\sum_{i=1}^{I_g} 2N_{ig}} - \sum_{i=1}^{I_g} \frac{\sum_{j=1}^{N_{ig}} \sum_{k=1}^{N_{ig}} \delta_{jk}^2}{2N_{ig}} \right) \tag{55b}$$

$$SSD_{AG} = \frac{\sum_{j=1}^{N} \sum_{k=1}^{N} \delta_{jk}^2}{2N} - \sum_{g=1}^{G} \frac{\sum_{i=1}^{I_g} \sum_{j=1}^{N_{ig}} \sum_{i'=1}^{I_g} \sum_{k=1}^{N_{i'g}} \delta_{jk}^2}{\sum_{i=1}^{I_g} 2N_{ig}} \tag{55c}$$

Here N_{ig} is the number of sequences in population i of group g, $N = \sum_{g=1}^{G} \sum_{i=1}^{Ig} N_{ig}$ the total number of individuals in the study, I_g the number of populations in group g, $\sum_{g=1}^{G} I_g$ the total number of populations and G the number of groups in the study.

The mean squared deviations (*MSD*) are obtained by dividing each *SSD* by the appropriate number of degrees of freedom, as reported in Table 3. The n-coefficients in Table 3 denote the average sample sizes of particular hierarchical levels, allowing for unequal sample sizes:

$$n = \frac{\sum_{g=1}^{G}\sum_{i=1}^{I_g} N_{ig} - \sum_{g=1}^{G}\left(\frac{\sum_{i=1}^{I_g} N_{ig}^2}{\sum_{i=1}^{I_g} N_{ig}}\right)}{\sum_{g=1}^{G} I_g}, \tag{56a}$$

$$n' = \frac{\sum_{g=1}^{G}\left(\frac{\sum_{i=1}^{I_g} N_{ig}^2}{\sum_{i=1}^{I_g} N_{ig}}\right) - \frac{\sum_{g=1}^{G}\sum_{i=1}^{I_g} N_{ig}^2}{\sum_{g=1}^{G}\sum_{i=1}^{I_g} N_{ig}}}{G-1}, \tag{56b}$$

$$n'' = \frac{\sum_{g=1}^{G}\sum_{i=1}^{I_g} N_{ig} - \frac{\sum_{g=1}^{G}\left(\sum_{i=1}^{I_g} N_{ig}\right)^2}{\sum_{g=1}^{G}\sum_{i=1}^{I_g} N_{ig}}}{G-1}. \tag{56c}$$

Table 3 AMOVA Table

Source of Variation	d. f.	MSD	Expected MSD
Among regions	$G-1$	MSD_{AG}	$\sigma_e^2 + n'\,\sigma_b^2 + n''\sigma_a^2$
Among populations within regions	$\sum_{g=1}^{G} I_g - G$	$MSD_{AP/WG}$	$\sigma_e^2 + n\sigma_b^2$
Among individuals within populations	$N - \sum_{g=1}^{G} I_g$	MSD_{WP}	σ_e^2
Total	$N-1$		

Taking the expected values of the mean squared deviations, we obtain the variance components for each hierarchical level. The structure of the analysis is akin to the F-statistics derived for the treatment of polymorphic systems (Cockerham, 1969; 1973). Here, it is also useful to employ an analogous array of sequence correlation measures, the Φ-statistics:

$$\sigma_e^2 = (1-\Phi_{ST})\sigma^2,\ \sigma_b^2 = (1-\Phi_{CT}),\ \sigma_a^2 = \Phi_{CT}\sigma^2 \tag{57}$$

where $\sigma^2 = \sigma_a^2 + \sigma_b^2 + \sigma_e^2$. Φ_{ST} represents the correlation of random sequences within populations, relative to that of random pairs of sequences drawn from the whole study. Φ_{CT} is the correlation of random sequences within a group of populations, relative to that of random pairs of sequences drawn from the whole study, and Φ_{SC} is the correlation of random sequences within populations, relative to that of random pairs of sequences from the group. We can also rewrite equations (57) in terms of the Φ-statistics:

$$\Phi_{ST} = \frac{\sigma_a^2 + \sigma_b^2}{\sigma^2}, \quad \Phi_{CT} = \frac{\sigma_a^2}{\sigma^2}, \quad \Phi_{SC} = \frac{\sigma_b^2}{\sigma_b^2 + \sigma_e^2} \tag{58}$$

4.2.3 Testing Significance of the Variance Components and Φ-Statistics

The AMOVA procedure requires very few assumptions. Under the null hypothesis, samples are regarded as drawn from a global population, with the observed variation due to random sampling. To obtain a null distribution, we allocate each sequence to a randomly chosen population (e.g., a specific subtype), while holding sample sizes constant at their realized values. This method is a permutation test like the one proposed by Mantel (1967). The rows (and corresponding columns) of the squared-distance matrix are permuted randomly. The variance components are estimated for each of a large number of permuted matrices. We thus obtain a null distribution for the test of significance of Φ_{ST} and σ_e^2.

Two different permutation schemes are utilized for the other variance components. The first permutation method generates the null distribution of Φ_{SC} and σ_b^2. It is assumed that the groups (e.g., different geographical areas or risk groups) are real but that the populations within them are arbitrary (it permutes individuals within groups without regard to populations). The second permutation method creates the null distribution of Φ_{CT} and σ_a^2. It is now assumed that while the populations are real, the groupings are artificial (it permutes whole populations across groups). In this case, the sizes of the groups may vary with each run.

5 CATEGORICAL ANALYSIS OF VARIANCE

Pinheiro and her colleagues (2000) developed a quasi-multivariate analysis of variance for categorical data when the response variable is not ordered. It is particularly well suited for the comparison of sets of sequences. Comparisons can be performed between and within groupings (e.g., clades) to see whether the variability is similar in each. Similarly, when we study several individuals and obtain a set of sequences from each at different time points, our interest lies in estimating the variability between and within individuals.

On the basis of a measure of variation similar to the Simpson Index of ecological diversity (1), Light and Margolin (1971; 1974) constructed a one-way analysis of variance for categorical variables (CATANOVA). Anderson and Landis (1980; 1982) extended this work to multidimensional contingency tables involving several factors. This technique can be used to contrast the variability of the response variable at a single position between and within groups. In the analysis of HIV sequences, however, a single position yields little information. Extending the methodology so that it can deal with genomic segments is the basic motivation of the work described in this section. Components of variation are derived from the fact that the sum of squares of deviations from the mean can be expressed as a function of the squares of the pairwise differences for all possible pairs. The sequences are not considered on an individual basis but only as contributing to the overall variability in the distribution of the categorical response (i. e., the nucleotide or amino-acid label). Measures of diversity are partitioned according to the factors considered assuming independence among all positions. The null hypothesis deals with homogeneity among groups.

5.1 Quantifying Sequence Variation

The data are summarized in Table 4. The interest is in assessing the homogeneity among groups: the null hypothesis is that $p_{cgk} = p_{ck}$ where p_{cgk} is the probability of belonging to category c in group g at position k (assuming independence among positions).

Table 4 Contingency Table (K positions)

Groups	Positions	1	2	…	C	Totals
1	1	n_{111}	n_{211}	…	n_{C11}	$n_{\cdot 11}=N$
1	2	n_{112}	n_{212}	…	n_{C12}	$n_{\cdot 12}=N$
⋮	⋮	⋮	⋮	…	⋮	⋮
1	K	n_{11K}	n_{21K}	…	n_{CK}	$n_{\cdot 1K}=N$
	Totals	$n_{11\cdot}$	$n_{21\cdot}$	…	$n_{C1\cdot}$	$n_{\cdot 1\cdot}=NK$
2	1	n_{121}	n_{221}	…	n_{C21}	$n_{\cdot 21}=N$
2	2	n_{122}	n_{222}	…	n_{C22}	$n_{\cdot 22}=N$
⋮	⋮	⋮	⋮	…	⋮	⋮
2	K	n_{12K}	n_{22K}	…	n_{C1K}	$n_{\cdot 2K}=N$
	Totals	$n_{12\cdot}$	$n_{22\cdot}$	…	$n_{C2\cdot}$	$n_{\cdot 2\cdot}=NK$
⋮	⋮	⋮	⋮	⋮	⋮	⋮
G	1	n_{1G1}	n_{2G1}	…	n_{CG1}	$n_{\cdot G1}=N$
G	2	n_{1G2}	n_{2G2}	…	n_{CG2}	$n_{\cdot G2}=N$
⋮	⋮	⋮	⋮	…	⋮	⋮
G	K	n_{1GK}	n_{2GK}	…	n_{CGK}	$n_{\cdot GK}=N$
	Totals	$n_{1G\cdot}$	$n_{2G\cdot}$	…	$n_{CG\cdot}$	$n_{\cdot G\cdot}=NK$
	TOTALS	$n_{1\cdot\cdot}$	$n_{2\cdot\cdot}$	…	$n_{C\cdot\cdot}$	$n_{\cdot\cdot\cdot}=NGK$

Consider Gini's measure of variation for categorical data (Gini, 1912). Let $X_1, X_2, \ldots, X_N$ denote measurements on N independent experimental units. The variance of X may be expressed as E $(\phi(X_1, X_2))$, where $\phi(a, b) = {}^1/_2(a-b)^2$ (Hoeffding, 1948). In a similar fashion, the sum of squares is

$$\sum_{i=1}^{N}(X_i-\overline{X})^2 = \frac{1}{2N}\sum_{i=1}^{N}\sum_{j=1}^{N}(X_i-X_j)^2 = \frac{1}{N}\sum_{1\le i\le j\le N} d_{ij}^2 \tag{59}$$

where $\overline{X} = \sum_{i=1}^{N} X_i / N$ and $d_{ij} = X_i - X_j$. In the present context, each X_i falls into one of C possible categories. Define $d(X_i, X_j) = d_{ij}$ as

$$d_{ij} = \begin{cases} 1 \text{ if } X_i \text{ and } X_j \text{ name different categories} \\ 0 \text{ if } X_i \text{ and } X_j \text{ name the same category.} \end{cases} \tag{60}$$

The variation for categorical responses $X_1,\ldots, X_N$ is

$$\frac{1}{2N}\sum_{j=1}^{N}\sum_{i=1}^{N} d_{ij}^2 = \frac{1}{2N}\sum_{j=1}^{N}\sum_{i=1}^{N} d_{ij}. \tag{61}$$

As each response assumes one and only one of the C possible categories, the data are summarized by the vector $(n_1,\ldots, n_C)$ where n_i is the number of responses in category i ($i = 1,\ldots, C$), so that $\sum_{i=1}^{C} n_i = N$.. The variation in the responses is defined as

$$D_G \equiv \frac{1}{2N}\sum_{i \neq j} n_i n_j = \frac{N}{2}\left\{1 - \sum_{i=1}^{C}\left(\frac{n_i}{N}\right)^2\right\} \tag{62}$$

If $p_1,\ldots, p_C$ stand for the probabilities of X belonging to these C categories, the sample counterpart of the Simpson Index of ecological diversity (1) is

$$h(\hat{\mathbf{p}}) = 1 - \hat{\mathbf{p}}'\,\hat{\mathbf{p}} = 1 - \sum_{i=1}^{C} \hat{p}_i^2, \tag{63}$$

where $\hat{p}_i = n_i / N, i = 1$ relate to the sample proportion. Therefore,

$$D_G = \frac{N}{2} h(\hat{\mathbf{p}}) \tag{64}$$

Definition (61) is motivated by two properties:

(d) the variation of N categorical responses is minimized if and only if they all belong to the same category, i.e., $p_i = 1$, $\forall\, i = 1,\ldots, C$

(e) the variation of N responses is maximized when the responses are evenly distributed among the available categories, i.e, $p_i = 1/C$, $\forall\, i = 1,.., C$.

5.2 Partitioning Diversity Measures

Let $\mathbf{X}^g_i = (X^g_{i1}, X^g_{i2},\ldots, X^g_{iK})'$ be a random vector representing sequence i of group g. Suppose $i = 1,\ldots, N$, $k = 1,\ldots, K$ and $g = 1\ldots, G$. So, X^g_{ik} represents position k of sequence i of group g. X^g_{ik} is a categorical variable assuming C (unordered) categories. For instance, if comparisons are made at the nucleotide level, $x^g_{ik} \in$ (A, C, T, G) and there are 4 categories.

First, assume there is only one position for each sequence. The data appear in Table 5. Now d_{ij} is defined as

$$d_{ij} = \begin{cases} 1 \text{ if } X_i^g \neq X_j^g \\ 0 \text{ if } X_i^g = X_j^g \end{cases}. \tag{65}$$

Table 5 Summary of the Data (one position)

	Groups				
Sequences	1	2	3	…	G
1	x_1^1	x_1^2	x_1^3	…	x_1^G
2	x_2^1	x_2^2	x_2^3	…	x_2^G
3	x_3^1	x_3^2	x_3^3	…	x_3^G
⋮	⋮	⋮	⋮	⋮	⋮
N	x_N^1	x_N^2	x_N^3	…	x_N^G

The total number of responses is

$$NG = \sum_{g=1}^{G} n_{.g} = \sum_{c=1}^{C} n_{c.} = \sum_{c=1}^{C}\sum_{g=1}^{G} n_{cg} \tag{66}$$

where n_{cg} is the number of responses in category c for group g and $N = n_{.g} = \sum_{c=1}^{C} n_{cg}$ is the number of responses for group g, which is simply the number of sequences in each group.

The Total Simpson Index (TSI) is

$$TSI = 1 - \sum_{c=1}^{C}\left(\frac{n_{c.}}{NG}\right)^2 \tag{67}$$

The dispersion within group g (i.e, within $\{x_1^g,\ x_2^g, \ldots,\ x_N^g\}$) is

$$1 - \sum_{c=1}^{C}\left(\frac{n_{cg}}{n_{.g}}\right)^2 \tag{68}$$

Therefore, the within-group Simpson Index (WSI) is found by averaging (67) over all g's:

$$WSI = \frac{1}{G}\sum_{g=1}^{G}\left\{1 - \sum_{c=1}^{C}\left(\frac{n_{cg}}{n_{.g}}\right)^2\right\} = 1 - G\sum_{g=1}^{G}\sum_{c=1}^{C}\left(\frac{n_{cg}}{NG}\right)^2 \tag{69}$$

The between-group Simpson Index (BSI) is

$$BSI = TSI - WSI = G\sum_{g=1}^{G}\sum_{c=1}^{C}\left(\frac{n_{cg}}{NG}\right)^2 - \sum_{c=1}^{C}\left(\frac{n_{c.}}{NG}\right)^2 \tag{70}$$

Now, assume there are K positions along each sequence. Let

$$\mathbf{X}_i^g = \left(X_{i1}^g, X_{i2}^g, \ldots, X_{iK}^g\right)' \text{ and } \left(\mathbf{X}_1^g, \mathbf{X}_2^g, \ldots, \mathbf{X}_N^g\right)'.$$

The total number of responses is

$$NGK = \sum_{g=1}^{G} n_{.g.} = \sum_{c=1}^{C} n_{c..} = \sum_{k=1}^{K} n_{..k} = \sum_{c=1}^{C}\sum_{g=1}^{G}\sum_{k=1}^{K} n_{cgk} \tag{71}$$

where n_{cgk} denotes the number of instances of category c at position k for group g. The variation within group g at position k is

$$1-\sum_{c=1}^{C}\left(\frac{n_{cgk}}{n_{\cdot gk}}\right)^2 = 1-\sum_{c=1}^{C}\left(\frac{n_{cgk}}{N}\right)^2 \tag{72}$$

since $n_{\cdot gk}, = N$. The variation within group g is

$$1-\sum_{c=1}^{C}\left(\frac{n_{cg\cdot}}{n_{\cdot g\cdot}}\right)^2 = 1-\sum_{c=1}^{C}\left(\frac{n_{cg\cdot}}{NK}\right)^2 \tag{73}$$

since $n_{\cdot g\cdot} = NK$. The measures of dispersions are

$$WSI = \frac{1}{G}\sum_{g=1}^{G}\left\{1-\sum_{c=1}^{C}\left(\frac{n_{cg\cdot}}{NK}\right)^2\right\} = 1-G\sum_{c=1}^{C}\left(\frac{n_{cg\cdot}}{NGK}\right)^2 \tag{74}$$

$$TSI = 1-\sum_{c=1}^{C}\left(\frac{n_{c\cdot\cdot}}{NGK}\right)^2, \tag{75}$$

and

$$BSI = TSI - WSI = G\sum_{g=1}^{G}\sum_{c=1}^{C}\left(\frac{n_{cg\cdot}}{NGK}\right)^2 - \sum_{c=1}^{C}\left(\frac{n_{c\cdot\cdot}}{NGK}\right)^2. \tag{76}$$

5.3 Probabilistic Model

Assuming that outcomes in different groups are independent, $(n_{1gk}, n_{2gk},\ldots, n_{Cgk})$ follows a multinomial distribution: let p_{cgk} stand for the probability of category c at position k for group g

$$\Pr\{n_{1gk}, n_{2gk},\ldots,n_{Cgk}\} = \binom{N}{n_{1gk}\ldots n_{Cgk}}\prod_{c=1}^{C}(p_{cgk})^{n_{cgk}} \tag{77}$$

where $\sum_{c=1}^{C} p_{cgk} = 1, p_{cgk} > 0, c = 1,\ldots, C, k = 1,\ldots, K$ and $g = 1,\ldots, G$,

$$E(n_{cgk}) = Np_{cgk}, Var(n_{cgk} = Np_{cgk}(1-p_{cgk}) \tag{78}$$

and $\mathrm{Cov}\left(n_{c_1g_1k_1}, n_{c_2g_2k_2}\right) = -\delta N p_{c_1g_1k_1} p_{c_2g_2k_2}$ with

$$\delta = \begin{cases} 1 \text{ if } g_1 = g_2 \text{ and } k_1 = k_2 \\ \quad 0 \text{ otherwise} \end{cases} \tag{79}$$

If we assume that all positions are independent, the model is

$$\prod_{g=1}^{G}\prod_{k=1}^{K}\Pr\{n_{1gk}, n_{2gk},\ldots,n_{Cgk}\} = \prod_{g=1}^{G}\prod_{k=1}^{K}\binom{N}{n_{1gk}\ldots n_{Cgk}}\prod_{c=1}^{C}(p_{cgk})^{n_{cgk}} \tag{80}$$

Then $\mathbf{V}_g \equiv (n_{1g1} \ldots n_{Cg1}\ n_{1g2} \ldots n_{Cg2} \ldots n_{1gK} \ldots n_{CgK})'$ is a $CK \times 1$ vector and $\mathbf{V} \equiv (\mathbf{V}_1\mathbf{V}_2 \ldots \mathbf{V}_G)'$ is a $GCK \times 1$ vector with

$$\mathrm{E}(\mathbf{V}) \equiv \mu = N(p_{111} \ldots p_{C11} \ldots p_{1GK\cdot} \ldots p_{CGK})'. \tag{81}$$

Let $\oplus$ denote the direct-sum operation. Then

$$\begin{aligned}\mathrm{Cov}(\mathbf{V}) &\equiv \Sigma \equiv N\Sigma^{\Diamond} \\ &= N(\Sigma_{11} \oplus \Sigma_{12} \oplus \ldots \oplus \Sigma_{1K} \oplus \Sigma_{21} \oplus \ldots \oplus \Sigma_{2K} \oplus \ldots \oplus \Sigma_{GK})\end{aligned} \tag{82}$$

where Σ_{gk} is a $C \times C$ matrix of the form

$$\Sigma_{gk} = \mathbf{D}_{gk} - \mu_{\Diamond gk}\mu'_{\Diamond gk} \tag{83}$$

with $\mathbf{D}_{gk}$ being a $C \times C$ diagonal matrix with elements $p_{1gk}, \ldots, p_{Cgk}$ and $\mu_{\Diamond gk} = (p_{1gk} \ldots p_{Cgk})'$. Now define the component matrices of $\Sigma^{\Diamond}$: $\Sigma^{\Diamond} = \Sigma_1 \oplus \Sigma_2 \oplus \ldots \oplus \Sigma_G$ with $\Sigma_g = \Sigma_{g1} \oplus \Sigma_{g2} \oplus \ldots \oplus \Sigma_{gK}$, $g = 1, \ldots, G$.

5.4 Test Statistics

Define the population variation within group g at position k as

$$h(\mathbf{p}_{gk}) = 1 - \sum_{c=1}^{C} p_{cgk}^2. \tag{84}$$

H_0: $p_{cgk} = p_{ck}$ for all g implies that

$$h(\mathbf{p}_{1k}) = h(\mathbf{p}_{2k}) = \ldots = h(\mathbf{p}_{Gk}) = h(\mathbf{p}_k), \tag{85}$$

i.e., within-group variation at position k is the same over all groups, and this entails

$$\|\mathbf{p}_{1k}\| = \|\mathbf{p}_{2k}\| = \ldots = \|\mathbf{p}_{Gk}\| \tag{86}$$

Under H_0,

$$E_0(TSI) = 1 - \frac{1}{NGK} + \frac{1}{NGK^2}\sum_{c=1}^{C}\left[\sum_{k=1}^{K} p_{ck}^2 - NGp_{c\cdot}^2\right] \tag{87a}$$

$$E_0(WSI) = 1 - \frac{1}{NK} + \frac{1}{NK^2}\sum_{c=1}^{C}\left[\sum_{k=1}^{K} p_{ck}^2 - Np_{c\cdot}^2\right] \tag{87b}$$

$$E_0(BSI) = \frac{G-1}{NGK}\left[1 - \frac{1}{K}\sum_{c=1}^{C}\sum_{k=1}^{K} p_{ck}^2\right] \equiv \theta_1 \tag{87c}$$

and, for any $g = 1, \ldots, G$, Σ_{gk} becomes Σ_{0k} and $\Sigma_g \Sigma_0^* = \Sigma_{01} \oplus \Sigma_{02} \oplus \cdots \oplus \Sigma_{0K}$ where Σ_{0k} is the $C \times C$ matrix

$$\Sigma_{0k} = \mathbf{D}_k - \mu_{\Diamond k}\mu'_{\Diamond k} \tag{88}$$

with $\mathbf{D}_k$ being a $C \times C$ diagonal matrix with elements $p_{1k}, \ldots, p_{Ck}$ and $\mu_{\Diamond k} = (p_{1k}, \ldots, p_{Ck})'$. Therefore, under H_0, Σ is now

$$N\Sigma_0^{\Diamond} = N(\mathbf{I}_G \otimes \Sigma_0^*) \tag{89}$$

where $\otimes$ denotes the Kronecker product and $\mathbf{I}_u$ the u-dimensional identity matrix. Let

$$\theta_2 \equiv 1 - \frac{1}{K^2}\sum_{c=1}^{C} p_{c\cdot}^2. \tag{90}$$

Asymptotically,

$$F \equiv N\left(\frac{BSI}{WSI}\right) \sim \frac{1}{\theta^2}\sum_{i=1}^{CGK} \lambda_i \left(\chi_1^2\right)_i \tag{91}$$

where $\{\lambda_i : 1, \ldots, KGC\}$ is the set of characteristic roots of $\frac{1}{NGK^2} B^{\Diamond}\Sigma_0^{\Diamond}$ where

$$B^{\Diamond} = \left((I_G \otimes U_K) - \frac{1}{G} U_{KG} \right) \otimes I_C \tag{92}$$

with $\mathbf{U}_u$, the u-dimensional square matrix whose entries are all 1's. Asymptotically,

$$E_0(F) = \frac{G-1}{GK\theta_2} \left[1 - \frac{1}{K} \sum_{c=1}^{C} \sum_{k=1}^{K} p_{ck}^2 \right] \tag{93}$$

$$\mathrm{Var}_0(F) = N^2 \frac{\mathrm{Var}_0(BSI)}{(\theta_2)^2} = \frac{2\mathrm{trace}(B^{\Diamond}\Sigma_0^{\Diamond})^2}{(GK^2\theta_2)^2} \tag{94}$$

Since p_{ck}'s are unknown, one can only get estimates for the λ_i's. To derive the distribution of F, estimate $\{p_{ck}\}$, then get the λ_i's based on those estimates.

Alternatively, since the terms of the R.H.S. of (91) are independent and identically distributed χ_1^2 random variables, if CGK is large and

$$\frac{\max(\lambda_i)}{\sqrt{\sum_{i=1}^{CGK} \lambda_i^2}} \to 0, \tag{95}$$

$$K^2 \left(F - N \frac{\theta_1}{\theta_2} \right) \sim N(0, \sigma^2) \tag{96}$$

where

$$\sigma^2 = \frac{2\mathrm{trace}(B^{\Diamond}\Sigma_0^{\Diamond})^2}{(G\theta_2)^2} \tag{97}$$

Extending the CATANOVA approach to several positions, the sum of squares are

$$WSS = \frac{NGK}{2} - \frac{1}{2NK} \sum_{g=1}^{G} \sum_{c=1}^{C} n_{cg.}^2 = \frac{NGK}{2} WSI \tag{98a}$$

$$TSS = \frac{NGK}{2} - \frac{1}{2NGK} \sum_{c=1}^{C} n_{c..}^2 = \frac{NGK}{2} TSI \tag{98b}$$

$$BSS = \frac{1}{2NGK} \left(G \sum_{g=1}^{G} \sum_{c=1}^{C} n_{cg.}^2 - \sum_{c=1}^{C} n_{c..}^2 \right) = \frac{NGK}{2} BSI. \tag{98c}$$

In this setup, the test statistic is

$$F^* = \frac{BSS/(G-1)}{WSS/(NGK-G)} = \frac{BSI/(G-1)}{WSI/(NGK-1)} = \frac{(NGK-G)}{N(G-1)} F \tag{99}$$

When CGK is large,

$$K \left(F^* - \frac{a_1}{a_2} \right) \sim N(0, \sigma_*^2) \tag{100}$$

where $\sigma_*^2 = \dfrac{\mathrm{trace}(B^{\Diamond}\Sigma_0^{\Diamond})^2}{2(a_2(G-1))^2}$, $a_1 = \dfrac{NGK}{G-1}\dfrac{\theta_1}{2}$ and $a_2 = \dfrac{1}{2}\theta_2$. For large N,

$$F^* = \frac{GK}{G-1} F. \tag{101}$$

Therefore,

$$\sigma_*^2 = \frac{G^2}{(G-1)^2}\sigma^2. \tag{102}$$

In order to look at the suitability of the asymptotic distribution of the test statistic F^*, Pinheiro *et al.* (2000) generated N sequences, with K positions each, in G groups and computed the test statistic F^* and the standardized F^* (i.e., $K\left(F^* - \frac{a_1}{a_2}\right)/\sigma_*$). On the basis of a simulation with 500 replications, the asymptotic distribution was found to be a good approximation when $N = 100$, $K = 10$ and $G = 10$. With half the number of groups, the standard normal distribution no longer applied. As the null hypothesis assumes homogeneity among groups, one pools all groups to estimate the parameters. Hence, the more groups, the more reliable the estimates are. The number of positions appears to have little influence on the validity of the asymptotics, nor does the composition of the sites.

When data sets are not large enough for the asymptotic results to apply, one needs to call upon a bootstrap procedure:

(a) Estimate p_{ck} from the data, i. e. $\hat{p}_{ck} = \frac{n_{c1k} + n_{c2k}}{2N}$ and compute the statistic F (called the realized value F_{obs}).
(b) Generate N sequences, with K positions each, in each of G groups, using $\hat{p}_{ck}$.
(c) Recompute the test statistic F from the bootstrap data and call it $F^\dagger$.
(d) Repeat steps 2 and 3 1,000 times.

The p-value is then $\frac{\# F^\dagger \geq F_{obs}}{1000}$.

6 CASE STUDY

When sequences within a group are not independent, for instance when they are sampled from a single individual, the reference distribution need to be altered to reflect the phylogenetic relationships. One then resorts to reproducing the evolution of the sequences and generating this distribution on the basis of the simulated sequences (Seillier-Moiseiwitsch *et al.*, 1999). The event of a mutation at a specific site is modeled as a two-step process. First, whether the position undergoes a change is governed by the overall rate for the genome under study. In the context of HIV sequences, this is the error rate for reverse transcriptase, i.e., 0.0005 per site per replication (Preston *et al.*, 1988). Next, in case of mutation, the specific substitution follows a transition matrix estimated from the observed frequencies.

The simulation starts with the consensus sequence as seed. It is subjected to the mutation process a random number of times between 100 and 2,400. This number represents the number of replications before transmission. This original sequence gives rise to offspring sequences: no offspring with probability 0.20 and 1 to 5 offspring with probability 0.16 each (Blower and McLean, 1994). The tree is grown by repeating this process many times (with the output sequences from the previous generation as seeds). A total of 10,000 to 20,000 sequences is thus generated. One samples without replacement as many sequences as there are in the original data set, and computes the test statistic. This sampling is performed a large number of times (here 1,000) to build up a reference distribution.

Table 6 Between-compartment Comparison of Variation.

Patient	Week	Blood	Semen	Statistics
1	0			15.51
	12		†	
	16			
	24			
2	0			55.90**
	8			
3	0			13.29
	4	†		
	12			
	40			
	56			
4	0			51.78**
	22	†		
	26			
5	0			52.46*
	3	†		
	4		†	
	8			
6	0			0.01
	8		†	
	13		†	
	25		†	
	29	†		
	49			
	58			
7	0			23.58**
	12			
	32			
8	0			12.06
	12			
	16			
	24	†		
	41			
	28			

† indicates visits for which data are missing.
* test statistic falls above the 99.5th percentile of the reference distribution.
** test statistic falls above the maximum of the reference distribution.

Pinheiro and colleagues (2000) applied this procedure to viral sequences from 8 individuals who were treated with a protease inhibitor. Sequences were sampled from blood and semen at the same or different time points. They are 1,041 nucleotides long and span the region coding for protease and reverse transcriptase. For each compartment, the data consist of the consensus sequence at each visit. The object of the study is to ascertain whether the variability is comparable in both compartments. There are thus 2 groups. Each patient provides a set of sequences on which the null hypothesis of homogeneity is tested (Table 6). For 4 of the 8 subjects, the null hypothesis is rejected.

ACKNOWLEDGEMENTS

This research was funded in part by Coordenação de Aperfeiçomento de Pessoal de Nível Superior, the National Science Foundation (DMS-9305588), the American Foundation for AIDS Research (70428-15-RF) and the National Institutes of Health (R2Y-GM49804 and P30-II D37260).

REFERENCES

Anderson, R. J. and Landis, J. R. 1980. CATANOVA for multidimensional contingency tables: Nominal-scale response. *Comm. Stat. Theory Meth.* **9:** 1191-1206.

Anderson, R. J. and Landis, J. R. 1982. CATANOVA for multidimensional contingency tables: Ordinal-scale response. *Comm. Stat. Theory Meth.* **11**: 257-270.

Bhattacharyya, A. 1946. On a measure of divergence between two multinomial populations. *Sankhya* **7**: 401-406.

Bickel, P., Cosman, P., Olshen, R., Spector, P., Rodrigo, A. and Mullins, J. 1996. Covariability of V3 loop amino acids. *AIDS Res. Hum. Retroviruses* **12**: 1401-1411.

Blower, S. M. and McLean, A. R. 1994. Prophylactic vaccines, risk-behavior-change and the probability of eradicating HIV in San Francisco. *Science* **265**: 1451-1454.

Bowman, E. O., Hutcheson, E., Odum, E. P. and Shenton, L. R. 1971. Comments on the distribution of indices of diversity, In *Statistical Ecology Vol. 3.* (Patil, G. P., Pielou, E. C. and Waters, W. E. eds.) Pennsylvania State University Press, University Park, PA.

Chakraborty, R. 1985. Genetic distance and gene diversity: Some statistical considerations, In *Multivariate Analysis - VI*, (Krishnaiah, P. ed.) Elsevier, Amsterdam.

Chakraborty, R. and Fuerst, P. A. 1979. Some sampling properties of selectively neutral alleles - effects of variability of mutation-rates. *Genet. Res.* **34**: 253- 267.

Chakraborty, R. and Griffths, R. C. 1982. Correlation of heterozygosity and the number of alleles in different frequency classes. *Theoret. Pop. Biol.* **21**: 205-218.

Chakraborty, R. and Rao, C. R. 1991. Measurement of genetic variation for evolutionary studies, In *Handbook of Statistics 8* (Rao, C. R. and Chakraborty, R. eds.) North-Holland, Amsterdam.

Cockerham, C. C. 1969. Variance of gene frequencies. *Evolution* **23**: 72-84.

Cockerham, C. C. 1973. Analyses of gene frequencies. *Genetics* **74**: 679-700.

Cockerham, C. C. 1984. Drift and mutation within a finite number of allelic states. *Proc. Natl. Acad. Sci. USA* **81**: 530-534.

Excoffier, L., Smouse, P. E. and Quattro, J. M. 1992. Analysis of molecular variance inferred from metric distances among DNA haplotypes: Application to human mitochondrial DNA restriction data. *Genetics* **131**: 479-491.

Felsenstein, J. 1981. Evolutionary trees from DNA sequences: A maximum likelihood approach. *J. Mol. Evol.* **17**: 368-376.

Fuerst, P., Chakraborty, R. and Nei, M. 1977. Statistical studies on protein polymorphism in natural populations: I. Distribution of single locus heterozygosity. *Genetics* **86**: 455-483.

Gini, C. W. 1912. Variabilita e mutabilita. Studi Economico-Giuridici della R. Universita di Cagliari **3**(2): 3-159.

Gojobori, T., Ishii, K. and Nei, M. 1982. Estimation of average number of nucleotide substitutions when the rate of substitution varies with nucleotide. *J. Mol. Evol.* **18**: 414-423.

Gojobori, T., Moriyana, E. N. and Kimura, M. 1990. Statistical methods for estimating sequence divergence. *Meth. Enzymol.* **183**: 531-550.

Hasegawa, M., Kishino, H. and Yano, T. 1985. Dating of the human-ape splitting by a molecular clock of mitochondrial DNA. *J. Mol. Evol* **22**: 160-174.

Hill, M. 1973. Diversity and evenness: A unifying notation and its consequences. *Ecology* **54**: 427-431.

Hoeffding, W. 1948. A class of statistics with asymptotically normal distribution. *Ann. Math. Stat.* **19**: 293-325.

Hutcheson, K. 1970. A test for comparing diversities based on the Shannon formula. *J. Theoret. Biol.* **29**: 151-154.

Jorde, L. 1980. The genetic structure of subdivided human populations: A review, In *Current Developments in Anthropological Genetics: Theory and Methods. Vol. 1* (Mielke, J. and Crawford, M. eds.) Plenum, New York.

Jukes, T. and Cantor, C. 1969. Evolution of protein molecules, In *Mammalian Protein Metabolism III*, (Munro, H. N. ed.) Academic Press, New York.

Karnoub, M., Seillier-Moiseiwitsch, F. and Sen, P. 1999. A conditional approach to the detection of correlated mutations, In *Statistics in Motecular Biology. Vol. 33* (Seillier-Moiseiwitsch, F. ed.) Lecture Notes Series, Institute of Mathematical Statistics, Hayward, CA.

Kimura, M. 1980. A simple method for estimating evolutionary rates of base substitutions through comparative studies of nucleotide sequences. *J. Mol. Evol.* **16**: 111-120.

Kimura, M. 1981. Estimation of evolutionary distances between homologous nucleotide sequences. *Proc. Natl. Acad. Sci. USA* **78**: 454-458.

Kimura, M. and Crow, J. F. 1964. The number of alleles that can be maintained in a finite population. *Genetics* **49**: 725-738.

Kimura, M. and Ohta, T. 1972. On a stochastic model for estimation of mutational distance between homologous proteins. *J. Mol. Evol.* **2**: 87-90.

Korber, B. T. M., Farber, R. M., Wolpert, D. H. and Lapedes, A. S. 1993. Covariation of mutations in the V3 loop of HIV-1: an information theoretic analysis. *Proc. Natl. Acad. Sci. USA* **90**: 7176-7180.

Lalouel, J. 1980. Distance analysis and multidimensional scaling, In *Current Developments in Anthropological Genetics: Theory and Methods. Vol. 1.* (Mielke, J. and Crawford, M. eds.) Plenum, New York.

Lewontin, R. C. 1972. The apportionment of human diversity. *Evol. Biol.* **6**: 381-398.

Li, W.-H. 1997. *Molecular Evolution.* Sinauer Associates, Sunderland MA.

Light, R. J. and Margolin, B. H. 1971. An analysis of variance for categorical data. *J. Amer. Stat. Assoc.* **66**: 534-544.

Light, R. J. and Margolin, B. H. 1974. An analysis of variance for categorical data II: Small sample comparisons with chi square and other competitors. *J. Amer. Stat. Assoc.* **69**: 755-764.

Lockhart, P., Steel, M., Hendy, M. and Penny D. 1994. Recovering evolutionary trees under a more realistic model of sequence evolution. *Mol. Biol. Evol.* **11**: 605-612.

Magurran, A. E. 1988. *Ecological Diversity and Its Measurement.* Princeton University Press.

Mahalanobis, P. 1936. On the generalized distance in statistics. *Proc. Natl. Inst. Sci. India* **2**: 49-55.

Manly, B. F. J. 1985. *The Statistics of Natural Selection on Animal Populations.* Chapman and Hall. London.

Mantel, N. 1967. The detection of disease clustering and a generalised regression approach. *Cancer Res.* **27**: 209-220.

Nayak, T. 1983. Applications of Entropy Functions in Measurement and Analysis of Diversity. PhD thesis. Department of Mathematics and Statistics, University of Pittsburgh, PA.

Nei, M. 1972. Genetic distance between populations. *Amer. Natur.* **106**: 283-292.

Nei, M. 1975. *Molecular Population Genetics and Evolution.* North-Holland, Amsterdam .

Nei, M. 1978. Estimation of average heterozygosity and genetic distance from a small number of individuals. *Genetics* **89**: 583-590.

Nei, M. 1987. *Molecular Evolutionary Genetics.* Columbia University Press, New York .

Nei, M. and Roychoudhury, A. 1974. Sampling variance of heterozygosity and genetic distance. *Genetics* **76**: 379-390.

Nei, M., Fuerst, P. and Chakraborty, R. 1976. Testing the neutral mutation hypothesis by distribution of single locus heterozygosity. *Nature* **262**: 491-493.

Peet, R. K. 1974. The measurement of species diversity. *Ann. Rev. Ecol. Syst.* **5**: 285-307.

Pinheiro, H. P., Seillier-Moiseiwitsch, F., Sen, P. K. and Eron, J. J. 2000. Genomic sequences and quasi-multivariate CATANOVA, In *Handbook of Statistics, Bioenvironmental and Public Health Statistics Vol. 18* (Rao, C. R. and Sen, P. K. eds.) Elsevier, Amsterdam.

Preston, B. D., Poiesz, B. J. and Loeb, L. A. 1988. Fidelity of HIV-1 reverse- transcriptase. *Science* **242**: 1168-1171.

Rao, C. R. 1982a. Diversity and dissimilarity coefficients: A unified approach. *Theor. Pop. Biol.* **21**: 24-43.

Rao, C. R. 1982b. Diversity: Its measurement, decomposition, apportionment and analysis. *Sankhya A* **44**: 1-21.

Rao, C. R. 1982c. Gini-Simpson index of diversity: A characterization, generalization and applications. *Util. Math.* **21**: 273-282.

Rao, C. R. 1984. Use of diversity and distance measures in the analysis of qualitative data, In *Multivariate Staistical Methods in Physical Anthropology* (Van Vark, G and Howell, W. eds.) Reidel, Dordrecht.

Rao, C. R. and Boudreau, R. 1984. Diversity and cluster analyses of blood group data on some human populations, In *Human Population Genetics: The Pittsburgh Symposium* (Chakravarti, A. ed.) Van Nostrand Reinhold, New York.

Sanghvi, L. D. 1953. Comparison of genetical and morphological methods for a study of biological differences. *Amer. J. Phys. Anthropol.* **11**: 385-404.

Searle, S. R. 1971. *Linear Models.* John Wiley and Sons, New York.

Seillier-Moiseiwitsch, F., Man, Z. M. and Swanstrom, R. 1999. Detecting linked genomic mutations, In *Statistics in Genetics. Vol. 112. IMA Volume in Mathematics and its Applications.* (Halloran, M. E. and Geisser, S. eds.) Springer-Verlag, New York.

Seillier-Moiseiwitsch, F., Margolin, B. H. and Swanstrom, R. 1994. Genetic variability of the human immunodeficiency virus: Statistical and biological issues. *Ann. Rev. Genet.* **28**: 559-596.

Simpson, E. H. 1949. The measurement of diversity. *Nature* **163**: 688.

Takahata, N. and Kimura, M. 1981. A model of evolutionary base substitutions and its application with special reference to rapid change of pseudogenes. *Genetics* **98**: 641-657.

Tavaré, S. 1986. Some probabilistic and statistical problems in the analysis of DNA sequences, In *Lectures on Mathematics in the Life Sciences* (Miura, R. M. ed.) American Mathematical Society, Providence, RI.

Tavaré, S. and Giddings, B. W. 1989. Some statistical aspects of the primary structure of nucleotide sequences, In *Mathematical Methods for DNA Sequences,* (Waterman, M. S. ed.) CRC Press, Boca Raton, FL.

Weir, B. S. and Basten, C. J. 1990. Sampling strategies for DNA sequence distances. *Biometrics* **46**: 551-582.

PHYLOGENETICS OF HIV

David Posada*, Keith A. Crandall* and David M. Hillis[§]

*Department of Zoology, Brigham Young University, Provo, UT 84602 USA
[§]Section of Integrative Biology and Institute of Cellular and Molecular Biology, University of Texas, Austin, TX 78712 USA

1. PHYLOGENIES AND HIV

A phylogeny is a set of relationships among groups of genes or organisms that reflects their evolutionary history. Inferring a phylogeny is an estimation procedure, a statistical inference of a true phylogenetic tree that is unknown. However, the aim of the phylogenetic analysis is not merely the reconstruction of a tree topology; the phylogeny provides a powerful framework in which several hypotheses can be tested and parameters of interest can be estimated from the data (Huelsenbeck and Crandall, 1997). Once a reliable estimate of the phylogeny of the sequences under study has been obtained, it can be used for testing diverse hypotheses about evolution. All phylogenetic methods are based on some set of assumptions. To understand the scope of the derived inferences, the assumptions of a method must be explained and delimited, and then tested and contrasted with the biological data at hand. Inferences derived from the phylogeny can be only as good as the phylogenetic estimate from which they were derived.

There are many reasons why a clear understanding of the genetic relationships among different strains of a virus is desirable. Such knowledge can provide information on the origins and geographic distributions of particular strains, on their routes of transmission, and for the development of vaccines (Leigh Brown and Holmes, 1994). In the case of HIV, its rapid evolution provides an ideal system for a successful application of a variety of phylogenetic approaches, as evidenced by the increasing number of studies on HIV using phylogenies. Phylogenetic analyses of HIV sequences have been used to investigate a variety of problems (see Seiller-Moiseiwitsch *et al.*, 1994; Crandall, 1999; Crandall *et al.*, 1999b). These problems include potential transmission of the HIV virus among individuals

(Krushkal and Li, 1999), cross-species transmissions (Sharp *et al.*, 1995; Sharp *et al.*, 1996), origins (Sharp *et al.*, 1994), epidemiology (Holmes *et al.*, 1999), subtyping (Louwagie *et al.*, 1993; Simon *et al.*, 1998) and drug resistance (Crandall *et al.*, 1999a). Phylogenetic studies have been critical for understanding the biology and evolution of HIV (Hillis, 1999). In fact, the wealth of data accumulated over the last few years has made the immunodeficiency viruses the most data-rich group of organisms for any evolutionary analyses (Leigh Brown, 1994). In this chapter, we introduce procedures for estimating phylogenies from DNA and protein sequences, including hypothesis testing and applications, and point out diverse special concerns about HIV. This chapter gives some simple guidelines for the phylogenetic analysis of HIV sequences, including references to more specific reviews and available software. Swofford *et al.* (1996a) provide the most comprehensive current review of the phylogenetic methodology.

2. PHYLOGENETIC RECONSTRUCTION

Phylogeny reconstruction is a complex process that requires several steps. Each step is equally important and should be completed carefully. In the next section we outline the main phases in phylogenetic analysis: alignment, selection of optimality criteria and search strategies for optimal trees, use of appropriate models of evolution, and confidence assessment. Also, a necessary discussion about consensus and ancestral sequences is included.

2.1 Alignment

The first step in any evolutionary study is to establish homology. In DNA sequence analysis one hypothesizes that the nucleotides observed at a given position came from the same position in the common ancestor of the taxa under study (Swofford *et al.*, 1996a). This statement of positional homology constitutes an alignment. In the alignment, positions inferred to be homologous are in the same column of the data matrix, so insertions or deletions (indels) are postulated by inserting gaps in one or several sequences. The quality of an alignment is measured as some cost resulting from different penalties. The insertion of a gap, its size, or position can each be penalized in different ways. In general, penalties are bigger for gaps than for mismatches, as indels are usually rarer than substitutions. The cost is also bigger for internal gaps than for leading or trailing gaps, as the latter usually represent different lengths of sequences rather than actual evolutionary changes (Swofford *et al.*, 1996a). Also, a matrix of change costs may be specified for the different nucleotide substitutions, thereby allowing, for example, the specification of different costs for transitions and transversions. In the case of protein-coding sequences, information about the protein reading frame or about the secondary structure of the protein can be incorporated in the alignment process (see Kjer, 1995). For example, gaps that are not multiples of three can be penalized more heavily than those that are multiples of three, because the former produce a shift in the protein reading frame.

Although alignment methods can be efficient, especially when the sequences are similar, they are not foolproof, and manual refinement of the alignment may still be required (Weiller *et al.*, 1995). Regions of the sequence alignment with substantial numbers of gaps, in which positional homology is too uncertain, should be omitted from the analysis (Swofford *et al.*, 1996a). However, deleting all the gapped columns—a procedure known as "gapstripping"—results in an unnecessary loss of information that neglects the reality of indels as evolutionary events. Some phylogeny methods (e. g., maximum parsimony) can treat gaps as a fifth state, acknowledging the evolutionary reality of indel evolution. Some models of indel evolution (Thorne *et al.*, 1991; Thorne *et al.*, 1992) have been proposed for use in likelihood or distance analyses, but no widely available software programs currently implement these models. Therefore, indels are often treated as missing data in maximum likelihood and distance analyses, thereby resulting in some loss of information. In the software section, some of the many programs for the alignment of DNA and amino acid sequences are described.

2.2 Consensus Sequences and Ancestral State Reconstruction

Consensus sequences can be constructed by examining each site or position in an alignment. When a site is invariant, the nucleotide is assigned to the consensus sequence in that position. When a site is variable, a certain nucleotide is assigned to the consensus sequence if its frequency reaches some predetermined value (e. g., 80%, 90%) called a consensus threshold. If no nucleotide reaches the threshold frequency, an ambiguity state is assigned to the consensus sequence. We would like to emphasize the artificial nature of consensus sequences. Consensus sequences are inappropriate representations of the variability in a group; this variation is better represented by using all the sequences in that group. Consensus sequences have often been used in the HIV literature as a surrogate for an ancestral sequence type. This is a serious mistake, for consensus sequences are not in any way ancestral sequences (see Figure 1).

Ancestral sequences can be readily reconstructed using various techniques, including parsimony and maximum likelihood methods. Parsimony reconstructs ancestral states by minimizing the number of steps in the tree (Fitch, 1971; Maddison and Maddison, 1994; Swofford, 1998). The problem is that the accuracy of the reconstruction is in doubt when there is a high level of sequence divergence, and parsimony often suggests many equally good reconstructions without a way to choose among them (Yang *et al.*, 1995b) (see Figure 1). Ancestral state reconstruction using parsimony is implemented in MACCLADE, PAUP* and COMPARE (see software section for software references and associated web sites). Yang *et al.* (1995b) proposed a statistical method for estimating ancestral states using maximum likelihood. In this method, a model of evolution is used to obtain maximum likelihood estimates of parameters such as branch lengths. These estimates are used to compare posterior probabilities of assignments of character states to interior nodes of the tree (see Figure 1). The best reconstruction at the site is the character-state assignment with the highest posterior probability. This likelihood-based method has been found to be superior to the parsimony method (Yang *et al.*, 1995b). Maximum likelihood ancestral state reconstruction is implemented in PAML and PAUP*.

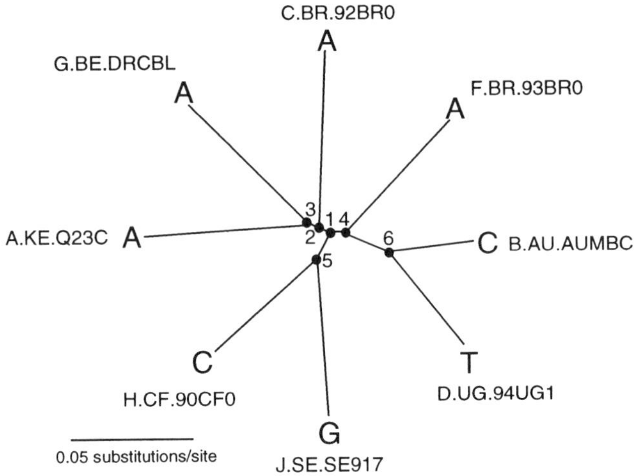

ANCESTRAL SEQUENCE CONSTRUCTION

Parsimony reconstruction of ancestral sequences.

	site 2865
Node 1	AC
Node 2	A
Node 3	A
Node 4	AC
Node 5	ACG
Node 6	ACT

Maximum likelihood reconstruction of ancestral sequences (posterior probabilities in parenthesis)

	site 2865
Node 1	A (0.949)
Node 2	A (0.969)
Node 3	A (0.978)
Node 4	A (0.925)
Node 5	A (0.846)
Node 6	C (0.651)

CONSENSUS SEQUENCE CONSTRUCTION

Threshold	Consensus Sequence at site 2865
0 %	A/C/G/T
20 %	A/C
30 %	A
50 %	A
70 %	?
100 %	?

Figure 1 Reconstruction of ancestral and consensus sequences. A maximum likelihood tree was estimated under the best-fit model of nucleotide substitution selected by MODELTEST (GTR+G) from a set of aligned *pol* sequences (3009 bp). Parsimony reconstruction was implemented in PAUP*. There were 10 most parsimonious possible reconstructions. The marginal likelihood ancestral reconstruction was implemented with the program *baseml* in PAML under the best- fit model of evolution.

2.3 Optimality Criteria and Searching Methods

Once an alignment has been proposed, several methods can be used for estimating the phylogeny of the sequences under study. All commonly used phylogenetic methods have two parts: the specification of an optimality criterion, and the specification of a search strategy to find optimal or near-optimal trees. The optimality criterion is a statement of how goodness-of-fit between data and alternative hypotheses is measured, whereas the search strategy is the means for looking for the best tree among the universe of possibilities. Given an optimality criterion, a score can be assigned to each possible phylogenetic hypothesis, so that all the different hypotheses can be ranked in order of preference. The main optimality criteria used in phylogenetics are maximum parsimony, maximum likelihood, and minimum evolution. For any of these criteria, searches for optimal solutions can be quick and approximate (e. g., neighbor joining, stepwise addition) or thorough and exact (e. g., branch-and-bound, exhaustive searches). A comparative review of optimality criteria and search methods as applied to HIV analyses is given in Hillis (1999), so this discussion will not be repeated here. We will only note that each of the three optimality criteria has advantages, and that thoroughly searching for optimal solutions is often of much greater importance than which of the three optimality criteria is selected. Parsimony and minimum evolution analyses of DNA and protein sequences can be implemented in programs such as PAUP*, PHYLIP, MEGA, and PHYLO_WIN. Maximum likelihood methods for DNA sequences are implemented in PAUP*, PHYLIP, PHYLO_WIN, fastDNAML, PAML, MOLPHY, and GAML. Maximum likelihood analyses of protein sequences are implemented in PAML and MOLPHY.

2.4 Models Of Evolution

All the phylogenetic methods make assumptions, whether implicit or explicit, about the process of DNA substitution or amino acid replacement (Felsenstein, 1973; Goldman, 1990; Penny *et al.*, 1992). This set of assumptions about the evolutionary process defines a DNA substitution or amino acid replacement model, respectively. Models are abstractions or simplifications of the real world, but they are intended to include the most important features and omit irrelevant detail (Penny *et al.*, 1994). Muse (1999) provides an extensive overview about modeling HIV evolution.

2.4.1 Models of Evolution of DNA Sequences

All the models of DNA substitution that we will discuss here share two basic assumptions:

- Substitution is described by a homogeneous Markov process, in which the probability of a change from nucleotide *i* to nucleotide *j* does not depend on the previous state of nucleotide *i*, and does not change in different parts of the tree.
- Substitution events are independent across sites, so that the probability of change in one site does not affect the probability of change in another site.

Other assumptions can be relaxed (or not) to allow a more realistic interpretation of the process that led to the data set at hand:

- Substitution events are reversible, meaning that the probability of change from nucleotide *i* to nucleotide *j* is the same as the probability of change from nucleotide *j* to nucleotide *i*.
- Rates of change are homogeneous among sites: all the sites along the DNA sequence evolve at the same rate.
- Base composition is stationary: the expected base frequencies do not change in different parts of the tree.

DNA substitution models are expressed as a 4 × 4 instantaneous rate matrix, **Q**, in which each element Q_{ij} represents the instantaneous rate of change from nucleotide *i* (rows) to nucleotide *j* (columns):

$$\mathbf{Q} = \{q_{ij}\} = \begin{pmatrix} -r_2\pi_C - r_4\pi_G - r_6\pi_T & r_2\pi_C & r_4\pi_G & r_6\pi_T \\ r_1\pi_A & -r_1\pi_A - r_8\pi_G - r_{10}\pi_T & r_8\pi_G & r_{10}\pi_T \\ r_3\pi_A & r_7\pi_C & -r_3\pi_A - r_7\pi_C - r_{12}\pi_T & r_{12}\pi_T \\ r_5\pi_A & r_9\pi_C & r_{11}\pi_G & -r_5\pi_A - r_9\pi_C - r_{11}\pi_G \end{pmatrix} \quad (1)$$

The rows and columns are ordered A, C, G, and T. The r_i's are the rate parameters that define the rates of change between nucleotides, and the π_i's are the base frequencies. If the matrix is made symmetric, so that $r_1 = r_2$, $r_3 = r_4$, $r_5 = r_6$, etc., then it corresponds to the general time-reversible model of DNA substitution (GTR or REV, Tavaré, 1986; Rodríguez *et al.*, 1990; Yang, 1994a), which is the most general model that we will consider here. Most of the commonly used models are special cases of the GTR model, and can be obtained by specifying different constraints in the values of the rate parameters or base frequencies. For instance, by restricting $r_3 = r_4 = r_9 = r_{10}$ (equal transition rates) and $r_1 = r_2 = r_5 = r_6 = r_7 = r_8 = r_{11} = r_{12}$ (equal transversion rates), one gets the Hasegawa-Kishino-Yano model (HKY, Hasegawa *et al.*, 1985). By restricting $r_1 = r_2 = r_3 = r_4 = r_5 = r_6 = r_7 = r_8 = r_9 = r_{10} = r_{11} = r_{12}$ (all rates equal) and $\pi_A = \pi_C = \pi_G = \pi_T$ (all base frequencies equal), the simplest model (JC, Jukes and Cantor, 1969) is specified. To calculate a likelihood score for a tree we need the probabilities of change from any state to any other along a branch of length *t*. This substitution probability matrix **P** is calculated as

$$\mathbf{P}(t) = e^{Qt} \quad (2)$$

In the JC model, these probabilities are:

$$P_{ij}(t) = \begin{cases} \frac{1}{4} + \frac{3}{4}e^{-\mu t} & (i = j) \\ \frac{1}{4} - \frac{1}{4}e^{-\mu t} & (i \neq j) \end{cases} \quad (3)$$

A very important extension of these models incorporates the possibility of rate heterogeneity across sites (Yang, 1993), which relaxes the assumption that all sites along the DNA sequence evolve at the same rate. This is accomplished by assigning to each site a certain probability of evolving at each rate contained in a discrete probability distribution. The simplest model assumes a proportion of invariable sites, while the rest of the sites evolve at the same rate. But the most commonly used distribution for modeling rate heterogeneity is the discrete gamma (Γ) distribution, in which a specific number of rate categories (e. g., four) are defined (Yang, 1994b). Yang (1996) reviewed the role of the incorporation of rate heterogeneity among sites and its impact on phylogenetic studies. In an attempt to relax the assumption of stationary base composition (i. e., that the base frequencies do not change in different parts of the tree), Galtier and Gouy (1998) recently proposed a nonhomogeneous, nonstationary model of DNA evolution that allows varying equilibrium G+C content among lineages. This model is implemented in the program NHML.

2.4.2 Protein Coding Sequences

DNA positions in a coding sequence can undergo either nonsynonymous substitutions, in which nucleotide substitutions correspond with amino acid replacements, or synonymous substitutions, in which nucleotide substitutions do not result in a change of amino acid. Under neutral evolution, synonymous substitutions occur at a higher rate than nonsynonymous substitutions. Since the number and type of nonsynonymous substitutions that can occur at a given site in a coding sequence can change over time, some assumptions do not hold as well in coding sequences as they do in noncoding sequences. In particular, nucleotides within a codon do not evolve independently, rates of substitution differ among the nucleotides within a codon, and rates at a specific site change over time (Muse, 1999). Some models of codon evolution that use 61 × 61 matrices have been developed in order to account for these problems (Goldman and Yang, 1994; Muse and Gaut, 1994).

Muse and Gaut (1994) define $P_{ij}(\alpha,\beta,dt)$ as the instantaneous probability of changing from codon *i* to codon *j* in a small amount of time *dt*. The parameters α and β are the synonymous and nonsynonymous substitution rates, respectively. Assuming that only one nucleotide substitution can occur in time *dt*, the substitution process among the 61 nonterminating codons is:

$$Pij(\alpha,\beta,dt) = \begin{cases} \alpha\pi_n dt & \text{when synonymous change} \\ \beta\pi_n dt & \text{when nonsynonymous change} \\ 0 & \text{when multiple substitutions needed} \end{cases} \quad (4)$$

where π_n is the frequency at equilibrium of the target nucleotide. Once again, transition probabilities for a given amount of time (t) need to be calculated in order to estimate branch lengths (or to calculate likelihoods given the branch length t). If we form a matrix **A** with the instantaneous probabilities, $P_{ij}(\alpha,\beta,dt)$, the transition probabilities are given by $e^{\mathbf{A}t}$. Once the matrix of transition probabilities **P** is approximated, the probability of observing the data, given values of α, β, and t, can be evaluated, and the parameters can be estimated using likelihood. Only the products of substitution rates and time, αt and βt, are estimable. This model is implemented in the program HYPHY.

The matrix of transition probabilities of Goldman and Yang (1994) is similar to that of Muse and Gaut (1994), but the elements of this matrix are calculated in a different way:

$$
Pij = \begin{cases} 0 & \text{if 2 or 3 of pair } (i_1, j_1), (i_2, j_2), (i_3, j_3) \\ & \text{are different} \\ \mu\pi_j \exp(-d_{aa_i,aa_j} / V) & \text{if exactly 1 of the pairs } (i_1, j_1), (i_2, j_2), (i_3, j_3) \\ & \text{differ by a transversion} \\ \mu\kappa\pi_j \exp(-d_{aa_i,aa_j} / V) & \text{if exactly 1 of the pairs } (i_1, j_1), (i_2, j_2), (i_3, j_3) \\ & \text{differ by a transition} \end{cases} \quad (5)
$$

where μ is the mutation rate, π_j is the frequency at equilibrium of the codon being changed to, κ is a parameter that allows the empirical finding that transitions often occur more frequently than do transversions (Brown *et al.*, 1982), $d_{aai,aaj}$ are physicochemical distances following Grantham (1974), and V is a parameter representing the variability of the gene or its tendency to undergo nonsynonymous substitutions. However, a simplified version is recommended (Yang, 1998; Yang and Nielsen, 1998; Yang *et al.*, 1998):

$$
q_{ij} = \begin{cases} 0 & \text{if the two codons differ at more than one position} \\ \pi_j & \text{for synonymous transversion} \\ \kappa\pi_j & \text{for synonymous transition} \\ \omega\pi_j & \text{for nonsynonymous transversion} \\ \omega\kappa\pi_j & \text{for nonsynonymous transition} \end{cases} \quad (6)
$$

where q_{ij} is the instantaneous substitution rate from codon i to codon j, κ is the transition/transversion ratio, ω is the nonsynonymous/synonymous rate ratio, and π_j is the equilibrium frequency of codon j. These models are implemented in the program *codeml* in the software package PAML.

2.4.3 *Amino Acid Sequences*

The simplest model of amino acid evolution is a Poisson model, analogous to the JC model for DNA, but with 20 possible states instead of four. The probability of change for this model is given by:

$$
P_{ij}(t) = \begin{cases} \dfrac{1}{20} + \dfrac{1}{19} e^{-\mu t} & (i = j) \\ \dfrac{1}{20} - \dfrac{1}{20} e^{-\mu t} & (i \neq j) \end{cases} \quad (7)
$$

As in the case of DNA, different constraints in the values of the relative rate parameters lead to the specification of distinct models. More complex empirically derived models have been developed taking into account amino acid physicochemical properties and protein secondary structure (Thorne *et al.*, 1996; Goldman *et al.*, 1998). Goldman, Thorne and Jone's model (1998) is implemented in the program PASSML. An empirical general reversible model of amino acid replacement (mtREV) analogous to the REV model of DNA substitution has been proposed by Adachi and Hasegawa (1996a), and it is implemented along with other models in MOLPHY (Adachi and Hasegawa, 1996b). Yang and colleagues (1998) transformed a Markov model of codon substitution into a mechanistic model of amino acid replacement that seems to provide a better fit to amino acid sequence data. A number of models of protein evolution are implemented in the program *aaml* in the software package PAML.

2.4.4 Estimating Model Parameters

The models described above contain a number of parameters that must be estimated from the data. This is best done in a maximum likelihood framework, simultaneously optimizing the different parameters, and finding the values that maximize the likelihood. As long as the parameter estimates are fairly consistent across tree topologies, a useful method is to estimate model parameters on some reasonably good tree (parsimony, neighbor joining, maximum likelihood using the JC model), and then use the resulting estimates in a search for a better tree under the adequate model of evolution. Yang *et al.* (1994; 1995a) have shown that in estimating some important parameters of molecular evolution, such as ti/tv ratio, κ, and the α parameter of the gamma distribution, knowledge of the true phylogeny is not very important as long as a reasonable model of evolution is adopted. How to decide if a model of evolution is reasonable is discussed in the next section.

2.4.5 Selecting a Model of Evolution

To have confidence in inferences, it is necessary to have confidence in the models on which these inferences are based (Goldman, 1993b). The advantage of methods that incorporate explicit models of DNA substitution, such as distance or maximum likelihood methods, is that confidence on the models can be assessed (Huelsenbeck and Crandall, 1997). Discrimination between models of DNA substitution will become more important as molecular sequence databases expand and interest in DNA analysis increases (Goldman, 1993b). One of the most widely used statistics for comparing the fit of two competing models is the likelihood ratio test (LRT) statistic δ:

$$\delta = 2(\ln L_1 - \ln L_0) \tag{8}$$

where L_1 is the maximum likelihood under the complex model (alternative hypothesis) and L_0 is the maximum likelihood under the simple model (null hypothesis). When the models compared are nested (the simple model is a special case of the complex model), the δ statistic is asymptotically distributed as χ^2 with q degrees of freedom, where q is the difference in number of free parameters between the two models. It is important to note that to preserve the nesting of the models, the likelihood scores must be estimated on the same tree topology.

Goldman (1993a) questioned the χ^2 approximation of the test statistic, but simulation study of Yang and coworkers (1995a) suggests that the χ^2 approximation is acceptable in most cases. However, the χ^2 distribution may not be reliable when the null model is equivalent to fixing some parameters at the boundary of the parameter space of the alternative model, e. g., rate homogeneity test, where the null hypothesis is a special case of the gamma-distribution model with shape parameter (α) equal to infinity (Yang, 1996). Whelan and Goldman (1999) showed that for comparisons of rate variation across sites and nucleotide frequencies estimated as the observed base frequencies, the observed distribution of the LRT statistic was significantly different from the χ^2 distribution. However, the likelihood differences when comparing models may be very large, and the inaccuracy of the χ^2 approximation should not change the conclusions of the tests in these cases. When the models compared are not nested, the δ statistic is not distributed as χ^2 anymore, and one must use alternative means of generating its null distribution, typically through Monte Carlo simulation (Goldman, 1993b) (see below and Figure 4).

A different approach for comparing models is to compare all competing models at the same time by calculating the minimum theoretical information criterion (Akaike, 1974), AIC= $-2\ln L + 2n$, where L is the maximum value of the likelihood function for a specific model using n independently adjusted parameters. Smaller values of AIC indicate better models (Hasegawa, 1990). The advantages of the AIC criterion are that it does not require the compared models to be nested, and it is very fast and easy to implement. Muse (1999) demonstrates the use this approach to model testing using HIV–1 sequence data. Rzhetsky and Nei (Rzhetsky and Nei, 1995) used linear invariants to develop several tests for the fit of a particular model to the data. They test whether the deviation for the expected invariant would be significant if the evaluated model were true. These tests do not require the use of an initial phylogeny, and are independent of evolutionary time, but they are model specific, and can only be applied to a small set of possible substitution models

Some other methods have been developed for assessing the fit of a single model to the data. These methods calculate the maximum value of the likelihood function under the multinomial distribution as an upper bound to which the likelihood of any model can be compared as a test for model fit (Goldman, 1993a). The likelihood function under the multinomial distribution refers to an unconstrained model of evolution, and for n aligned DNA sequences of length N sites (excluding gapped sites) it has the form

$$L = \prod_{b \in \Re} (p_b)^{n_b} \qquad (9)$$

where $\Re$ is a set of 4^n possible nucleotide patterns that may be observed at each site, p_b is the probability that any site exhibits the pattern b in $\Re$ given the tree and a substitution model, and n_b is the number of times the pattern b is observed out of the N sites.

One tool for choosing a model of evolution using likelihood-ratio tests or the AIC is the program MODELTEST. This program compares the likelihood scores (obtained through PAUP*) corresponding with different models of evolution for a given tree topology using LRTs and the AIC criterion. This approach tests a number of hypotheses concerning the sequence data, including 1) Are nucleotide frequencies equal? 2) Are transition rates equal to transversion rates? 3) Are transition rates and transversion rates equal within these classes? 4) Is there rate heterogeneity within the data set? (Γ) and 5) Is there a significant proportion of invariable sites (I) ?

It has been shown that the use of one model of evolution or another can change the results of the analysis (Sullivan and Swofford, 1997; Kelsey *et al.*, 1999). The methodology described in Huelsenbeck and Crandall (1997) and implemented and extended in Posada and Crandall (1998) provides a justification for the use of a specific model of DNA substitution. Empirical tests support the idea that best-fit models identified using LRT tests seem to be a conservative choice, but their relative performance seems to be the greatest when the choice of models is most important (Cunningham *et al.*, 1998). A model does not have to be perfect to be useful (Swofford *et al.*, 1996a). All the current models of evolution are often rejected when compared against the multinomial distribution (Goldman, 1993a; Yang *et al.*, 1994), but this means only that actual models do not completely describe the underlying process of evolution, not that they are inadequate to lead to a reasonable estimation of the phylogeny. The use of adequate models of evolution (even current ones) improves the accuracy of the phylogenetic inference (Leitner *et al.*, 1997; Sullivan and Swofford, 1997).

Example 1. The subtype reference *pol* alignment was downloaded from Los Alamos database at *http://hiv-web.lanl.gov/ALIGN_98/subtype_alignments.html*. A block of commands (included in the package MODELTEST) is executed in PAUP* to obtain likelihood scores for 24 different models of evolution, given the data (the *pol* alignment) and a neighbor-joining (NJ) tree constructed using the JC model of evolution. The output of this execution of PAUP* is the input for the program MODELTEST, which indicates that the best-fitting model for this data set is the GTR + Γ (see above) model, after rejecting the null hypotheses of equal base frequencies, equal transition and transversion rates, equal transition rates and equal transversion rates, and rate homogeneity among sites, and failing to reject the null hypothesis of no invariable sites (Table 1). For the *pol* data set we have estimated two trees using the neighbor-joining algorithm. One tree has been estimated using the Kimura two-parameter model, (K2P or K80, Kimura, 1980) (Figure 2A); the other tree was estimated using the best fitting model, the GTR + Γ model (Figure 2B). The topology of these trees is different (see Figure 2, position of subtype A), although, in this specific case, this difference is not significant (Kishino-Hasegawa test; P-value = 0.9693).

Table 1 Likelihood ratio tests of models of molecular evolution (Huelsenbeck and Crandall, 1997; Posada and Crandall, 1998). *P*-value were corrected using the Bonferroni correction (Miller, 1966). $\delta = 2\ (\ln L_1 - \ln L_0)$

Null Hypothesis	Models Compared	$-\ln L_0$	$-\ln L_1$	δ	df	*P*
Equal base frequencies	H_0: JC69 H_1: F81	17650.73	17422.55	456.36	3	<0.0001*
Equal ti/tv rates	H_0: F81 H_1: HKY85	17422.55	16708.51	1428.08	1	<0.0001*
Equal ti and equal tv rates	H_0: HKY85 H_1: GTR	16708.51	16636.97	143.08	3	<0.0001*
Equal rates among sites	H_0: GTR H_1: GTR+Γ	16636.97	15902.03	1469.88	1	<0.0001*
Proportion of invariable sites	H_0: GTR+Γ H_1:GTR+Γ+I	15902.03	15898.80	6.46	1	0.0554

* Null hypothesis rejected

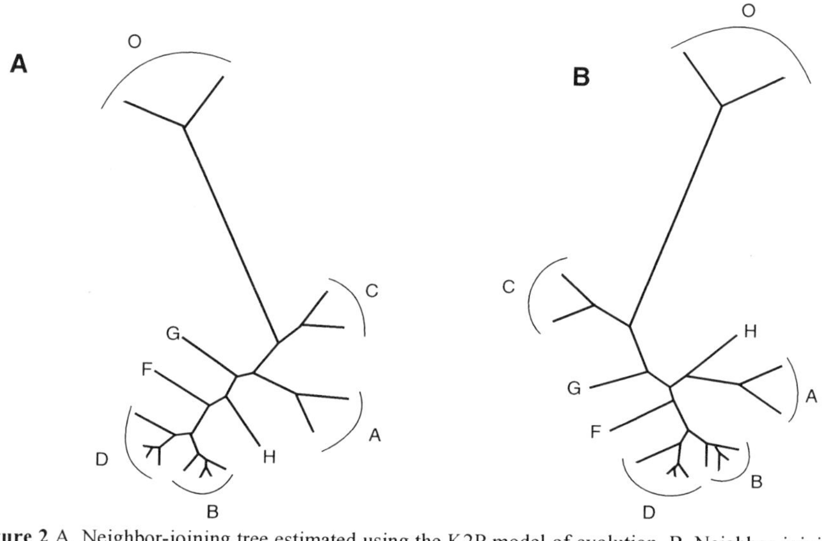

Figure 2 A. Neighbor-joining tree estimated using the K2P model of evolution. B. Neighbor-joining tree estimated using the GTR + Γ model of evolution. Observe the differences in the position of subtype A.

2.4.6 *A General Model of HIV–1 Evolution*

It is now well known that the substitution matrix in HIV–1 is highly asymmetric. The most common type of change observed in HIV–1 sequences is the transition A to G, A to C transversions are more common than C to T transitions, and all of these types are several times more common than C to G transversions (Hillis *et al.*, 1994). Simple models such as K2P cannot explain the complexity of HIV–1 evolution (Moriyama *et al.*, 1991), and usually more complex models such as GTR fit the data better (Leitner *et al.*, 1997). One possible way to incorporate more specific information on HIV evolution would be to use the large HIV database to estimate important parameters for use in implementing different general models of evolution for different parts of the HIV genome (Hillis, 1999). This approach would be expected to increase the accuracy and power of phylogenetic analysis of HIV sequences. Indeed, a first step in this direction is the codon-based model of Pedersen *et al.* (1998), that incorporates unequal base compositions in the three codon positions and selection against the CpG dinucleotide.

2.5 Confidence Assessment

Without some assessment of reliability, a phylogenetic estimate has limited value. A phylogenetic estimate based on data should normally be accompanied by an assessment of the estimate's reliability (Penny and Hendy, 1986). There are several methods of assessing the reliability of the individual internal branches of an estimated tree.

The decay index or Bremer support (Bremer, 1988) is the difference in length between the shortest tree (the tree that implies the smallest number of nucleotide substitutions) that contains that branch and the shortest tree that does not contain that branch. The significance of the different values of Bremer support is not clear, because there is no defined range of values, and it can be applied only for parsimony trees. Bremer support can be calculated using the program AUTODECAY.

The a priori T-PTP (topology-dependent permutation tail probability) test (Faith, 1991) calculates the proportion of the time that a particular Bremer support value is matched or exceeded when calculated from permuted data sets. The a posteriori T-PTP (Faith, 1991) test uses a different method for generating the permuted data sets. Swofford *et al.* (1996b) have argued that, because the permutation procedure destroys all phylogenetic structure in the data, the null hypothesis tested by T-PTP is that of no phylogenetic structure, rather than that a particular group is non-monophyletic. They used computer simulation to support their arguments. On the other hand, Faith and Trueman (1996) suggest that when the T-PTP test is significant, it fails to falsify a hypothesis of monophyly. The T-PTP test is implemented in PAUP*.

The interior branch test (Rzhetsky and Nei, 1992; Sitnikova *et al.*, 1995) is a t-test that assesses whether the length of the branch separating the hypothesized monophyletic group from the remaining taxa is significantly greater than zero. If the branch length is not significantly greater than zero, then it is not considered significantly supported. The interior branch test is implemented in METREE.

Resampling techniques such as bootstrapping and jackknifing (Efron and Tibshirani, 1993) are used to estimate the variance of a statistic from which the underlying distribution is either unknown or difficult to derive analytically. The variance of the statistic of interest is approximated by the variance of a sampling distribution obtained by repeatedly resampling data from the original data set. Each new sample obtained by resampling is called a pseudosample. When the resampling is made with replacement and the size of the pseudosample is the same as the size of the original sample, the technique is called bootstrapping or nonparametric bootstrapping (see Figure 3). Consequently, in the bootstrap pseudosamples, some data points are lost and others are repeated. When the resampling is made without replacement and the size of the pseudosamples is smaller than the size of the original sample, the technique is called jackknifing. In the jackknifed pseudosamples, some data points are lost but none are repeated. In a phylogenetic context, the resampled data points are the columns of the alignment (characters), because the statements of homology (the columns) must be preserved. From each pseudosample a new tree is estimated, and the number of times that a specific internal branch appears in the whole set of trees is recorded as the bootstrap or jackknife proportion for that branch (Felsenstein, 1985; Felsenstein, 1988). In general, the bootstrap or jackknife pseudosamples are summarized by computing a consensus tree of all bootstrap or jackknife replicates.

Bootstrapping is used more in phylogenetics than jackknifing or any of the other techniques described above (Swofford *et al.*, 1996a). However, interpretation of bootstrap proportions is often problematic. A bootstrap proportion is not the probability that a branch (a grouping) is correct. Analytical (Zharkikh and Li, 1992b; Zharkikh and Li, 1992a) and empirical (Hillis and Bull, 1993) studies have shown that bootstrap values are biased estimates of accuracy. Under a consistent method, high bootstrap values underestimate accuracy, while low bootstrap values overestimate accuracy. The extent of the bias depends on the data set at hand, so it is incorrect to assume that bootstrap proportions provide a direct measure of accuracy. Two methods, iterated bootstrapping (Hall and Martin, 1988; Rodrigo *et al.*, 1993) and the complete-and-partial (C-P) bootstrap technique (Zharkikh and Li, 1995), have been proposed to correct for this bias. In addition, it has been recommended that the number of bootstrap replicates should be greater than 400 to reduce the variance of the estimate (Li, 1997). Although the interpretation of bootstrap values is still in debate, what is clear is that high values (>90%) are very likely to indicate correct branches if the method is consistent. It also should be clear that the bootstrap proportions are no better than the phylogenetic method used. If the phylogenetic method used is inconsistent, it will converge repeatedly to the same wrong topology, providing high bootstrap values that have no correspondence to phylogenetic accuracy. One common unjustified use of bootstrap values compares them across different trees for establishing different levels of support for different hypotheses. This procedure lacks any statistical or logical basis. Nonparametric bootstrapping is designed to provide a general measure of support, and is not a method for testing specific a priori hypotheses. For testing specific hypotheses, appropriate statistical tests are available that make efficient use of the available information (all the information in the data that is relevant to the desired inference is contained in the statistic). Some of these tests are described below. Another subtle issue is how to present bootstrap values. Commonly, bootstrap values are presented in a consensus tree of the trees estimated from the pseudosamples. However, this tree does not represent our best estimate of relationships. Because of that, it is often desirable to present the bootstrap values on the best estimate of topology and branch lengths. Bootstrap and jackknife techniques are implemented in general phylogenetics packages, such as PAUP*, PHYLIP, and MEGA.

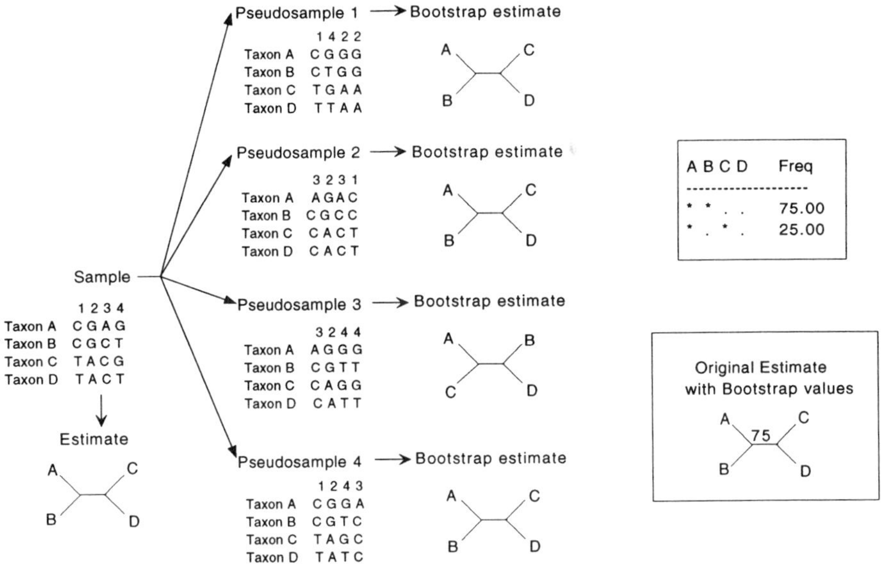

Figure 3 The bootstrap process

3. HYPOTHESIS TESTING IN A PHYLOGENETIC FRAMEWORK

Whereas nonparametric bootstrapping provides a rough measure of support for various branches in an estimated tree, it is often desirable to test specific a priori hypotheses of phylogeny. Huelsenbeck and Rannala (1997) provided a review of testing hypotheses in an evolutionary context, using likelihood-ratio tests. Tests of this type can be generalized for any optimality criterion (Hillis *et al.*, 1996), with the test statistics generated through parametric bootstrapping (also called Monte Carlo simulation). In this procedure, many data sets of the same size as the original are simulated under an explicit model of evolution and an explicit phylogenetic hypothesis. In Figure 4, the process of parametric bootstrapping is described for testing the monophyly of a group (see also Huelsenbeck *et al.*, 1996c). In the phylogeny problem, the parameters used in the simulation would include the tree topology, branch lengths, and substitution parameters (e. g., transition:transversion ratio, shape parameter of the gamma distribution). Huelsenbeck *et al.* (1996b) provides a review of performance and applications of parametric bootstrapping in phylogenetics. Some of these applications are discussed below. Programs for simulating DNA and protein sequences given a tree and a specified model of evolution are THE SIMINATOR, SEQGEN and TREEVOLVE.

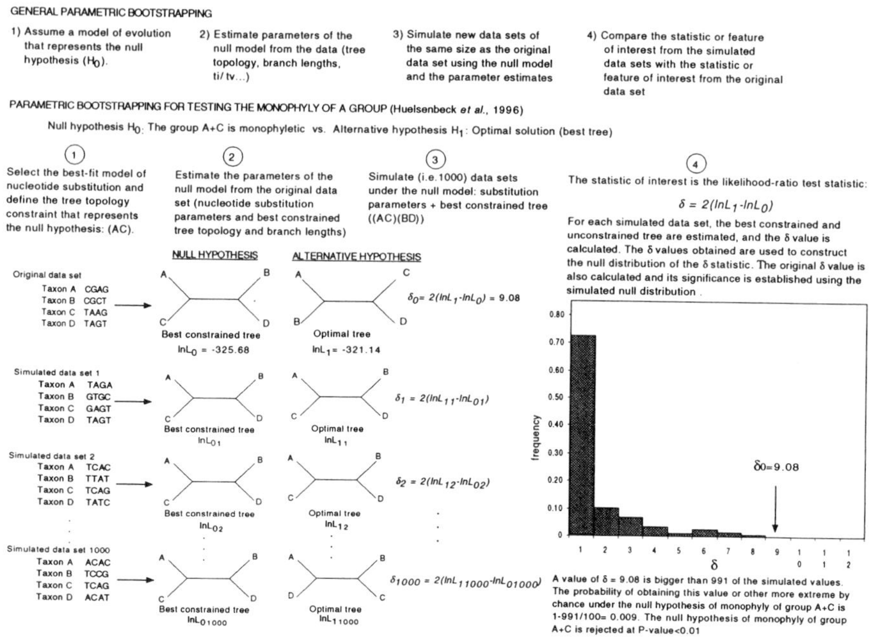

Figure 4 Parametric bootstrapping description and its application for testing the monophyly of a given group.

Obtaining a phylogeny should not be the end of the analysis when testing phylogenetic hypotheses. Because the phylogeny can provide answers to many biological questions, a number of statistical tests have been developed that take into account the phylogeny of the group of interest. In this section, we also describe some of these extensions of phylogenetic tests.

3.1 Comparing Two Trees: Is Tree A Different than Tree B?

Often it is of interest to test alternative trees that represent different hypotheses, for example, monophyly of a group versus nonmonophyly of that group, or congruent partitions of the data versus incongruent partitions. When testing phylogenetic hypotheses, it is essential to designate the two alternative trees to be tested before the estimation procedure; i. e., the hypotheses must be declared *a priori*; the comparison considered most appropriate should not be selected after evaluating the results of the test.

In a parsimony framework, several tests have been proposed for testing the null hypothesis that the number of substitutions is not significantly different in the two trees. The Templeton test (Templeton, 1983b) is a one-tailed Wilcoxon signed-ranks test that compares the differences at each site in the number of substitutions required for each tree. The winning sites test (Prager and Wilson, 1988) is a signed test (approximated to a binomial) for the departure from one-half of the proportion of sites that support one of the trees. Both of these tests can be easily implemented in MACCLADE using the “Compare two trees” option or in PAUP* under “Tree Scores” for parsimony. The Kishino-Hasegawa test (Kishino and Hasegawa, 1989) uses the variance of the difference in steps (substitutions) in single sites between tree topologies; a t-test is then used to compare the observed differences in number of steps between the two trees. These three tests are explained in Figure 5.

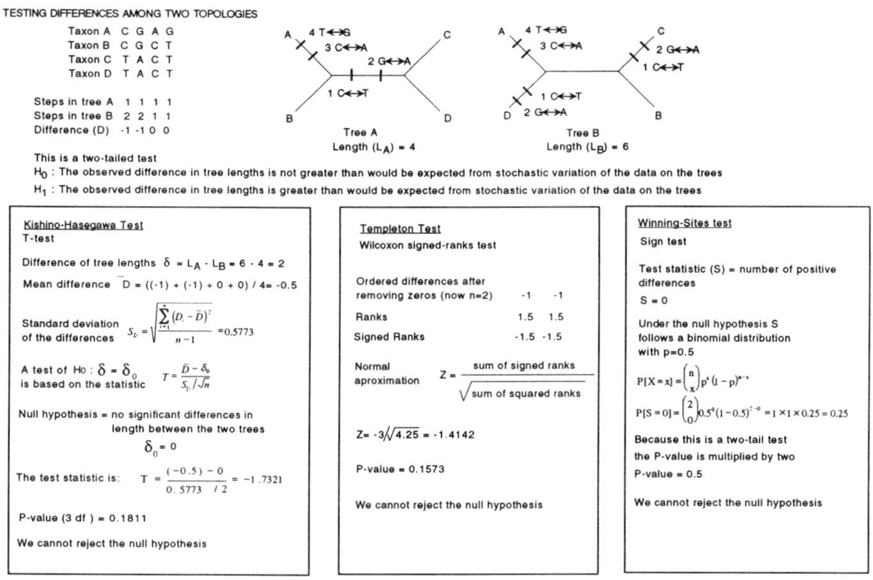

Figure 5 Tests for comparing tree topologies.

Rzhestky and Nei (1992) developed a t-test for testing the difference in the sums of branch lengths between two topologies. This test is equivalent to testing whether the lengths of the interior branches at which the two topologies differ are statistically greater than zero (Nei, 1996). The interior branch test can be implemented in METREE.

In a maximum likelihood framework, other tests have been proposed to compare two trees. Kishino and Hasegawa (1989) proposed the estimation of the variance of the difference in single-site log-likelihood scores between tree topologies for performing a simple t-test. It is important to note that this test compares not only the topology of the trees but also their branch lengths. The Kishino-Hasegawa test for parsimony or maximum likelihood trees can be implemented in PAUP* or PHYLIP. Huelsenbeck and Bull (1996) designed a test for comparing two trees derived from different data partitions (e. g., different genes). They proposed the use of a likelihood-ratio test (LRT) for testing the null hypothesis that the same phylogenetic tree underlies all of the data partitions. Monte Carlo simulation is used to establish the significance of the LRT test statistic (as in Figure 4). This test estimates whether the difference in likelihood between the best solution that supports each hypothesis is significantly larger than expected if the null hypothesis is true.

Another approach to compare trees is through the use of *tree comparison metrics* that quantify differences in topology. The most widely used tree comparison metric is the symmetric-difference distance, or partition metric (Robinson and Foulds, 1981), which is the number of groups that appear in one of the trees or the other but not in both. It is easy to calculate and its probability distribution is known (Steel and Penny, 1993), which allows for the calculation of its significance (i. e., whether the value observed could have arisen by chance) (Penny and Hendy, 1985). The calculation of this metric is implemented in PAUP*.

Example 2. Simon *et al.* (1998) identified a highly divergent new HIV–1 isolate from Cameroon (YBF30), proposing it as the prototype strain of a new human immunodeficiency virus (group N). In their analysis, the neighbor-joining tree based on the *env* gene indicated clustering of YBF30 with a chimpanzee lentivirus from Gabon (SIVcpz-gab). However, the hypothesis of interest here is whether this strain falls within the M group, O group, or neither group; i. e., can we reject the null hypothesis of YBF30 clustering with the M or O group? The point estimate of the phylogenetic relationships does not constitute a test of this hypothesis. To test this hypothesis we also need to estimate two trees with the constraint of YBF30 being a member of the M and O groups, respectively. Then we compare these constrained trees with the original point estimate of the phylogeny (the tree clustering YBF30 with SIVcpz-gab). The authors provided the accession numbers of the *env* (gp 160) sequences used in the analysis, and we tested this hypothesis (Figure 6). Given the data and the best-fit model of evolution, we cannot reject the null hypothesis of YBF30 falling into the M group, but strongly reject the hypothesis of YBF30 be-

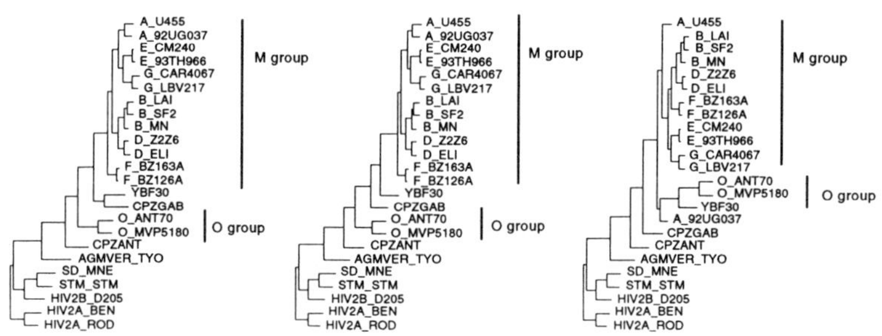

Point Estimate (Unconstrained NJ) | YBF30 - group M constraint (NJ) | YBF30 - group O constraint (NJ)

Trees	Length	Templeton Test P-value	-ln Likelihood	KH test P-value
Unconstrained NJ	9025		38776.396	
YBF30 - M	9044	0.2040^{NS}	38789.292	0.0830^{NS}
YBF30 - O	9232	<0.0001***	39007.463	<0.0001***

Figure 6 Analysis of the YBF30 sequence. *P*-values were adjusted using sequential Bonferroni.

longing to the O group. The best-fit model (GTR + I + Γ) was different from the model used by the authors (K2P). Even using the K2P model, the results were the same. Therefore, the conclusion drawn by Simon *et al.* is not supported by our statistical analysis of these sequences.

3.2 Comparing Rates of Evolution

Several tests have been proposed for comparing rates of nucleotide substitution between homologous sequences. Some of these tests have been designed for comparing two lineages (relative rate tests), while others test for overall heterogeneity of rates in a given phylogeny (rate heterogeneity tests or molecular clock tests). The relative rate tests use a third taxon as a reference point, and compare the relative rates from the reference to each of the test sequences. Other tests are based on variance estimates for performing simple t-tests (Sarich and Wilson, 1973; Wu and Li, 1985; Pamilo and Bianchi, 1993). Muse and Weir (1992) and Muse and Gaut (1994) proposed the use of a likelihood-ratio test for the same purpose. These methods are implemented in HYPHY. Templeton (1983a), Gu and Li (1992), and Tajima (1993) proposed a nonparametric approach using a sign test for comparing the number of sites that the two taxa of interest have in common with the reference taxon. The likelihood-ratio tests are the more powerful approaches. The codon-based models of sequence evolution provide powerful likelihood-based tests that compare nonsynonymous and synonymous substitution rates between lineages (Muse and Gaut, 1994). Relative-rate tests may be generalized to compare substitution rates between more than two sequences (Robinson *et al.*, 1998). Steel, Cooper and Penny (1996) also described a relative rate test that uses the variation within two monophyletic groups to estimate their divergence time. The tests of Wu and Li, Tajima, and Steel and colleagues are implemented in the program R8S.

The first test of the molecular clock was the maximum likelihood approach of Langley and Fitch (1974), which tests whether the estimated branch lengths are consistent with a Poisson process under a constant rate. This method and an extended version are implemented in the program R8S. Other molecular clock tests are based on least squares approaches (Felsenstein, 1984; Uyenoyama, 1995). The rate constancy hypothesis can also be tested by calculating the likelihood values of the best trees with and without enforcing the molecular clock and then performing a likelihood-ratio test (Felsenstein, 1988). A likelihood-ratio test can also be applied for testing assertions about rates of evolution in different parts of a molecule, such third codon positions or different genes (Felsenstein, 1988; Gaut and Weir, 1994). Steel *et al.* (1996) developed a t-test of the molecular clock that does not rely on any specific model of evolution and is based on relative distances. Hartmann and Golding (1998) also proposed a permutation method for detecting regional substitution rate heterogeneity based on maximum likelihood—a method they claim is more accurate statistically than the likelihood-ratio methods.

Ideally, molecular clocks should be calibrated using independent lineages in the phylogeny (Hillis *et al.*, 1996). The common calibration using pairwise differences among taxa within a group inflates the correlation between divergence and time, because many pairwise differences are based on the same portions of the phylogeny, and therefore are not independent (Hillis *et al.*, 1996). This lack of independence makes the regression analysis of genetic distance on time inadequate. Moreover, estimates of HIV–1 divergence rates vary depending on the region of the genome under study, alignment, amount of recombination, different selection pressure among individuals, and phylogenetic accuracy (Korber *et al.*, 1998). The existence of a molecular clock in HIV cannot be assumed unless it is tested statistically with techniques other than regression.

The assumption of a molecular clock in HIV is controversial. The rate of evolution of HIV has been estimated at 10^{-2}–10^{-3} nucleotide substitutions per site per year (Li *et al.*, 1988) using a molecular clock. This rate correlates well with epidemiological data for the branching points within the HIV phylogeny (Sharp *et al.*, 1994). Estimates of the age of the M group are in accord with the assumed age given the existence of a well-dated sequence (ZR59; Zhu *et al.*, 1998) at the base of this group (Korber *et al.*, 1998), although the confidence limits of the estimate for the age of the M group are very large. However, the molecular clock is rejected when analyzing subtype A and B *env*, *gag* and *pol* gene sequences (Holmes *et al.*, 1999). Given the relevance of the molecular clock assumption for parameter estimation, more research is needed in this direction. An alternative and powerful approach is the use of coalescent theory for estimating divergence times. A review of these techniques is included in Chapter 10 in this book by Vasco and Fu.

Recently, Sanderson (1997) and Thorne and colleagues (1998) have developed two promising approaches for estimating divergence times in the absence of a molecular clock, based on the idea of autocorrelation of rates in time. Sanderson's method is implemented in the program R8S. Thorne, Kishino and Painter's method (1998) is implemented in a C program available from Jeffrey Thorne at the Department of Statistics at North Carolina State University.

Example 3. A simple likelihood-ratio test of the molecular clock can be implemented in PAUP*. The sequences used are those in the subtype reference *pol* alignment from Los Alamos HIV database at *http://hiv-web.lanl.gov/ALIGN_98/subtype_alignments.html*. The first step is to obtain the best estimate of the phylogenetic relationships. A likelihood-ratio test is implemented that compares the likelihood of the estimated tree constrained with the null hypothesis of a molecular clock (L_0) versus the likelihood of same tree but allowing each lineage to have different rates (L_1). These likelihood scores can be obtained in PAUP* by enforcing the molecular clock under Analysis–Likelihood settings–Miscellaneous, and obtaining the maximum likelihood scores in Trees–Tree Scores–Likelihood under the best-fit model, and repeating the same steps without the molecular clock enforcement. The likelihood-ratio test statistic is calculated as twice the difference between the log-likelihood scores of the two models being contrasted. When the model representing the null hypothesis is a special case of the alternative model, as in this situation, this statistic fits a chi-square distribution, with n-2 degrees of freedom ($2n - 3$ rates for the non-clock model minus n-1 rates for the clock-like model), n being the number of taxa.

- The best-fit model is selected using PAUP* and MODELTEST (see above). This model is GTR + Γ, and these were the parameter estimates:

rAC	rAG	rAT	rCG	rCT	rGT	Γ shape
3.0409	10.4617	1.3261	1.4239	14.9662	1.0000	0.3314

- A neighbor-joining tree (or an ML tree) is estimated from the data using the GTR + Γ model of evolution with the parameter estimates
- The likelihood of this estimated tree is calculated in PAUP* with the molecular clock constraint
  L_0 = -ln likelihood tree with the molecular clock assumption = 15918.9027
- The likelihood of this estimated tree is calculated in PAUP* without the molecular clock constraint
  L_1 = -ln likelihood tree without molecular clock assumption = 15904.5491
- The ratio likelihood test statistic is $\delta = 2\ (\ln L_1 - \ln L_0) = 2\ (-15904.5491 + 15918.9027) = 28.7072$

The significance of this test is calculated comparing the test statistic to a chi-square distribution with (17 – 2) = 15 degrees of freedom. *P*-value = 0.0175, which is significant ($P < 0.05$). In this case we would reject the molecular clock at the 95% level, although not at the 99% level of confidence.

3.3 Detecting Selection in Protein Coding Sequences: Synonymous and Nonsynonymous Substitution Rates

HIV–1 is known to exhibit high levels of genetic variation even within a single patient (Hahn *et al.*, 1986; Fisher *et al.*, 1988). Since RNA viruses have high mutation rates (Holland *et al.*, 1992), one possible explanation for the HIV–1 polymorphism is that the variation is selectively neutral and a consequence of the high mutation rates. An alternative hypothesis is that HIV–1 genetic variation is maintained by positive natural selection by the immune system (Holmes *et al.*, 1992; Seibert *et al.*, 1995). The nonsynonymous/synonymous substitution ratio (*dn*/*ds*) has been proposed as an indicator for discriminating between the neutral and the selective hypotheses. Under purifying selection (neutral theory), the *dn*/*ds* substitution ratio is smaller than one (Kimura, 1977), as synonymous mutations are much more likely to become fixed than are nonsynonymous mutations, since the latter have a negative effect on gene function. Under positive Darwinian selection the *dn*/*ds* ratio is greater than one, because advantageous nonsynonymous substitutions are fixed at a higher rate than synonymous substitutions (Hughes and Nei, 1988; Messier and Stewart, 1997). Several methods have been proposed for estimating *dn* and *ds* substitution rates. The Nei and Gojobori (1986) method and related methods (Miyata and Yasunaga, 1980; Lewontin, 1989; Li, 1993; Pamilo and Bianchi, 1993; Ina, 1995) start by counting the number of silent and replacement sites. Next they compare homologous codons site by site and infer the number of silent and replacement differences, using the shortest pathways between the codons. Finally, they adjust these counts for multiple substitutions (for example, using the JC model). But there are problems with this approach, as variation at most sites does not result exclusively from synonymous or nonsynonymous substitutions, and the parameter being estimated (expected number of silent substitutions per silent site) is not clearly defined. Ina (1995) pointed out that these methods give underestimates of the *dn* rate and overestimates of the *ds* rate, because use of a simple model such as JC does not allow for unequal nucleotide change probabilities, unequal base frequencies, or heterogeneity of the rate of substitution among sites. Consequently, these methods are conservative tests of positive selection. Moreover, these pairwise approaches have problems because they count the substitutions occurring on internal branches multiple times (Crandall *et al.*, 1999a) (Figure 7). Nei and Gojobori's method is implemented in MEGA, SITES, and DNASP.

By using a 61×61 model of codon evolution, the problems found in standard methods can be repaired (Goldman and Yang, 1994; Muse and Gaut, 1994), although the cost is a higher computational demand. Muse (1996) proposed a maximum-likelihood estimation of the *dn*/*ds* ratio based on the latter model. This method is implemented in the program HYPHY. Yang (1998) and Nielsen and Yang (1998) used a modification of Goldman and Yang's (1994) model for constructing a likelihood-ratio test of neutral evolution. This test can be implemented in PAML by estimating likelihood scores under a neutral and a positive selective model. Some other tests of neutrality exist (Hudson *et al.*, 1987; Tajima, 1989; McDonald and Kreitman, 1991; Fu and Li, 1993) based on

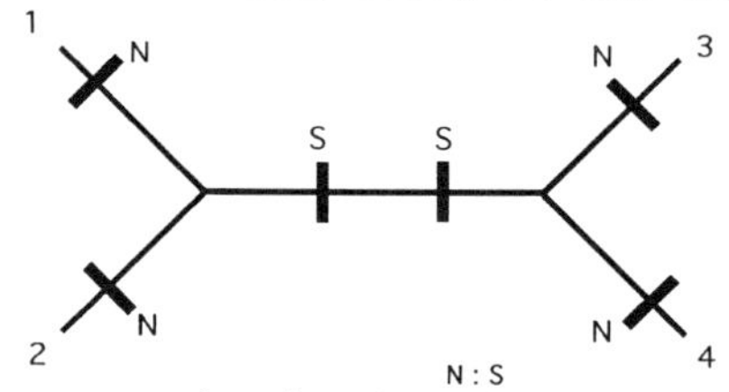

	N : S
Comparison 1-2 =	2 : 0
Comparison 1-3 =	2 : 2
Comparison 1-4 =	2 : 2
Comparison 2-3 =	2 : 2
Comparison 2-4 =	2 : 2
Comparison 3-4 =	2 : 0

Estimated dn/ds = 12 : 8 (3 : 2)
True dn/ds = 4 : 2 (2 : 1)

Figure 7 Unrooted tree with four terminal taxa (1-4) and four nonsynonymous (N) and two synonymous (S) changes along the branches. Pairwise estimates of dn/ds overestimate changes on the internal branch and therefore can lead to biased estimates.

population genetics theory, and the reader is referred to Chapter 11 in this book. These tests can be implemented in SITES and DNAsp.

Using the *dn*/*ds* criterion, positive selection has been identified in HIV–1, especially in the hypervariable V3 loop of the envelope gene (Holmes *et al.*, 1992; Bonhoeffer *et al.*, 1995; Seibert *et al.*, 1995; Mindell, 1996; Yamaguchi and Gojobori, 1997; Nielsen and Yang, 1998). Positive selection can be acting in a region where the *dn*/*ds* ratio is smaller than one, because typically only a few amino acids are responsible for adaptive evolution (Hughes and Nei, 1988; Yokoyama *et al.*, 1988) and because variation in selection intensity leads to underestimation of *dn* rates (Nielsen, 1997). Crandall *et al.* (1999a) provide an example of this situation in the case of the evolution of drug resistance.

It is also important to note the difference between neutral and random evolution. Under neutral evolution, the *dn/ds* substitution ratio is smaller than one because functional constraints in proteins do not allow the fixation of certain mutations (Kimura, 1977). In contrast, under random evolution there are no functional constraints, and any substitution can be fixed with equal probability. Because only one–third of the possible changes in a codon are synonymous, under random evolution, we expect approximately two times as many nonsynonymous substitutions as synonymous substitutions.

4. SPECIAL CONCERNS WITH HIV

Different organisms present different characteristics and these should be taken into account in the phylogenetic analysis. Because of the rapid increase in the size of the HIV sequence database, it is now common to use large data sets containing several genes. The high rate of substitution and the possibility of recombination should be also taken into account when designing an HIV phylogenetic study.

4.1 Large Data Sets

The number of possible bifurcating topologies increases rapidly with the addition of taxa. For unrooted trees this number is

$$B(T) = \prod_{i=3}^{T} (2i-5), \tag{10}$$

T being the number of taxa, while the number of possible rooted trees is increased by a factor of $2T-3$. For example, for 10 taxa, there are 2,027,025 possible unrooted and 34,459,425 rooted trees; for 20 taxa there are 2.216431×10^{20} possible unrooted and 8.200794×10^{21} rooted trees; for 100 taxa there are 1.700459×10^{182} possible unrooted and 1.649445×10^{184} rooted trees. Methods based on an optimality criterion (i. e., maximum parsimony, minimum evolution, and maximum likelihood) search in the tree space for the tree with the best score given the specific optimality criterion. In an exact search the best tree (global optimum) is guaranteed to be found. This can be done by evaluating all possible trees (exhaustive search) or by using some exact algorithms (e. g., branch and bound) that do not explore the complete tree space. However, with more than 10 taxa for maximum likelihood, more than 15 for minimum evolution, or more than 20 for maximum parsimony, these exact algorithms often require an impractical amount of computation. In these cases, a heuristic search is performed, where the tree space is explored partially and the tree obtained is not guaranteed to be the best possible tree (i. e., it may be only locally optimal). In this situation, it is a good idea to perform several replicates of the heuristic search with random addition of taxa to obtain a starting tree. In this way the search is started several times at different points in the tree landscape, thereby reducing the possibility of entrapment in local optima.

Many phylogenetic analyses of HIV include more than 20 sequences of several genes or even complete genomes. When making an effort to optimize the phylogenetic solution, parsimony analyses are much faster than distance analyses (neighbor joining does not optimize the solution; it gives a simple point estimate), which are in turn faster than maximum likelihood analyses (Hillis, 1999). With large data sets, then, maximum likelihood analyses may be prohibitive. But some of the advantages offered by the maximum likelihood approach, like the definition of models, can still be incorporated into parsimony and distance analyses (see section "Implementing a phylogenetic study of HIV sequences" in Hillis (1999)). In addition, new methods are being developed for reducing the amount of computation of maximum likelihood methods. One of such strategies is the quartet puzzling method (Strimmer and Haeseler, 1996). This method reconstructs the maximum-likelihood tree for each possible quartet (groups of 4 sequences). The resulting quartet trees are combined in an overall tree during the puzzling step. The quartet-puzzling tree is obtained as a majority-rule consensus of all trees that result from multiple runs of the puzzling step. The number of times a group is reconstructed during the puzzling steps is converted into reliability values for each internal branch. The quartet puzzling method is implemented in PUZZLE. The use of genetic algorithms (Lewis, 1998) is a new avenue for a faster tree search. Genetic algorithms imitate natural processes such as natural selection to find an optimal or near-optimal tree. In the case of Lewis' method, an initial population of trees is evolved during several generations by mutation and recombination under selective pressure for improving the likelihood score. The genetic algorithm search strategy is implemented in GAML. Another promising (and not exclusive) approach is the parallelization of search strategies, where the different search paths are split among many processors, making the process much faster. Some programs, such as fastDNAML and GAML, can be executed in parallel. Several groups are starting to actively develop parallel versions of phylogenetics programs, creating the possibility of analyzing very large data sets in the near future.

4.2 Data Partitions

There are good reasons for thinking that different evolutionary processes occur in different genes or even within distinct parts of the genes (e. g., 1st, 2nd and 3rd positions of a codon). It is not unusual to have HIV nucleotide sequences from several different genes, and several ways of partitioning HIV data sets can be considered. One of the biggest debates in molecular phylogenetics is how partitioned data should be analyzed (Nixon and Carpenter, 1996). Kluge (1989) proposed that all data sets should be combined when performing a phylogenetic analysis. On the other hand, Miyamoto and Fitch (1995) argued that each tree should be estimated independently from each data partition, and the different estimates compared for congruence. Congruence among data partitions should provide strong evidence that the proposed phylogeny is accurate (Penny and Hendy, 1986; Swofford, 1991). The third approach is to subject the data partitions to a statistical test of homogeneity

(Bull *et al.*, 1993; de Queiroz, 1993; Rodrigo *et al.*, 1993; Huelsenbeck *et al.*, 1996a). If the data partitions result in significantly different estimates of phylogeny, they are considered heterogeneous and the results of the different analyses are considered separately. If the estimates of phylogeny are not significantly different, the data partitions are then combined.

There are several ways of testing for data heterogeneity. De Queiroz (1993) suggested that if there is high bootstrap support for conflicting clades, the data should not be combined. Rodrigo *et al.* (1993) proposed the use of the distance between the shortest trees for each partition as a test statistic, whose null distribution could be constructed by bootstrapping. The incongruence length difference (*ILD*) (Mickevich and Farris, 1981; Farris *et al.*, 1994) is the difference between the length of the shortest tree from the combined data set (L_C) and the sum of lengths of the shortest trees (L_i) from each one of the n partitions:

$$ILD = L_C - \sum_{i=1}^{n} L_i \tag{11}$$

The null distribution of the *ILD* statistic is generated by randomly partitioning the combined data set into subsets of the same size as the original partitions. If the original value of the *ILD* statistic is greater than 95% of the *ILD* values in the null distribution, the null hypothesis of congruence is rejected. The *ILD* test is implemented in PAUP* in the partition homogeneity test in the Analysis menu. Huelsenbeck and Bull (1996) proposed a likelihood-ratio test for data heterogeneity. The null hypothesis (homogeneity) is represented by the likelihood of the tree when the same tree is assumed to underlie all data partitions, whereas the alternative hypothesis (heterogeneity) is represented by the likelihood of the tree when different trees can underlie each data partition. The null distribution of the statistic is calculated using parametric bootstrapping. Topology tests (see above) can also be used to detect partition incongruence when the partitions support significantly different trees. Cunningham (1997) compared the ILD test, Templeton's topology test, and Rodrigo and coworkers' test by applying them to well-corroborated vertebrate phylogenies, and showed the ILD test to be the most useful and accurate.

4.3 Recombination

Genetic recombination can result in a direct violation of one of the fundamental assumptions of most methods of phylogenetic reconstruction—namely, that there is a single history common to all the sequences under study. Therefore, recombination can cause incorrect phylogenetic inference (Sanderson and Doyle, 1992). Recombination clearly plays a role in the evolution of RNA viruses (Lai *et al.*, 1995). Recombination has been shown to be a common phenomenon within HIV–1 subtypes (Groenink *et al.*, 1991; Vartanian *et al.*, 1991; Zhu *et al.*, 1995) and among subtypes (Sabino *et al.*, 1994; Leitner *et al.*, 1995; Robertson *et al.*, 1995a; Robertson *et al.*, 1995b; Cornelissen *et al.*, 1996; Gao *et al.*, 1996; Lole *et al.*, 1998). Recombination in HIV therefore may be widespread (Sharp *et al.*, 1996). Although several statistical tests have been proposed for testing the occurrence of recombination within a gene region and for delimiting its boundaries (reviewed in Crandall and Templeton, 1999), very little is known about how well they work and under which conditions. Often methods of the "bootscanning family" (Robertson *et al.*, 1995a; Siepel *et al.*, 1995; Siepel and Korber, 1995; Salminen *et al.*, 1996; Lole *et al.*, 1998) are applied to HIV data sets. In these methods, a sliding window is used over and over until character partitions that result in different tree estimates are found. However, these methods have severe limitations, for they will never identify overlapping recombinant regions, do not correct for multiple comparison, and their use of bootstrap values for comparing different topologies is inappropriate (see above). Crandall and Templeton (1999) proposed two new methods that capitalize on the strengths and correct the weaknesses of existing methods. Any HIV phylogeny estimated without exploring the possibility of recombination can be erroneous, thereby compromising the inferences derived. Moreover, recombination in HIV has immediate consequences for the understanding of HIV pathogenesis and for vaccine development (Sharp *et al.*, 1996). Therefore, testing for recombination should be a common practice in any HIV phylogenetic study. Several programs have been developed for detecting recombination (see software section).

Most methods of phylogenetic analysis are designed to build trees in which genes or species always diverge and never recombine or hybridize to form new lineages. However, in situations where genetic recombination among sequences is likely, it is more appropriate to build a network rather than a tree, as the former can depict both divergence as well as recombination of sequences. One method that can be used to produce networks of HIV sequences is the method of statistical parsimony (Templeton *et al.*, 1992; Crandall, 1994; Crandall *et al.*, 1994; Crandall and Templeton, 1996). This method makes all pairwise connections between sequences that differ minimally and whose connections are justified by a statistical criterion—namely, that the probability is highest that a specific site difference between a pair of haplotypes corresponds with a single substitution. Recombination may be discovered by examining the distribution of homoplasy on the resulting tree or network (Crandall and Templeton, 1999). Recombinants are either eliminated from the analysis or the recombinational history is depicted as a network of genealogical relationships. There are five advantages to the statistical parsimony approach over traditional analysis: (1) a more accurate estimation of phylogenetic relationships for data with low levels of divergence is possible (Crandall *et al.*, 1994); (2) a rigorous hypothesis-testing framework is introduced, which in turn provides a quantitative partitioning of population phenomena across evolutionary time (Templeton and Sing, 1993); (3) the method allows for (and calculates) uncertainty in the phylogenetic estimate, rather than relying on a single estimate of phylogenetic relationships (Templeton *et al.*, 1992); (4) the approach allows for the potential of recombination within the data set (Crandall and Templeton, 1999); and (5) a probabilistic determination of appropriate rooting is produced (Crandall and Templeton, 1993; Castelloe and Templeton, 1994). This method complements other methods of phylogenetic reconstruction, because it allows greater statistical resolution when differences are few and similarities are many (Crandall, 1994), whereas most methods are more powerful when there are many differences. Statistical parsimony has been used successfully in HIV studies, including transmission identification (Crandall, 1995) and drug resistance studies (Crandall *et al.*, 1999a; Crandall *et al.*, 1999b). A computer program to implement this method (TCS) is available at *http://bioag.byu.edu/zoology/crandall_lab/programs.html.*

4.4 Summary

Phylogenetic analysis is a complex field of study that embraces a variety of techniques that can be applied to a wide range of evolutionary questions. However, the complete understanding of all the assumptions involved in the analysis is essential for a correct interpretation of the results. Although computation still represents a boundary to the application of advanced phylogenetic theory, recent improvements in computer science methodology enhance the application of more sophisticated techniques. The HIV community can take advantage of the rich phylogenetic methodology, as has been shown throughout this chapter.

5. PHYLOGENETIC SOFTWARE

A large amount of software is available for performing phylogenetic analysis. The most comprehensive list of software is compiled by Joe Felsenstein on the WWW at *http://evolution.genetics.washington.edu/phylip/software.html.* An interesting link where diverse analyses (conversion, alignment, phylogeny) can be performed online is *http://bioweb.pasteur.fr/intro-uk.html.* Some tools for the analysis of HIV sequences can also be found at the HIV sequence database at Los Alamos at *http://hiv-web.lanl.gov/HTML/tools.html.*

5.1 General Phylogeny Packages

- PAUP* (Swofford, 1998) is the most sophisticated and user–friendly program for phylogenetic analysis, with many options (e. g., bootstrapping, ancestral reconstruction) and close compatibility with MACCLADE (see below). It includes parsimony, distance matrix, invariants, and maximum likelihood methods. PAUP* 4.0 beta is distributed as Macintosh, DOS, and Unix versions. It is described in its web page at *http://www.lms.si.edu/PAUP* with information on bugs, commands, frequently asked questions and ordering.
- PHYLIP (Felsenstein, 1993) includes programs to carry out parsimony, distance matrix methods, and maximum likelihood, including bootstrapping and consensus trees. It accepts a variety of types of data, including DNA and RNA, proteins, restriction sites, 0/1 discrete character data, gene frequencies, continuous characters, and distance matrices. It is distributed free of charge in C source code, or as executables for DOS, 386/486/Pentium Windows, Macintosh, or PowerMac. It is available at its web site: *http://evolution.genetics.washington.edu/phylip.html*
- MEGA (Kumar *et al.*, 1993) is for analysis of data from DNA, RNA, and protein sequences, and distance matrices produced from other kind of data. It includes the neighbor-joining method, a branch-and-bound parsimony method, and bootstrapping. It is distributed as an executable program for DOS ma-

chines. It also runs under Windows in a DOS window. The program costs $20 (for the documentation, mailing and handling), and can be ordered from *http://www.bio.psu.edu/People/Faculty/Nei/Lab/Programs.html*

- PHYLO_WIN (Galtier *et al.*, 1996) performs neighbor-joining, parsimony, and maximum likelihood methods and can bootstrap with any of them. It runs under X Windows on many Unix workstations, including Sun (SunOS and Solaris), Silicon Graphics, IBM, DEC Alpha, and HP. You also need the NCBI Vibrant toolkit. The program can be downloaded at *http://pbil.univ-lyon1.fr/software/phylowin.html*
- TREEALIGN (Hein, 1990) builds trees as it aligns DNA or protein sequences. It uses a combination of distance matrix and approximate parsimony methods. It is available by anonymous ftp at the European Bioinformatics Institute molecular biology software distribution site *ftp://ftp.ebi.ac.uk* in directories *pub/software/unix* and *pub/software/vms*
- CLUSTALX (Thompson *et al.*, 1997) is another multisequence alignment program that estimates trees as it aligns multiple sequences. It is probably the easiest alignment program to use given the current implementations. It provides an integrated environment for performing multiple sequence and profile alignments. It is distributed as C source code and executables for DOS, Macintosh, and some Unix systems. It is available by anonymous ftp at *ftp://ftp-igbmc.u-strasbg.fr*. There is a description on its web page at *http://www-igbmc.u-strasbg.fr/BioInfo/ClustalX/Top.htm*
- MALIGN (Wheeler, 1996) is a parsimony-based alignment program for molecular sequences. It implements the idea that alignment and phylogenies can be done at the same time by finding the tree that minimizes the total alignment score along the tree. It is distributed as a DOS or Macintosh executable or as C source code (with a Makefile) for Unix workstations. It is available by anonymous ftp from the American Museum of Natural History's anonymous ftp site, *ftp://ftp.amnh.org*, in directory *pub/molecular*
- HMMER (Eddy, 1998) uses Profile Hidden Markov Models for the alignment of DNA or protein sequences. It is distributed as source code and also as executables for Solaris, SGI and Linux. It is available from its web page at *http://hmmer.wustl.edu/*

5.3 Maximum Likelihood Programs

- PAML (Yang, 1997) is a program for the maximum likelihood analysis of nucleotide or protein sequences. The program can be used to test evolutionary models, to calculate substitution rates at particular sites, to reconstruct ancestral nucleotide or amino acid sequences, to perform codon-based likelihood analysis (for estimating synonymous and nonsynonymous rates, testing hypotheses concerning *dn/ds* rate ratios), to do amino acid likelihood analysis with rate variation among sites, and for phylogenetic tree reconstruction by maximum likelihood and Bayesian methods. The package is distributed as ANSI C source code and executables for Macintosh, Windows, and Unix systems, and it is available at the its web page at *http://abacus.gene.ucl.ac.uk/software/paml.html*

- MOLPHY (Adachi and Hasegawa, 1996b) carries out maximum likelihood inference of phylogenies for either nucleotide sequences or protein sequences. The package is distributed free as C source code. It is available for Unix machines by anonymous ftp from *ftp://sunmh.ism.ac.jp* in directory *pub/molphy*. An executable version for Windows95 or Windows NT on Intel processors, and also one that works on Windows NT on DEC Alpha processors, is available at *http://dogwood.botany.uga.edu/malmberg/software.html*
- fastDNAML (Olsen *et al.*, 1994) it is an enhanced replacement for the PHYLIP program DNAML. The C program and PowerMac executables are also available by anonymous ftp from the Indiana University Biology ftp server at *ftp://ftp.bio.indiana.edu* in directory *molbio/evolve*
- GAML (Lewis, 1998) uses a genetic algorithm for finding the maximum likelihood trees. It is quite fast and allows the analysis of a large number of taxa (100). It is available for Macintosh, Windows, and Unix machines at *http://biology001.unm.edu/~lewisp/gaml.html*
- PASSML (Lio *et al.*, 1998) has been developed to implement an evolutionary model that combines protein secondary structure and amino acid replacement and permits analysis of phylogeny and secondary structure from aligned amino acid sequences. It is distributed as a C source for Unix *at http://ng-dec1.gen.cam.ac.uk/hmm/Passml.htm.l.*
- NMHL (Galtier and Gouy, 1998) is an implementation of a nonhomogeneous, nonstationary model of DNA evolution for performing maximum likelihood analyses. It is available for Unix machines by anonymous ftp from *ftp://pbil.univ-lyon1.fr* in directory *pub/mol_phylogeny/nhml*

5.4 Parsimony Programs

- A list of software for parsimony analysis is maintained by the Willi Hennig Society at *http://www.vims.edu/~mes/hennig/software.html*
- AUTODECAY (Eriksson, 1998) generates decay indices from an existing PAUP treefile. Its intent is to simplify the task of creating reverse constraint trees in PAUP and subsequent generation of Bremer support values. It is distributed as a Macintosh executable at *http://www.bergianska.se/personal/TorstenE/*
- NONAME (Goloboff, 1997) searches for most parsimonious trees according to character weights defined by the user a priori. with versions available for both 386-486-Pentium machines and earlier 16-bit machines. The demo version and the documentation are available from the Willi Hennig Society's software pages at *http://www.vims.edu/~mes/hennig/software.html*

5.5 Distance Programs

- METREE (Rzhetsky and Nei, 1993) computes minimum evolution trees from DNA and amino acid sequence data, and tests the statistical significance of topological differences and of the branch lengths of the minimum evolution tree. Different distance measures may be used. It is distributed as a PC program free of charge from *http://www.bio.psu.edu/People/Faculty/Nei/Lab/Programs.html*

5.6 Character Evolution

- MACCLADE (Maddison and Maddison, 1994) has its analytical strength in studies of character evolution. It also provides many tools for entering and editing data and phylogenies, and for producing tree diagrams and charts. It runs on Macintosh and is available at *http://phylogeny.arizona.edu/MACCLADE/MACCLADE.html.*
- COMPARE (Martins, 1997) includes various programs for conducting statistical analyses of comparative data in a phylogenetic context. It includes programs to compute independent contrasts, spatial autocorrelation analyses, sum of squares parsimony, random data, and trees and/or branch lengths. It is distributed as C source code and as Windows95, Windows 3.1, and Unix applications. The program is available from its web page at *http://evolution.uoregon.edu/~COMPARE/indexV3.html*

5.7 Simulation Software

- THE SIMINATOR (Huelsenbeck, 1995) simulates the evolution of nucleotide sequences along a given tree or trees. It allows for gamma-distributed rate variation among sites, and the Hasegawa-Kishino-Yano 1985 model of nucleotide substitution. It is distributed as C source code, with examples of input files. It can be obtained from the Slatkin Lab's software Web page at *http://ib.berkeley.edu/labs/slatkin/software.html*
- SEQGEN (Rambaut and Grassly, 1997) will simulate the evolution of nucleotide sequences along a phylogeny, using common models of the substitution process. A range of models of molecular evolution is implemented including the general reversible model. Nucleotide frequencies and other parameters of the model may be given and site-specific rate heterogeneity may also be incorporated in a number of ways. It is distributed as C source code and Macintosh executable from its Web site at *http://evolve.zoo.ox.ac.uk/software.html*
- TREEVOLVE and PTREEVOLVE (Grassly and Rambaut, 1997) simulate the evolution of DNA and protein sequences respectively. The molecular sequences are simulated under coalescent models with constant population size, or with exponential population size growth. In addition, different levels of recombination can be specified. They are distributed as C source code and Macintosh executable from their Web site at *http://evolve.zoo.ox.ac.uk/software.html*

5.8 Selecting Models of Evolution

- MODELTEST (Posada and Crandall, 1998) implements a hierarchical likelihood-ratio test procedure for choosing the model of DNA substitution that best fits the data, as well as AIC estimates. It is distributed as C source code and as Macintosh, Windows and Unix executable from its web site at *http://bioag.byu.edu/zoology/crandall_lab/modeltest.htm*

5.9 Rates of Evolution

- HYPHY (Muse, 2000) is a free multiplatform (Mac, Windows and UNIX) software package intended to perform maximum likelihood analyses of genetic sequence data and equipped with tools to test various statistical hypotheses. HYPHY was designed with maximum flexibility in mind and to that end it incorporates a simple high level programming language which enables the user to tailor the analyses precisely to his or her needs. It is available from *http://peppercat.stat.ncsu.edu/~hyphy/*
- R8S(Sanderson, 1997) is designed to perform miscellaneous analyses of rates of molecular evolution, estimation of divergence times under clock and non-clock models, estimation of birth-death parameters of the branching process, and miscellaneous functions, including the construction of phylogenetic super-trees. It is distributed as C source code from *http://phylo.ucdavis.edu/r8s/r8s.html.*

5.10 Population Genetics Programs

- DNASP (Rozas and Rozas, 1999) is a software package that performs extensive population genetic analyses on DNA sequence data. DNASP estimates several measures of DNA sequence variation within and between populations, as well as estimating linkage disequilibrium, recombination, gene flow, and gene conversion. DNASP can also carry out several tests of neutrality, including those of Hudson *et al.* (1987), Tajima (1989), McDonald and Kreitman (1991), and Fu and Li (1993). It is distributed as a Windows application from its web site at *http://www.bio.ub.es/~julio/DnaSP.html*
- ARLEQUIN (Schneider *et al.*, 1997) is population genetics software environment able to analyze RFLPs, DNA sequences, microsatellites, standard multi-locus data or allele frequency data. It implements a variety of population genetics methods either at the intra-population or at the inter-population level. It is distributed as PC executable from its web site at *http://anthropologie.unige.ch/arlequin/software/.* A Java version that works in Windows, Unix and Macintosh environments can be requested from the authors
- SITES (Hey and Wakeley, 1997) is a computer program for the analysis of comparative DNA sequence data. It is intended primarily for data sets with multiple closely related sequences. SITES is written in ANSI C. Precompiled versions are available for DOS and Macintosh. It is available from its web page at *http://heylab.rutgers.edu/#software*

5.11 Tree Analysis

- COMPONENT (Page, 1993) is a computer program for analyzing evolutionary trees and is intended for use in studies of phylogeny, tree-shape distribution, gene trees/species trees, host-parasite cospeciation, and biogeography. It runs on PC-DOS 286 or 386 systems under Windows 3.0 or higher. It costs 40 pounds U.K., and an order form can be filled at its web site at *http://taxonomy.zoology.gla.ac.uk/rod/cpw.html*

5.12 Detecting Recombination

- David Robertson has created a web page with links for several recombination analysis programs at *http://grinch.zoo.ox.ac.uk/RAP_links.html*

ACKNOWLEDGEMENTS

This work was supported by a BYU Graduate Studies Award (DP), NIH grant number RO1-HD34350 and the Alfred P. Sloan Foundation (KAC).

REFERENCES

Adachi J., Hasegawa, M. 1996a. Model of amino acid substitution in proteins encoded by mitochondrial DNA. *J. Mol. Evol.* **42**: 459-468.

Adachi J., Hasegawa, M. 1996b. MOLPHY version 2.3: programs for molecular phylogenetics based in maximum likelihood. *Comput. Sci. Monogr.* **28**: 1-150.

Akaike, H. 1974. A new look at the statistical model identification. *IEEE Trans. Autom. Contr.* **19**: 716-723.

Bonhoeffer, S., Holmes, E. C. and Nowak, M. A. 1995. Causes of HIV diversity. *Nature* **376**: 125.

Bremer, K. 1988. The limits of amino acid sequence data in angiosperm phylogenetic reconstruction. *Evolution* **42**: 795-803.

Brown, W. M., Prager, E. M., Wang, A. and Wilson, A. C. 1982. Mitochondrial sequences of primates: tempo and mode of evolution. *J. Mol. Evol.* **18**: 225-239.

Bull, J. J., Huelsenbeck, J. P., Cunningham, C. W., Swofford, D. L. and Waddell, P. J. 1993. Partitioning and combining data in phylogenetic analysis. *Syst. Biol.* **42**: 384-397.

Castelloe, J. and Templeton, A. R. 1994. Root probabilities for intraspecific gene trees under neutral coalescent theory. *Mol. Phylogenet. Evol.* **3**: 102-113.

Cornelissen, M., Kampingam, G., Zorgdrager, F. and Goudsmit, J. 1996. Human immunodeficiency virus type I subtypes defined by *env* show high frequency of recombinant *gag* genes. The UNAIDS Network for HIV Isolation and Characterization. *J. Virol.* **70**: 8209-8212.

Crandall, K. A. 1994. Intraspecific cladogram estimation: Accuracy at higher levels of divergence. *Syst. Biol.* **43**: 222-235.

Crandall, K. A. 1995. Intraspecific phylogenetics: Support for dental transmission of human immunodeficiency virus. *J. Virol.* **69**: 2351-2356.

Crandall, K. A. (ed.). 1999. *Molecular Evolution of HIV*. The Johns Hopkins University Press, Baltimore, MD.

Crandall, K. A., Kelsey, C. R., Imamichi, H., Lane, C. H. and Salzman, N. P. 1999a. Parallel evolution of drug resistance in HIV: failure of nonsynonymous/synonymous substitution rate ratio to detect selection. *Mol. Biol. Evol.* **16**: 372-382.

Crandall, K. A. and Templeton, A. R. 1993. Empirical tests of some predictions from coalescent theory with applications to intraspecific phylogeny reconstruction. *Genetics* **134**: 959-969.

Crandall, K. A. and Templeton, A. R. 1996. Applications of intraspecific phylogenetics, In *New Uses for New Phylogenies* (Harvey, P. H., Leigh Brown, A. J., Maynard Smith, J. and Nee, S., eds). Oxford University Press, Oxford, England.

Crandall, K. A. and Templeton, A. R. 1999. Statistical methods for detecting recombination, In *The Evolution of HIV* (Crandall, K. A., ed.) The Johns Hopkins University Press, Baltimore, MD.

Crandall, K. A., Templeton, A. R. and Sing, C. F. 1994. Intraspecific phylogenetics: problems and solutions, In *Models in Phylogeny Reconstruction* (Scotland, R. W., Siebert, D. J. and Williams, D. M., eds.) Clarendon Press, Oxford, England.

Crandall, K. A., Vasco, D., Posada, D. and Imamichi, H. 1999b. Advances in understanding the evolution of HIV. *AIDS* **13**: S39-S47.

Cunningham, C. W. 1997. Can three incongruence tests predict when data should be combined? *Mol. Biol. Evol.* **14**: 733-740.

Cunningham, C. W., Zhu, H. and Hillis, D. M. 1998. Best-fit maximum-likelihood models for phylogenetic inference: empirical tests with known phylogenies. *Evolution* **52**: 978-987.

de Queiroz, A. 1993. For consensus (sometimes). *Syst. Biol.* **42**: 368-372.

Eddy, S. 1998. *HMMER: profile hidden Markov models for biological sequence analysis. 2.1.1.* Department of Genetics, Washington University, St. Louis.

Efron, B., Tibshirani, R. J. 1993. *An Introduction to the Bootstrap.* Chapman and Hall, New York.

Eriksson, T. 1998. *AUTODECAY. 4.0.* Bergius Foundation, Royal Swedish Academy of Sciences, Stockholm.

Faith, D. P. 1991. Cladistic permutation tests for monophyly and nonmonophyly. *Syst. Zool.* **40**: 366-375

Faith, D. P. and Trueman, J. W. H. 1996. When the topology-dependent permutation test (T-PTP) for monophyly returns significant support for monophyly, should that be equated with (a) rejecting a null hypothesis of nonmonophyly, (b) rejecting a null hypothesis of "no structure", (c) failing to falsify a hypothesis of monophyly, or (d) none of the above? *Syst. Biol.* **45**: 580-586.

Farris, J. S., Källersjö, M., Kluge, A. G. and Bult, C. 1994. Testing significance of incongruence. *Cladistics* **10**: 315-320.

Felsenstein, J. 1973. Maximum likelihood and minimum-steps methods for estimating evolutionary trees from data on discrete characters. *Syst. Zool.* **22**: 240-249.

Felsenstein, J. 1984. Distance methods for inferring phylogenies: a justification. *Evolution* **38**: 16-24

Felsenstein, J. 1985. Confidence limits on phylogenies: an approach using the bootstrap. *Evolution* **39**: 783-791.

Felsenstein, J. 1988. Phylogenies from molecular sequences: inference and reliability. *Annu. Rev. Genet.* **22**: 521-565.

Felsenstein, J. 1993. *PHYLIP (Phylogeny Inference Package).* 3.5c. Department of Genetics, University of Washington, Seattle.

Fisher, A. G., Ensoli, B., Looney, D., Rose, A., Gallo, R. C., Saag, M. S., Shaw, G. M., Hahn, B. H. and Wong-Staal, F. 1988. Biologically diverse molecular variants within a single HIV-1 isolate. *Nature* **334**: 444-447.

Fitch, W. 1971. Toward defining the course of evolution: minimal change for a specific tree topology. *Syst. Zool.* **20**: 406-416.

Fu, Y. X. and Li, W. H. 1993. Statistical tests of neutrality of mutations. *Genetics* **133**: 693-709.

Galtier, N. and Gouy, M. 1998. Inferring pattern and process: maximum likelihood implementation of a nonhomogeneous model of DNA sequence evolution for phylogenetic analysis. *Mol. Biol. Evol.* **15**: 871-879.

Galtier, N., Gouy, M. and Gautier, C. 1996. SeaView and Phylo_win, two graphic tools for sequence alignment and molecular phylogeny. *Computer Applications in Biosciences* **12**: 543-548.

Gao, F., Robertson, D. L., Morrison, S. G., Hui, H., Craig, S., Decke, J., Fultz, P. N., Girard, M., Shaw, G. M., Hahn, B. H. and Sharp, P. M. 1996. The heterosexual human immunodeficiency virus type 1 epidemic in Thailand is caused by an intersubtype (A/E) recombinant of African origin. *J. Virol.* **70**: 7013-7029.

Gaut, B. S. and Weir, B. S. 1994. Detecting substitution-rate heterogeneity among regions of a nucleotide sequence. *Mol. Biol. Evol.* **11**: 620-629.

Goldman, N. 1990. Maximum likelihood inference of phylogenetic trees, with special reference to a Poisson process model of DNA substitution and to parsimony analyses. *Syst. Zool.* **39**: 345-361.

Goldman, N. 1993a. Simple diagnostic statistical tests of models for DNA substitution. *J. Mol. Evol.* **37**: 650-661.

Goldman, N. 1993b. Statistical tests of models of DNA substitution. *J. Mol. Evol.* **36**: 182-198.

Goldman, N., Thorne, J. L. and Jones, D. T. 1998. Assessing the impact of secondary structure and solvent accessibility on protein evolution. *Genetics* **149**: 445-458.

Goldman, N. and Yang, Z. 1994. A codon-based model of nucleotide substitution for protein-coding DNA sequences. *Mol. Biol. Evol.* **11**: 725-736.

Goloboff, P. A. 1997. *NONA*. 1.5. S. M. de Tucumón: Fundación e Instituto Miguel Lillo, Argentina.

Grantham, R. 1974. Amino acid difference formula to help explain protein evolution. *Science* **185**: 862-864.

Grassly, N. C. and Rambaut, A. 1997. *Treevolve: a program to simulate the evolution of DNA sequences under different population dynamic scenarios.* 1.3. Wellcome Centre for Infectious Disease, Department of Zoology, Oxford University, Oxford, UK.

Groenink, M., Fouchier, R. A. M., de Goede, R. E. Y., de Wolf, F., Gruters, R. A., Cupyers, H. T. M., Hisman, H. G. and Tersmette, M. 1991. Phenotypic heterogeneity in a panel of infectious molecular human immunodeficiency virus type 1 clones derived from a single individual. *J. Virol.* **65**: 1968-1975.

Gu, X. and Li, W-H. 1992. Higher rates of amino acid substitution in rodents that in humans. *Mol. Phylogenet. Evol.* **1**: 211-214.

Hahn, B. H., Shaw, G. M., Taylor, M. E., Redfield, R. R., Markham, P. D., Salahuddin, S. Z., Wong-Staal, F., Gallo, R. C., Parks, E. S. and Parks, W. P. 1986. Genetic variation in HTLV-III/LAV over time in patients with AIDS or at risk for AIDS. *Science* **232**: 1548-1553.

Hall, P. and Martin, M. A. 1988. On bootstrap resampling and iteration. *Biometrika* **75**: 661-671.

Hartmann, M. and Golding, B. G. 1998. Searching for substitution rate heterogeneity. *Mol. Phy. Evol.* **9**: 64-71.

Hasegawa, M. 1990. Phylogeny and molecular evolution in primates. *Jpn. J. Genet.* **65**: 243-266.

Hasegawa, M., Kishino, K. and Yano, T. 1985. Dating the human-ape splitting by a molecular clock of mitochondrial DNA. *J. Mol. Evol.* **22**: 160-174.

Hein, J. 1990. Reconstructing evolution of sequences subject to recombination using parsimony. *Math. Biosci.* **98**: 185-200.

Hey, J., and Wakeley, J. 1997. A coalescent estimator of the population recombination rate. *Genetics* **145**: 833-846.

Hillis, D. M. 1999. Phylogenetics and the study of HIV, In *The Evolution of HIV* (Crandall, K. A., ed.) Johns Hopkins University Press, Baltimore, MD.

Hillis, D. M., Bull, J. J. 1993. An empirical test of bootstrapping as a method for assessing confidence in phylogenetic analysis. *Syst. Biol.* **42**: 182-192.

Hillis, D. M., Huelsenbeck, J. P. and Cunningham, C. W. 1994. Application and accuracy of molecular phylogenies. *Science* **264**: 671-677.

Hillis, D. M., Mable, B. K. and Moritz, C. 1996. Applications of molecular systematics: The state of the field and a look to the future, In *Molecular Systematics*, (Hillis, D. M., Moritz, C. and Mable, B. K., eds.) Sinauer Associates, Sunderland, MA.

Holland, J. J., De la Torre, J. C. and Steinhauer, D. A. 1992. RNA virus populations as quasispecies. *Curr. Top. Microbiol. Immunol.* **176**: 1-20.

Holmes, E. C., Pybus, O. G. and Harvey, P. H. 1999. The molecular population dynamics of HIV-1, In *The Evolution of HIV*, (Crandall, K. A., ed.) The Johns Hopkins University Press , Baltimore, MD.

Holmes, E. C., Zhang, L. Q., Simmonds, P., Ludlam, C. A. and Leigh Brown, A. J. 1992. Convergent and divergent sequence evolution in the surface envelope glycoprotein of human immunodeficiency virus type 1 within a single infected patient. *Proc. Natl Acad. Sci. USA* **89**: 4835-4839.

Hudson, R. R., Kreitman, M. and Aguade, M. 1987. A test of neutral molecular evolution based on nucleotide data. *Genetics* **116**: 153-159.

Huelsenbeck, J. 1995. *The Siminator: a program for simulating data under the HKY85 model of DNA substitution with gamma distributed rates among sites. 2.0.* Department of Integrative Biology, University of California at Berkeley, Berkely, CA.

Huelsenbeck, J. P. and Bull, J. J. 1996. A likelihood ratio test to detect conflicting phylogenetic signal. *Syst. Biol.* **45**: 92-98.

Huelsenbeck, J. P., Bull, J. J. and Cunningham, W. 1996a. Combining data in phylogenetic analysis. *Trend Ecol. Evol.* **11**: 152-158.

Huelsenbeck, J. P. and Crandall, K. A. 1997. Phylogeny estimation and hypothesis testing using maximum likelihood. *Annu. Rev. Ecol. Syst.* **28**: 437-466.

Huelsenbeck, J. P., Hillis, D. M. and Jones, R. 1996b. Parametric bootstrapping in molecular phylogenetics: applications and performance, In *Molecular Zoology: Advances, Strategies, and Protocols*, (Ferraris, J. D. and Palumbi, S. R., eds.) Wiley-Liss, New York, NY.

Huelsenbeck, J. P. and Rannala, B. 1997. Phylogenetic methods come of age: testing hypothesis in a evolutionary context. *Science* **276**: 227-232.

Hughes, A. L. and Nei, M. 1988. Pattern of nucleotide substitution at major histocompatibility complex class I loci reveals overdominant selection. *Nature* **335**: 167-170.

Ina, Y. 1995. New methods for estimating the numbers of synonymous and nonsynonymous substitutions. *J. Mol. Evol.* **40**: 190-226.

Jukes, T. H. and Cantor, C. R. 1969. Evolution of protein molecules, In *Mammalian Protein Metabolism*, (Munro, H. M., ed.) Academic Press, New York, NY.

Kelsey, C. R., Crandall, K. A. and Voevodin, A. F. 1999. Different models, different trees: The geographic origin of PTLV-I. *Mol. Phylogenet. Evol.* **10**:336-347.

Kimura, M. 1977. Preponderance of synonymous changes as evidence for the neutral theory of molecular evolution. *Nature* **267**: 275-276.

Kimura, M. 1980. A simple method for estimating evolutionary rate of base substitutions through comparative studies of nucleotide sequences. *J. Mol. Evol.* **16**: 111-120.

Kishino, H. and Hasegawa, M. 1989. Evaluation of the maximum likelihood estimate of the evolutionary tree topologies from DNA sequence data, and the branching order in Hominoidea. *J. Mol. Evol.* **29**: 170-179.

Kjer, K. M. 1995. Use of rRNA secondary structure in phylogenetic studies to identify homologous positions: An example of alignment and data presentation from the frogs. *Mol. Phylogenet. Evol.* **4**: 314-330.

Kluge, A. G. 1989. A concern for evidence and a phylogenetic hypothesis of relationships among *Epicrates* (Boidae, Serpentes). *Syst. Zool.* **38**: 7-25.

Korber, B., Theiler, J. and Wolinsky, S. 1998. Limitations of a molecular clock applied to the considerations of the origin of HIV-1. *Science* **280**: 1868-1871.

Krushkal, J. and Li, W.-H. 1999. Use of phylogenetic inference to test an HIV transmission hypothesis, In *Molecular Evolution of HIV*, (Crandall, K. A., ed.) The Johns Hopkins University Press, Baltimore, MD.

Kumar, S., Tamura, K. and Nei, M. 1993. *MEGA: Molecular Evolutionary Genetics Analysis. 1.01.* The Pennsylvania State University, University Park, PA

Lai, S., Page, J. B. and Lai, H. 1995. HIV results in the frame. Paradox remains [letter]. *Nature* **375**: 196-197; discussion 198.

Langley, C. H. and Fitch, W. 1974. An estimation of the constancy of the rate of molecular evolution. *J. Mol. Evol.* **3**: 161-177.

Leigh Brown, A. 1994. Methods of evolutionary analysis of viral sequences, In *The Evolutionary Biology of Viruses*, (Morse, S. S., ed.) Raven Press, Ltd., New York.

Leigh Brown, A. J. and Holmes, E. C. 1994. Evolutionary biology of human inmunodeficiency virus. *Annu. Rev. Ecol. Syst.* **25**: 127-165.

Leitner, T., Escanilla, D., Marquina, S., Wahlberg, J., Brostrom, C., Hansson, H. B., Uhlen, M. and Albert, J. 1995. Biological and molecular characterization of subtype D, G, and A/D recombinant HIV-1 transmissions in Sweden. *Virology* **209**: 136-146.

Leitner, T., Kumar, S. and Albert, J. 1997. Tempo and mode of nucleotide substitutions in *gag* and *env* gene fragments in human immunodeficiency virus type 1 populations with a known transmission history. *J. Virol.* **71**: 4761-4770.

Lewis, P. O. 1998. A genetic algorithm for maximum-likelihood phylogeny inference using nucleotide sequence data. *Mol. Biol. Evol.* **15**: 277-283.

Lewontin, R. C. 1989. Inferring the number of evolutionary events from DNA coding sequences. *Mol. Biol. Evol.* **6**: 15-32.

Li, W.-H. 1993. Unbiased estimation of the rates of synonymous and nonsynonymous substitution. *J. Mol. Evol.* **36**: 96-99.

Li, W.-H. 1997. *Molecular Evolution*. Sinauer Associates, Inc., Sunderland, MA.

Li, W.-H., Tanimura, M. and Sharp, P. M. 1988. Rates and dates of divergence between AIDS virus nucleotide sequences. *Mol. Biol. Evol.* **5**: 313-330.

Lio, P., Goldman, N., Thorne, J. L. and Jones, D. T. 1998. PASSML: combining evolutionary inference and protein secondary structure prediction. *Bioinformatics* **14**: 726-733.

Lole, K. S., Bollinger, R. C., Paranjape, R. S., Gadkari, D., Kulkarni, S. S., Novak, N. G., Ingersoll, R., Sheppard, H. W. and Ray, S. C. 1998. Full-length human immunodeficiency virus type 1 genomes from subtype C-infected seroconverters in India, with evidence of intersubtype recombination. *J. Virol.* **73**: 152-160.

Louwagie, J. J., McCutchan, F., Brennan, T., Peeters, M., Brennan, T., Sanders-Buell, E., Eddy, G., van der Groen, G., Fransen, K., Bershy-Damet, M., Deleys, R. and Burke, D. 1993. Phylogenetic analysis of gag genes from seventy international HIV-1 isolates provides evidence for multiple genotypes. *AIDS* **7**: 769-780.

Maddison, W. P. and Maddison, D. R. 1994. *MacClade: Analysis of phylogeny and character evolution.* Sinauer Associates, Sunderland, MA.

Martins, E. 1997. *COMPARE: phylogenetic analysis of comparative data. 4.1.* Department of Biology, University of Oregon, Eugene, OR.

McDonald, J. H. and Kreitman, M. 1991. Adaptive protein evolution at the Adh locus in *Drosophila. Nature* **351**: 652-654.

Messier, W. and Stewart, C.-B. 1997. Episodic adaptive evolution of primate lysozymes. *Nature* **385**: 151-154.

Mickevich, M. F. and Farris, J. S. 1981. The implications of congruence in *Menidia. Syst Zool* **30**: 351-370.

Miller, R. G. 1966 *Simultaneous Statistical Inference.* McGraw-Hill, New York.

Mindell, D. P. 1996. Positive selection and rates of evolution in immunodeficiency viruses from humans and chimpanzees. *Proc. Natl Acad. Sci. USA* **93**: 3284-3288.

Miyamoto, M. M. and Fitch, W. M. 1995. Testing species phylogenies and phylogenetic methods with congruence. *Syst. Biol.* **44**: 64-76.

Miyata, T. and Yasunaga, T. 1980. Molecular evolution of mRNA: A method for estimating evolutionary rates of synonymous and amino acid substitution from homologous nucleotide sequences and its application. *J. Mol. Evol.* **16**: 23-36.

Moriyama, E. N., Ina, Y., Ikeo, K., Shimizu, M. and Gojobori, T. 1991. Mutation pattern of human immunodeficiency virus gene. *J. Mol. Evol.* **32**: 360-363.

Muse, S. 1999. Modeling the molecular evolution of HIV sequences, In *The Evolution of HIV*, (Crandall, K. A., ed.) Johns Hopkins University Press, Baltimore, MD.

Muse, S. V. 1996. Estimating synonymous and nonsynonymous substitution rates. *Mol. Biol. Evol.* **13**: 105-114.

Muse, S. V. 2000. *HYPHY: hypothesis testing using phylogenies. Beta 1.0.* Program in Statistical Genetics, Department of Statistics, North Carolina State University, Raleigh, NC

Muse, S. V. and Gaut, B. S. 1994. A likelihood approach for comparing synonymous and nonsynonymous nucleotide substitution rates, with application to the chloroplast genome. *Mol. Biol. Evol.* **11**: 715-724.

Muse, S. V. and Weir, B. S. 1992. Testing for equality of evolutionary rates. *Genetics* **132**: 269-276.

Nei, M. 1996. Phylogenetic analysis in molecular evolutionary genetics. *Ann. Rev. Genet.* **30**: 371-403.

Nei, M. and Gojobori, T. 1986. Simple method for estimating the numbers of synonymous and nonsynonymous nucleotide substitutions. *Mol. Biol. Evol.* **3**: 418-426.

Nielsen, R. 1997. The ratio of replacement to silent divergence and tests of neutrality. *J. Evol. Biol.* **10**: 217-231.

Nielsen, R. and Yang, Z. 1998. Likelihood methods for detecting positively selected amino acid sites and applications to the HIV-1 envelope gene. *Genetics* **148**: 929-936.

Nixon, K. C. and Carpenter, J. M. 1996. On simultaneous analysis. *Cladistics* **12**: 221-241.

Nowak, M. and Bangham, C. R. M. 1996. Population dynamics of inmune responses to persistent viruses. *Science* **272**: 74-79.

Olsen, G. J., Matsuda, H., Hagstrom, R. and Overbeek, R. 1994. Fast DNAml: a tool for construction of phylogenetic trees of DNA sequences using maximum likelihood. *Comput. Appl. Biosciences* **10**: 41-48.

Page, R. D. M. 1993. *COMPONENT. 2.0.* Natural History Museum, London, UK

Pamilo, P. and Bianchi, N. O. 1993. Evolution of the Zfx and Zfy genes: rates and interdependence between genes. *Mol. Biol. Evol.* **10**: 271-281.

Pedersen, A.-M. K., Wiuf, C. and Christiansen, F. B. 1998. A codon-based model designed to describe lentiviral evolution. *Mol. Biol. Evol.* **15**: 1069-1081.

Penny, D. and Hendy, M. D. 1985. The use of tree comparison metrics. *Syst. Zool.* **34**: 75-82.

Penny, D. and Hendy, M. D. 1986. Estimating the reliability of evolutionary trees. *Mol. Biol. Evol.* **3**: 403-417.

Penny, D., Hendy, M. D. and Steel, M. A. 1992. Progress with methods for constructing evolutionary trees. *Trends Ecol Evol* **7**: 73-79.

Penny, D, Lockhart, P. J., Steel, M. A. and Hendy, M. D. 1994. The role of models in reconstructing evolutionary trees, In *Models in Phylogenetic Reconstruction*, (Scotland, R. W., Siebert, D. J. and Williams, D. M., eds.) Clarendon Press, Oxford.

Posada, D. and Crandall, K. A. 1998. Modeltest: Testing the model of DNA substitution. *Bioinformatics* **14**: 817-818.

Prager, E. M. and Wilson, A. C. 1988. Ancient origin of lactalbumin from lysozyme: Analysis of DNA and amino acid sequences. *J. Mol. Evol.* **27**: 326-335.

Rambaut, A. and Grassly, N. C. 1997. Seq-Gen: an application for the Monte Carlo simulation of DNA sequence evolution along phylogenetic trees. *Comput. Appl. Biosciences* **13**: 235-238.

Robertson, D. L., Hahn, B. H. and Sharp, P. M. 1995a. Recombination in AIDS viruses. *J. Mol. Evol.* **40**: 249-259.

Robertson, D. L., Sharp, P. M., McCutchan, F. E. and Hahn, B. H. 1995b. Recombination in HIV-1. *Nature* **374**: 124-126.

Robinson, D. F. and Foulds, L. R. 1981. Comparison of phylogenetic trees. *Math. Biosci.* **53**: 131-147.

Robinson, M., Gouy, M., Gautier, C. and Mouchiroud, D. Sensitivity of the relative-rate test to taxonomic sampling. *Mol. Biol. Evol.* **15**: 1091-1098.

Rodrigo, A. G., Kelly-Borges, M., Bergquist, P. R. and Bergquist, P. L. 1993. A randomisation test of the null hypothesis that two cladograms are sample estimates of a parametric phylogenetic tree. *New Zealand J. Bot.* **31**: 257-268.

Rodríguez, F., Oliver, J. F., Marín, A. and Medina, J. R. 1990. The general stochastic model of nucleotide substitution. *J. Theor. Biol.* **142**: 485-501.

Rozas, J. and Rozas, R. 1999. DnaSP version 3: an integrated program for molecular population genetics and molecular evolution analysis. *Bioinformatics* **15**: 174-175.

Rzhetsky, A. and Nei, M. 1992. A simple method for estimating and testing minimum-evolution trees. *Mol. Biol. Evol.* **9**: 945-967.

Rzhetsky, A. and Nei, M. 1993. *METREE: program package for inferring and testing minimum evolution trees. 1.2.* Institute of Molecular Evolutionary Genetics and Department of Biology, The Pennsylvania State University, University Park, PA

Rzhetsky, A. and Nei, M. 1995. Tests of applicability of several substitution models for DNA sequence data. *Mol. Biol. Evol.* **12**: 131-151.

Sabino, E. C., Shpaer, E. G., Morgado, M. G., Korber, B. T. M., Diaz, R., Bongertz, V., Cavalcante, S., Galvao-Castro, B., Mullins, J. I. and Mayer, A. 1994 Identification of human immunodeficiency virus type 1 envelope genes recombinant between subtypes B and F in two epidemiologically linked individuals from Brazil. *J. Virol.* **68**: 6340-6346.

Salminen, M. O., Carr, J. K., Burke, D. S. and McCutchan, F. E. 1996. Identification of breakpoints in intergenotypic recombinants of HIV-1 by bootscanning. *AIDS Res. Hum. Retroviruses* **11**: 1423-1425.

Sanderson, M. 1997. A nonparametric approach to estimating divergence times in the absence of rate constancy. *Mol. Biol. Evol.* **14**: 1218-1231.

Sanderson, M. J. and Doyle, J. J. 1992. Reconstruction of organismal and gene phylogenies from data on multigene families: Concerted evolution, homoplasy, and confidence. *Syst. Biol.* **41**: 4-17.

Sarich, V. M. and Wilson, A. C. 1973. Generation time and genomic evolution in primates. *Science* **179**: 1144-1447.

Schneider, S., Kueffer, J.-M., Roessli, D. and Excofier, L. 1997. *Arlequin: A software for population genetic data analysis. 1.1.* Genetics and Biometry Lab, Dept. of Anthropology, University of Geneva.

Seibert, S. A., Howell, C. Y., Hughes, M. K. and Hughes, A. L. 1995. Natural selection on the *gag*, *pol*, and *env*, genes of human immunodeficiency virus 1 (HIV-1). *Mol. Biol. Evol.* **12**: 803-813.

Seiller-Moiseiwitsch, F., Margolin, B. H. and Swanstrom, R. 1994. Genetic variability of the human immunodeficiency virus: statistical and biological issues. *Annu. Rev. Genet.* **28**: 559-596.

Sharp, P. M., Robertson, D. L., Gao, F. and Hahn, B. H. 1994. Origins and diversity of human immunodeficiency viruses. *AIDS* **8**: S27-S42.

Sharp, P. M., Robertson, D. L. and Hahn, B. H. 1995. Cross-species transmission and recombination of AIDS viruses. *Phil. Trans. R. Soc. Lond. B* **349**: 41-47.

Sharp, P. M., Robertson, D. L. and Hahn, B. H. 1996. Cross-species transmission and recombination of 'AIDS' viruses, In *New Uses for New Phylogenies*, (Harvey. P. H., Leigh Brown, A. J., Smith, J. M. and Nee, S, eds.) Oxford University Press, Oxford.

Siepel, A. C., Halpern, A. L., Macken, C. and Korber, B. T. M. 1995. A computer program designed to screen rapidly for HIV type 1 intersubtype recombinant sequences. *AIDS Res. Hum. Retroviruses* **11**: 1413-1416.

Siepel, A. C. and Korber, B. K. 1995. Scanning the data base for recombinant HIV-1 genomes, In *Human Retroviruses and AIDS 1995: A Compilation and Analysis of Nucleic Acid and Amino Acid Sequences.*, (Myers, G., Korber, B., Hahn, B., Jeang, K.-T., Mellors, J., McCutchan, F., Henderson, L., Pavlakis, G. and Theoretical Biology and Biophysics Group LANL, Los Alamos, NM., eds.) Theoretical Biology and Biophysics Group, Los Alamos National Laboratory, Los Alamos, NM.

Simon, F., Mauclere, P., Roques, P., Loussert-Ajaka, I., Müller-Trutwin, M. C., Saragosti, S., Georges-Courbot, M. C., Barre-Sinoussi, F. and Brun-Vezinet, F. 1998. Identification of a new human immunodeficiency virus type 1 distinct from group M and group O. *Nature Medicine* **4**: 1032-1037.

Sitnikova, T., Rzhetsky, A. and Nei, M. 1995. Interior-branch and bootstrap tests of phylogenetic trees. *Mol. Biol. Evol.* **12**: 319-333.

Steel, M. A., Cooper, A. C. and Penny, D. 1996. Confidence intervals for the divergence time of two clades. *Syst. Biol.* **45**: 127-134.

Steel, M. A. and Penny, D. 1993. Distributions of tree comparison metrics — some new results. *Syst. Biol.* **42**: 126-141.

Strimmer, K. and Haeseler, Av. 1996. Quartet puzzling: a quartet maximum-likelihood method for reconstructing tree topologies. *Mol. Biol. Evol.* **13**: 964-969.

Sullivan, J. and Swofford, D. L. 1997. Are guinea pigs rodents? The importance of adequate models in molecular phylogenies. *Journal of Mammalian Evolution* **4**: 77-86.

Swofford, D. L. 1991. When are phylogeny estimates from molecular and morphological data incongruent? In *Phylogenetic analysis of DNA sequences*, (Miyamoto, M. M. and Cracraft, J., eds.) Oxford University Press, New York, Oxford.

Swofford, D. L. 1998. *PAUP* Phylogenetic analysis using parsimony and other methods. 4.0 beta.* Sinauer Associates, Sunderland, MA

Swofford, D. L., Olsen, G. J., Waddell, P. J. and Hillis, D. M. 1996a. Phylogenetic Inference, In *Molecular Systematics*, (Hillis, D. M., Moritz, C. and Mable, B. K., eds.) Sinauer Associates, Inc., Sunderland, MA.

Swofford, D. L., Thorne, J. L., Felsenstein, J. and Wiegmann, B. M. 1996b. The topology-dependent permutation test for monophyly does not test for monophyly. *Syst. Biol.* **45**: 575-579.

Tajima, F. 1989. Statistical method for testing the neutral mutation hypothesis by DNA polymorphism. *Genetics* **123**: 585-595.

Tajima, F. 1993. Simple methods for testing the molecular evolutionary clock hypothesis. *Genetics* **135**: 599-607.

Tavaré, S. 1986. Some probabilistic and statistical problems in the analysis of DNA sequences, In *Lectures on Mathematics in the Life Sciences*, (Miura, R. M., ed.) Amer. Math. Soc., Providence, RI.

Templeton, A. R. 1983a. Convergent evolution and nonparametric inferences from restriction data and DNA sequences, In *Statistical Analysis of DNA Sequence Data*, (Weir, B. S., ed.) Marcel Dekker, Inc., New York.

Templeton, A. R. 1983b. Phylogenetic inference from restriction endonuclease cleavage site maps with particular reference to the evolution of humans and the apes. *Evolution* **37**: 221-244.

Templeton, A. R., Crandall, K. A. and Sing, C. F. 1992. A cladistic analysis of phenotypic associations with haplotypes inferred from restriction endonuclease mapping and DNA sequence data. III. Cladogram estimation. *Genetics* **132**: 619-633.

Templeton, A. R. and Sing, C. F. 1993. A cladistic analysis of phenotypic associations with haplotypes inferred from restriction endonuclease mapping. IV. Nested analyses with cladogram uncertainty and recombination. *Genetics* **134**: 659-669.

Thompson, J. D., Gibson, T. J., Plewniak, F., Jeanmougin, F. and Higgins, D. G. 1997. The ClustalX windows interface: flexible strategies for multiple sequence alignment aided by quality analysis tools. *Nucleic Acids Res* **24**: 4876-4882.

Thorne, J., Kishino, H. and Painter, I. S. 1998. Estimating the rate of evolution of the rate of molecular evolution. *Mol. Biol. Evol.* **15**: 1647-1657.

Thorne, J. L., Goldman, N. and Jones, D. T. 1996. Combining protein evolution and secondary structure. *Mol. Biol. Evol.* **13**: 666-673.

Thorne, J. L., Kishino, H. and Felsenstein, J. 1991. An evolutionary model for the maximum likelihood alignment of sequence evolution. *J. Mol. Evol.* **33**: 114-124.

Thorne, J. L., Kishino, H. and Felsenstein, J. 1992. Inching toward reality: an improved likelihood model of sequence evolution. *J. Mol. Evol.* **34**: 3-16.

Uyenoyama, M. K. 1995. A generalized least-squares estimate for the origin of sporophytic self-incompatibility. *Genetics* 139: 975-992.

Vartanian, J.-P., Meyerhans, A., Åsjo, B. and Wain-Hobson, S. 1991. Selection, recombination, and GÆA hypermutation of human immunodeficiency virus type 1 genomes. *J. Virol.* **65**: 1779-1788.

Weiller, G. F., McClure, M. A. and Gibbs, A. J. 1995. Molecular phylogenetic analysis, In *Molecular Basis of Virus Evolution*, (Gibbs, A., Calisher, C. H. and García Arenal, F., eds.) Cambridge University Press, Cambridge.

Wheeler, W. 1996. Optimization alignment: the end of multiple sequence alignment in phylogenetics? *Cladistics* **12**: 1-9.

Whelan, S. and Goldman, N. 1999. Distributions of statistics used for the comparison of models of sequence evolution in phylogenetics. *Mol. Biol. Evol.* **16**: 1292-1299.

Wu, C.-I. and Li, W.-H. 1985. Evidence for higher rates of nucleotide substitution in rodents than in man. *Proc. Natl Acad. Sci. USA* **82**: 1741-1745.

Yamaguchi, Y. and Gojobori, T. 1997. Evolutionary mechanisms and population dynamics of the third variable envelope region of HIV within single hosts. *Proc. Natl Acad. Sci. USA* **94**: 1264-1269.

Yang, Z. 1993. Maximum likelihood estimation of phylogeny from DNA sequences when substitution rates differ over sites. *Mol. Biol. Evol.* **10**: 1396-1401.

Yang, Z. 1994a. Estimating the pattern of nucleotide substitution. *J. Mol. Evol.* **39**: 105-111.

Yang, Z. 1994b. Maximum likelihood phylogenetic estimation from DNA sequences with variable rates over sites: approximate methods. *J. Mol. Evol.* **39**: 306-314.

Yang, Z. 1996. Maximum-likelihood models for combined analyses of multiple sequence data. *J. Mol. Evol.* **42**: 587-596.

Yang, Z. 1997. Applications Note: PAML: a program package for phylogenetic analysis by maximum likelihood. *Comput. Appl. Biosciences* **13**: 555-556.

Yang, Z. 1998. Likelihood ratio tests for detecting positive selection and application to primate lysozyme evolution. *Mol. Biol. Evol.* **15**: 568-573.

Yang, Z., Goldman, N. and Friday, A. 1994. Comparison of models for nucleotide substitution used in maximum-likelihood phylogenetic estimation. *Mol. Biol. Evol.* **11**: 316-324.

Yang, Z., Goldman, N. and Friday, A. 1995a. Maximum likelihood trees from DNA sequences: a peculiar statistical estimation problem. *Syst. Biol.* **44**: 384-399.

Yang, Z., Kumar, S. and Nei, M. 1995b. A new method of inference of ancestral nucleotide and amino acid sequences. *Genetics* **141**: 1641-1650.

Yang, Z. and Nielsen, R. 1998. Synonymous and nonsynonymous rate variation in nuclear genes of mammals. *J. Mol. Evol.* **46**: 409-418.

Yang, Z., Nielsen, R. and Masami, H. 1998. Models of amino acid substitution and applications to mitochondrial protein evolution. *Mol. Biol. Evol.* **15**: 1600-1611.

Yokoyama, S., Chung, L. and Gojobori, T. 1988. Molecular evolution of the human immunodeficiency and related viruses. *Mol. Biol. Evol.* **5**: 237-251.

Zharkikh, A. and Li, W.-H. 1992a. Statistical properties of bootstrap estimation of phylogenetic variability from nucleotide sequences. I. Four taxa with a molecular clock. *Mol. Biol. Evol.* **9**: 1119-1147.

Zharkikh, A. and Li, W.-H. 1992b. Statistical properties of bootstrap estimation of phylogenetic variability from nucleotide sequences: II. Four taxa without a molecular clock. *J. Mol. Evol.* **35**: 356-366.

Zharkikh, A. and Li, W.-H. 1995. Estimation of confidence in phylogeny: The complete-and-partial bootstrap technique. *Mol. Phylogenet. Evol.* **4**: 44-63.

Zhu, T., Wang, N., Carr, A., Wolinsky, S. and Ho, D. D. 1995. Evidence for coinfection by multiple strains of human immunodeficiency virus type 1 subtype B in an acute seroconvertor. *J. Virol.* **69**: 1324-1327

GOALS AND STRATEGIES FOR ANALYSIS OF RECOMBINATION AMONG HIV MOLECULAR SEQUENCES

J. Claiborne Stephens
Genaissance Pharmaceuticals, New Haven CT 06511 USA

1. INTRODUCTION

It is a fascinating paradox that the widespread phenomenon of sexual reproduction, and consequent biparental inheritance, is totally abandoned in virtually all phylogenetic models and treatments of molecular evolution. Hence although the number of an individual diploid organism's ancestors *increases* as one goes backward in time, the number of ancestors for a given DNA or protein sequence is only one. So, for instance, while a collection of n individuals may in general have $2n$ ancestors in the previous generation, a collection of n allelic sequences may have n ancestors, or even n-1 ancestors when two of the sequences have a common ancestor. The latter concept is expressed more formally in the well-developed branch of population genetics known as "coalescent theory" (e.g., Hudson, 1990).

The standard phylogenetic depiction of a sample of n molecular sequences is thus a "tree" in which at some time point T_n in the past, two of the sequences share a common ancestor, reducing the number of lineages to n-1. Similarly, T_{n-1} is the time at which two of these n-1 lineages share a common ancestor, and so on back to the root of the tree (Figure 1). This model fails when recombination is present, because a sequence is no longer limited to a single ancestor in the previous generation — it may in fact have two. Hudson's work (1983, 1990) on coalescent theory has elegantly incorporated recombination by essentially recognizing that while each nucleotide has a tree-like phylogeny, an adjacent nucleotide has a different phylogeny if recombination has occurred between them. Below, this concept will be extended for the purpose of detecting historical recombination events.

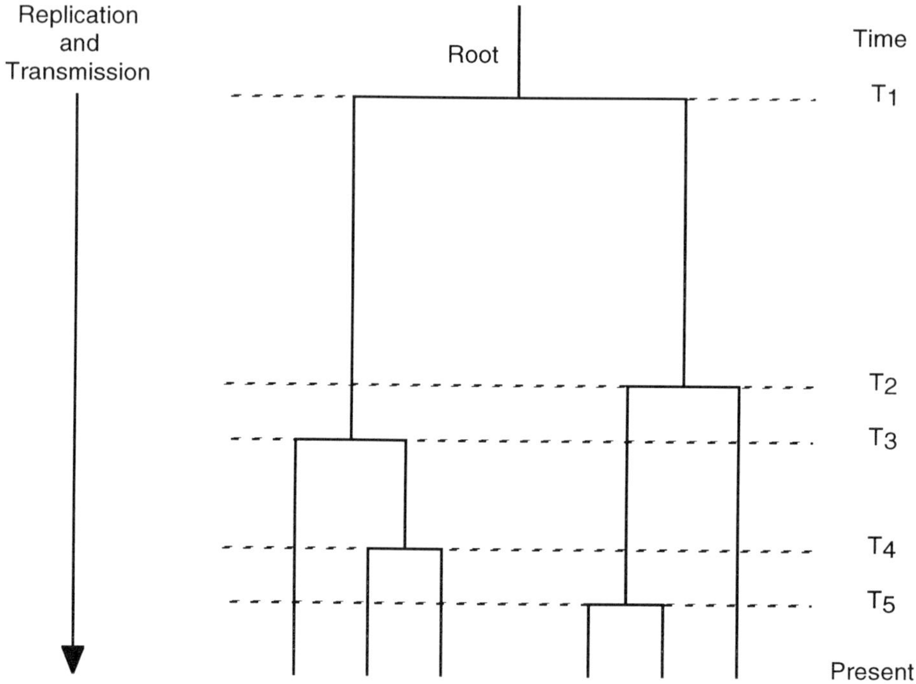

Figure 1. Standard depiction of phylogeny or genealogy of sequences. Sequences descend, with divergence (not shown), in a tree-like fashion with no further "communication" or interaction between sequences. This model is violated by recombination, in which case two sequences can interact to produce a distinct, third sequence.

For viruses such as HIV, one would think that the reservations voiced above are less relevant, in that "reproduction" is actually replication. However, recent molecular analyses (Sabino *et al.,* 1994; Leitner *et al.,* 1995; Robertson *et al.,* 1995; Siepel *et al.,* 1995; Salminen *et al.,* 1995) have shown that a fraction of HIV sequences are indeed recombinant, in spite of the virus's asexual reproduction. Apparently, instead of the sort of evolutionarily inert transmission from generation to generation depicted in Figure 1, HIV sequences are capable of interacting with each other in ways that alter the molecular sequence. The interaction is in fact recombination, presumably mediated by an individual being infected with distinct viral types. Indeed, Hu and Temin (1990) have shown the recombinogenic potential of retroviruses when different viral types coexist in the same cell. In what follows, recombination among viral sequences will be addressed in much the same way recombination among allelic sequences has been addressed, simply because most published algorithms are grounded in that mode of thought. One should note here, at the outset, that perhaps recombination as a biparental (i.e., binary) interaction may be too restrictive for HIV, and that more complex models may need to be entertained as the natural history of HIV super-infection is elucidated.

2. POTENTIAL GOALS

The focus in this paper is on the inference and characterization of historical recombination events from a series of molecular sequences. As such, it is intrinsically an evolutionary analysis and stands in contrast to more empirical inferences and characterization of recombination, such as that done for the human major histocompatibility Class II region (Cullen *et al.,* 1997). Therefore, since we may in general have very little useful ancillary information, it is best to focus discussion on what types of information are potentially discernible purely from the inspection and analysis of the pattern of variation within a sample of molecular sequences. For nuclear or mitochondrial DNA sequences, the frequencies of distinct molecular types (alleles) is an important piece of information, but seldom is frequency an issue for HIV sequences.

The initial goal of a recombination analysis should always be the determination of whether or not recombination is a plausible source of the variation seen among the sampled sequences. Ideally, this could be simply a yes or no answer, but in practice has often been confounded with detailed characterization of specific recombination events. Furthermore, while it would be useful to have a statistical assessment of the requirement for recombination, in practice statistical rigor may require knowledge of additional parameters or data that are simply not available (e.g., site-specific mutation rates or frequency and availability of specific sequences). In such cases, it may be sufficient to investigate the plausibility of recombination for the sample, rather than a rigorous evaluation of relative likelihoods (e.g., Hoelzel *et al.,* 1999).

Along the above lines, a very useful but relatively difficult (Hudson and Kaplan, 1985; Stephens, 1985b; but see Hey and Wakeley, 1997) parameter to extract from sampled sequences is the recombination rate. This would seem to be especially true for HIV sequences, since the opportunity for recombination depends on the frequency of co-infection, which itself would be expected to be highly variable. A useful fallback strategy is to focus on the impact recombination has had, especially relative to other sources of mutation (e.g., nucleotide substitution, insertion/deletion) in terms of accounting for the variety of observed sequences. Pragmatically, we are often most interested in how large a factor recombination has been in the evolution of a sample. For instance, in a sample of 211 HIV *env* sequences, Siepel *et al.* (1995) concluded that 9 (about 4%) were produced by recombination between subtypes. Similarly, Robertson *et al.* (1995) concluded at least 10 of 114 (9%) HIV-1 sequences were recombinant. However, a recombination analysis will usually be quite ambitious and attempt to determine the who, where, and when of specific recombination events (Table 1). That is, goals include the identification of recombinant and non-recombinant sequences, and for each recombinant sequence the details of the event regarding the sequences involved in recombination and the breakpoint(s) between them. Having identified a recombinant sequence and its likely progenitors, there is still the phylogenetic question of where and when the progenitors arose on the phylogenetic tree of the rest of the sample. It is noteworthy that most current methods of recombination analysis (detailed below) assume that the recombinant sequence will be a coarse mosaic of a small number, usually two, blocks of sequence matching one or the other progenitor sequence (R_1 in Figure 2). This need not be the case for viral sequences, in that multiple recombination events could conceivably occur between rounds of viral replication. Hence virus produced

A

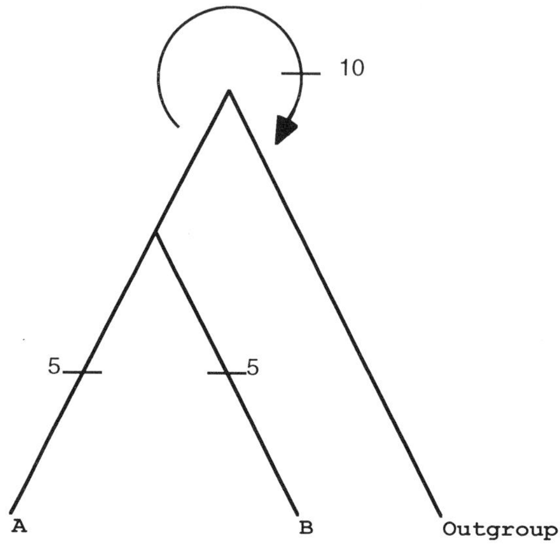


```
                    1       1       2
1       5           0       5       0
G C T A T C A G G A A A T C T C A G T T   Reference (AB Anc.)
A . A G . T . C . T . G C . . G . . C .   OUTGROUP
. . . . . . . . . . . . . . . . . . . .   AB ANCESTOR
. A . . A . . . C . . . . T . . . . . A   A
. . . . . . G . . . T . . . C . G C . .   B

. A . . A . . . C . T . . . C . G C . .   R1 (coarse mosaic)
. A . . . . G . C . . . . . C . . C . A   R2 (fine mosaic)
```

B

```
. . . . . . . . . . * . . * * . * * . *   R1 vs A
. * . . * . * . * . . . . . . . . . . .   R1 vs B

. . . . * . * . . . . . . * * . . * . .   R2 vs A
. * . . . . . . * . * . . . . . * . . *   R2 vs B
```

Figure 2. A) Two hypothetical sequences (A and B) diverge from each other and from an outgroup. DNA sequences shown are restricted to just the variable sites, with dots to indicate a match to the reference sequence. R_1 and R_2 are hypothetical recombinants between A and B. R_1 is a coarse mosaic, reflecting a single crossover: a tract matching A (sites 1-10) is followed by a tract matching B (sites 11-20). In contrast, R_2 is a finer mosaic, in which it switches back and forth several times between A and B. B) Sequence similarity between A, B, and the hypothetical recombinants R_1 and R_2. Matching is indicated by a dot, mismatch by *.

Table 1. Questions for the characterization of specific recombination events.

Who:	Which sequences are recombinant? Which sequences were they derived from or are most like the ancestral parental sequences?
Where:	Within a recombinant sequence, where is the transition point (breakpoint) from one parental type to the next?
When:	Is the event recent or ancient? If ancient, are there sequences in the sample that are closely related to the sequences involved in recombination?

by an individual infected with two distinct viruses could, in theory, be a heavily scrambled mixture of the two original types (R_2 in Figure 2). In fact, Robertson *et al.* (1995) have characterized many of the recombinants they identified as being the product of multiple crossovers.

Whether coarse or fine, a complicating effect of recombination is that the original phylogenetic signal of divergence between the parental sequences is replaced by a phylogenetic signal that is a hybrid of the two parental sequences. While this can be an opportunity to detect the recombination event, it also wreaks havoc on conventional phylogenetic analyses. For instance, in Fig. 2, note that while both R_1 and R_2 are about as divergent from the AB ancestor as are either A or B, each recombinant sequence has specific variants that would suggest clustering with A (e.g. site 2) *and* with B (e.g. site 18). This conflicting signal is similar to the homoplasy generated by recurrent mutation, such as reverse or parallel changes. By not being fully diverged in either A or B's direction such recombinant sequences often mimic an ancestral intermediate between the two progenitor sequences.

Since most of the methods so far proposed for recombination analysis attempt to provide as much detail as possible for each recombination event, attention should now be given to the more general considerations of detection of recombination from this type of data. It will be useful to work through some possibilities pertinent to potential sampled data sets.

3. PHYLOGENETIC CONSIDERATIONS

The two key aspects of recombination identification and characterization are the underlying phylogeny (actually *phylogenies*) of the sampled sequences and the level of variation available to diagnose recombination events. Each of these is determined by many factors. For example, the sample may be one of global coverage, so that extreme types of sequence are recognized but perhaps few opportunities to recognize individual recombining sequences. The level of variation available depends on mutation rate, length of sequence, and depth of the sample. Conceivably, one could simply sequence a longer region to identify additional variation for diagnosing recombination, but this practice could be confounded if the extended region contains many additional recombination events.

In Figure 3A, a general scenario of biparental recombination is presented. Our focus is on the ability to detect this event, which will depend on which

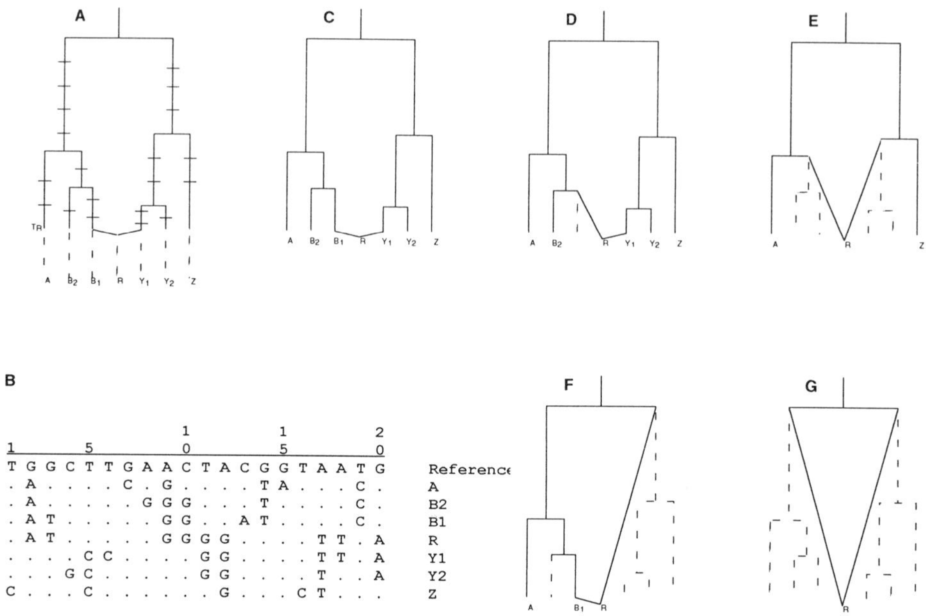

Figure 3. A) Actual pattern of divergence and differentiation in a hypothetical set of seven sequences. Two main lineages exist: (A, B1, B2) and (Z, Y1, Y2). A recombination event between B1 and Y1 produces sequence R. B) Nucleotide sequences at the variable sites among A-Z. C) – G) Various subsamples from sequences A-Z, used to visualize the potential of determining that R is indeed recombinant.

sequences are actually sampled. The recombinant sequence, R, is assumed to have been created by a single crossover between sequences B1 and Y1 at a time T_R in the past. Other sequences are assumed to have been sampled as well (not shown), but we only need to focus on those most closely related to R and its progenitors. Furthermore, we can restrict our attention to the regions of sequence immediately flanking the recombination event and assume that it is the only recombination event in this region. Without loss of generality, the sequences as they existed at time T_R (Fig. 3 B) are assumed to be available at present – our interest is in what can be said if only a subset are sampled. For instance, in the ideal situation in which all sequences are sampled (A, B2, B1, Y1, Y2, and Z along with R), it is reasonable to conclude (Fig. 3C) an AB lineage and a YZ lineage as in figure 3A, and that R is directly derived from B1 (left) and Y1 (right) through an ancestral recombination event. Note that even if B1 were not available, the left part of R would still be recognizably close to B2 (Fig. 3D). Similarly, even if both B1 and B2 were missing, the left part of R would still be recognizably closer to A than to the YZ lineage. In fact, even if only A, Z, and R were sampled, R could still be recognized as a recombinant sequence, being more closely related to A on the left, but much closer to Z on the right (Fig. 3B, 3E).

Now consider what happens if one entire side of the tree is missing (e.g., only A, B1, and R are sampled, Fig. 3F). In this case, R is identical to B1 on the left, but has numerous (8) differences from B1 on the right. There is no clear phylogenetic signal as to where these differences came from, since neither Y1 nor any other members of that lineage were in the sample. Still, it is possible to achieve a diagnosis of a recombination event. In this case, the diagnosis depends on this juxtaposition of identity on the left with differentiation on the right. In this scenario, one could conclude that R belongs in the AB lineage, but the level of difference observed for the right side is suspicious. If additional sequence data outside the A-Z lineages suggests that there is comparable variability between the right and left, then it is reasonable to conclude that the disparity within R is indeed due to recombination. Similarly, if only A and R were sampled, A and R would cluster together but again the disparity of differentiation between the right and left sides would be suspicious. What if only R were sampled (Fig. 3G)? Clearly, without any sequences related to R's progenitors, R could not be recognized as a recombinant. Thus, if HIV existed as distinct subpopulations and recombination only occurred within each subpopulation, a sample of one sequence from each subpopulation would not be expected to show evidence of recombination.

4. EXISTING ALGORITHMS

Algorithms that are currently available for the detection or characterization of historical recombination events capitalize on some of the principles outlined above. Because the goals for such an analysis can be diverse, the various algorithms are differentiated primarily by which aspects of the data they emphasize in the overall evaluation of historical recombination. Readers are encouraged to evaluate each algorithm in the context of their own applications, and to consult the original literature for further detail.

The algorithm introduced by Stephens (1985) is a good place to begin, in that it was an explicit attempt to capture variation in the phylogenetic signal in a sample of sequences, and to relate that pattern of variation to a simple model of recombination. One of the primary considerations is that if there is no recombination reflected in the sample of sequences, then there will be one true tree underlying the sample and there should be no particular pattern to the physical distribution of variable sites corresponding to the different branches of this tree. For instance, if a particular pair of sequences have each other as closest relatives in the sample, then the sites reflecting this relationship should be randomly distributed across the entire length of sequence. On the other hand, if the relationship holds only for the first three quarters of the sequence, because one of the pair is recombinant, then the physical distribution of sites pairing the two should be restricted to the first three quarters of sequence, and a different pairing for the recombinant sequence is restricted to the final quarter. Stephens' algorithms (1985) based on this concept appeal to the Bose-Einstein statistics (Feller, 1968) for evaluation of the statistical significance of any departures from randomness observed in the physical distribution of the sites with specific phylogenetic signals.

Stephens' original algorithm was applied in four basic ways. First, since a given set of sequences may cover a wide spectrum of sites in terms of variability (e.g., highly conserved vs. hyper-variable), application was made to both the complete sequence and to the list of variable sites (i.e., suppressing sites that did not vary in the sample). Second, both methods, "complete" and "list", were applied to each set of sites with identical phylogenetic signal in two ways. First, clustering of the entire set of such sites was tested to see if it was too compact relative to the length sequenced (complete method) or to the number of variable positions (list method). Second, the distribution of sites within a set was inspected to see whether any gaps existed that were statistically significantly too long. Interpretation of the latter was that positions in the gap reflected an alternative phylogenetic tree corresponding to positions to either side of the gap, potentially indicating a gene conversion event or a short double crossover. Alternatively, two independent single crossovers occurring at different times could conceivably produce such a pattern.

The main problem with Stephens' approach is that it depends on the existence of multiple sites reflecting any given branch in a phylogenetic tree. This concept breaks down in three distinct ways. First, too little variability in the sequence leads to failure to diagnose a recombination event. This effect has been quantified for nuclear allelic sequences (Stephens, 1985b). Second, hypervariability can also lead to failure with Stephens' algorithms in that the phylogenetic signal at a variable site can be reversed or obscured by additional mutation. This effect can be overcome by applying the algorithm to subsets of the data (e.g., DuBose *et al.*, 1988). In fact, it is often useful to simply inspect the distribution of differences in each pairwise comparison between sequences and to test each distribution by the four methods outlined above (e.g., Hoelzel *et al.*, 1999). Third, if the sample has endured many recombination events in its history, the inherent scrambling may generate so many alternative phylogenetic signals that no clusters of distinct signals exist.

In relation to HIV sequences, it is very clear that the hypervariability problem mentioned above confounds Stephens' method of analysis. It is reasonable to suppose that the identification and removal of hypervariable sites from a dataset would have value when applying algorithms of this type. Many of the strategies

Maynard Smith (1992) has developed a method, the maximum chi-squared method, for detecting mosaicism in the phylogeny of genes and assessing its statistical significance, and applied it to the question of horizontal gene transfer among prokaryotic species (e.g., Spratt *et al.,* 1992). In its original incarnation, it was similar in spirit to Stephens' list method applied to pairwise differences between sequences, but would be expected to be less sensitive to hypervariable sites. The gist of Maynard Smith's method is to compare levels of differentiation on both sides of each variable position: statistically significant differences suggest that the nucleotide is near a recombination breakpoint, so the nucleotide with the maximum differentiation is the most likely location for a breakpoint. Recall that in Figure 3F, the only signal of recombination was differences in the level of divergence.

Numerous authors have investigated sequence exchange in another complex system, among the genes of the mammalian major histocompatibility complex (*Mhc*). Takahata (1994) has proposed a method for detecting such exchanges that is relevant to detecting recombination among HIV sequences. His method is based on the statistical "runs test", in which sites are classified into two groups, a and b, and then the randomness of the joint distribution of these two types of sites is measured by the number of runs. Conceptually, this has an advantage over Stephens' method when the data are somewhat noisy. That is, there can be clear evidence of clustering, suggesting recombination, that would not be detected by the more demanding statistics Stephens suggested (1985). The runs test has very low power when one of the two classifications is far more frequent than the other, but this is expected to be less of a problem for HIV sequences.

Several direct applications of the above lines of thought have been made recently to HIV sequences (Robertson *et al.*, 1995; Siepel *et al.*, 1995; Salminen *et al.*, 1995). For instance, Robertson *et al.* (1995) looked for a transition point from one phylogenetic signal to another (compare Fig. 3C-E) among various regions of *gag* and *env* genes from a series of viral isolates. They then adapted Maynard Smith's method (1992) for localizing the recombination breakpoints. Siepel *et al.* (1995) and Salminen *et al.* (1995) describe computer programs aimed specifically at detecting recombinant HIV sequences. The Recombinant Identification Program (RIP) of Siepel *et al.* (1995) makes use of a list of HIV sequences classified as to their subtype and then takes additional test sequences as input. The extent to which each test sequence matches the consensus sequence of a given subtype is monitored by sliding a user-defined window the length of the alignment, noting when the test sequence switches over from one subtype to another as evidence of the test sequence being recombinant. As such, it is an exploratory tool that capitalizes on the wealth of knowledge captured in the classification of HIV sequences into know subtypes, and it utilizes the phylogenetic signal of switching from one subtype to another as the basis for detection. Obviously, in this mode of analysis it is not capable of detecting intra-subtype recombination but could be reasonably modified to do so. A vulnerability in this type of analysis is in the correctness of the input list of classified sequences. Similarly, Salminen *et al.* (1995) describe a computer program they call "bootscanning" that makes use of the phylogenetic evaluation procedure known as "bootstrapping", whereby a given branch in a phylogenetic tree is evaluated for overall consistency and confidence. Here the input data are again a reference set and a test sequence, and phylogenetic algorithms are used to pair the test sequence with one or more reference sequences. The strength of that pairing is evaluated by bootstrapping, and as a window slides along the aligned length of sequence, changes in the bootstrap values and the actual phylogeny are taken as evidence that the test sequence is recombinant with the breakpoint at the position of lowest bootstrap values. The power of this method would seem to depend heavily on choice of reference sequences (consider Fig. 3F-G, for instance).

Finally, several recent publications (Hein, 1993; Jakobsen and Easteal, 1996; Grassly and Holmes, 1997; McGuire *et al.*, 1997; Maynard Smith, 1999) have improved and extended the previous methods in apparently productive ways. Grassly and Holmes (1997), for instance, have written a program "PLATO" (for partial likelihoods assessed through optimization) that evaluates the spatial variation of phylogenetic signal, analogous to many of the algorithms mentioned above. However, their program is also useful for detecting certain types of selection and incorporates a maximum likelihood framework.

5. CONCLUSIONS

Detection of historical recombination events from a sample of molecular sequences depends heavily on the sample itself. The ideal signal of recombination comes from large shifts in the phylogenetic signal, such as that seen when an HIV strain has clear-cut affinity to one subtype when one gene is used to build the phylogeny, but clear affinity to another subtype when a different gene is used. However, it is still possible to detect recombination when the level of differentiation between sequences undergoes a substantial shift, and that shift can be separated from other phenomena, such as selective constraints. An important future direction that may depend on external estimation of evolutionary parameters will be the evaluation of the relative effects of mutation and recombination in shaping the pattern of variation among molecular sequences.

REFERENCES

Cullen, M., Noble, J., Erlich, H., Thorpe, K., Beck, S., Klitz, W., Trowsdale, J. and Carrington, M. 1997. Characterization of recombination in the HLA Class II region. *Am.. J. Hum. Genet.* **60**:397-407.

Dubose, R.F., Dykhuizen, D.E. and Hartl, D.L. 1988. Genetic exchange among natural isolates of bacteria: recombination within the phoA gene of *Escherichia coli*. *Proc. Natl Acad. Sci. USA* **85**:7036-7040.

Feller, W. 1968. *An Introduction to Probability Theory and Its Applications*. VOL I, Third ed. Wiley & Sons, New York.

Grassly, N.C. and Holmes, E.C. 1997. A likelihood method for the detection of selection and recombination using nucleotide sequences. *Mol. Biol. Evol.* **14**:239-247.

Hein, J. 1993. A heuristic method to reconstruct the history of sequences subject to recombination. *J. Mol. Evol.* **36**:396-405.

Hey, J. and Wakeley, J. 1997. A coalescent estimator of the population recombination rate. *Genetics* **145**:833-846.

Hoelzel, A.R., Stephens, J.C. and O'Brien, S.J. 1999. Molecular genetic diversity and evolution at the MHC DQB locus in four species of pinnipeds. *Mol. Biol. Evol.* **16**:611-618.

Hu, W.S. and Temin, H.M. 1990. Retroviral recombination and reverse transcription. *Science* **250**:1227-1233.

Hudson, R.R. 1983. Properties of a neutral allele model with intragenic recombination. *Theor. Popul. Biol.* **23**:183-201.

Hudson, R.R. 1990. Gene genealogies and the coalescent process. *Oxf. Surv. Evol. Biol.* **7**:1-44.

Hudson, R.R. and Kaplan, N.L. 1985. Statistical properties of the number of recombination events in the history of a sample of DNA sequences. *Genetics* **111**:147-164.

Jakobsen, I.B. and Easteal, S. 1996. A program for calculating and displaying compatibility matrices as an aid in determining reticulate evolution in molecular sequences. *CABIOS* **12**: 291-295.

Leitner, T., Escanilla, D., Marquina, S., Wahlberg, J., Brostrom, C.,Hansson, H.B., Uhlen, M. and Albert, J. 1995. Biological and molecular characterization of subtype D, G, and A/D recombinant HIV-1 transmissions in Sweden. *Virology* **209**:136-146.

Maynard Smith, J. 1992. Analyzing the mosaic structure of genes. *J. Mol. Evol.* **34**:126-129.

Maynard Smith, J. 1999. The detection and measurement of recombination from sequence data. *Genetics* **153**:1021-1027.

McGuire, G., Wright, F. and Prentice, M.J. 1997. A graphical method for detecting recombination in phylogenetic data sets. *Mol. Biol. Evol.***14**:1125-1131.

Robertson, D.L., Hahn, B.H. and Sharp, P.M. 1995a. Recombination in AIDS viruses. *J. Mol. Evol.* **40**:249-259.

Robertson, D.L., Sharp, P.M., McCutchan, F.E. and Hahn, B.H. 1995b. Recombination in HIV-1. *Nature* **374**:124-126.

Sabino, E.C., Shpaer, E.G., Morgado, M.G., Korber, B.T.M., Diaz, R.S., Bongertz, V., Cavalcante, S., Galvao-Castro, B., Mullins, J.I. and Mayer, A. 1994. Identification of human immunodeficiency virus type 1 envelope genes recombinant between subtypes B and F in two epidemiologically linked individuals from Brazil. *J. Virol.* **68**:6340-6346.

Salminen, M.O., Carr, J.K., Burke, D.S., McCutchan, F.E. 1995. Identification of breakpoints in intergenotypic recombinants of HIV type 1 by bootscanning. *AIDS Res. and Hum. Retroviruses* **11**:1423-1425.

Siepel, A.C., Halpern, A.L., Macken, C. and Korber, B.T.M. 1995. A computer program designed to screen rapidly for HIV type 1 intersubtype recombinant sequences. *AIDS Res. and Hum. Retroviruses* **11**:1413-1416.

Spratt, B.G., Bowler, L.D., Zhang, Q.Y., Zhou, J. and Maynard Smith, J. 1992. Role of interspecies transfer of chromosomal genes in the evolution of penicillin resistance in pathogenic and commensal *Neisseria* species. *J. Mol. Evol.* **34**:115-125.

Stephens, J.C. 1985. Statistical methods of DNA sequence analysis: detection of intragenic recombination or gene conversion. *Mol. Biol. Evol.* **2**:539-256.

Stephens, J.C. 1986. On the frequency of undetectable recombination events. *Genetics* **112**:923-926.

Takahata, N. 1994. Comments on the detection of reciprocal recombination or gene conversation. *Immunogenetics* **39**:146-149.

MOLECULAR POPULATION GENETICS: COALESCENT METHODS BASED ON SUMMARY STATISTICS

Daniel A. Vasco*, Keith A. Crandall* and Yun-Xin Fu[§]

*Department of Zoology, Brigham Young University, Provo, UT 84602, USA
[§]Human Genetics Center, School of Public Health, University of Texas at Houston, Houston TX 77030 USA

1. INTRODUCTION

Population genetics theory has recently undergone a renaissance of sorts with the application of coalescent methods to estimation of population parameters using sequence data. Even though this branch of theoretical population genetics only started in the early 1980's, it is already proving to be a fundamental tool in the development of computational and statistical methods for studying evolution. Applications of these methods to viral population dynamics, especially RNA viruses such as HIV, will provide an important interface between theory and empiricism. This interface could provide a fundamentally new understanding of the role of mutation, natural selection and genetic drift in the origins of the HIV epidemic as well as the rapid evolution of resistance to therapy.

In this chapter we have three goals. First, we will demonstrate that very fast, accessible, and useful coalescent methods exist to analyze DNA sequence data. Second, we will develop, in detail, application of the methods for a hypothetical data set of five haplotypes. We will also apply these methods to an actual HIV data set taken from Holmes and his coworkers (1992). Lastly, we discuss present and future developments for applications of coalescent theory to several parameter estimation problems in genetic epidemiology that require analysis of large number of sequences.

Although coalescent theory is a relatively new development in population genetics theory, several recent reviews have appeared that the reader may refer to (Tavaré 1984; Takahata 1991; Hudson 1990, 1993; Donnelly and Tavaré 1995; Li and Fu, 1999). Perhaps the most important issues we address in this review are the integration of coalescent theory and statistical principles and the ease of application of methods to large-scale data sets. However, when possible, we also attempt to discuss estimation methods that allow answering questions of a more practical nature in analyzing large data sets such as: Can a given set of estimators be computed in minutes? Hours? Or days? How does an estimator behave when new data are added to the sample, such as more sites, more sequences, or data from independent loci? As the field of statistical inference using coalescent methods is still in its infancy much research remains before practical answers to these types of questions are obtained. For the case of methods based upon summary statistics much recent progress has been made that we will elaborate on in this chapter.

Section 2 introduces the essential concepts from coalescent theory that we will use in this chapter. We define the concepts of coalescent times and neutral mutation models as they will be used in the chapter. First, we focus on showing how branch length statistics can be computed over coalescent trees. Second, we develop neutral mutation models and show how they can be related to the coalescent time distribution of a model. Lastly, we show how a genealogy with known topology can be used, along with coalescent statistics based upon Monte Carlo simulation of many thousands of genealogies, to estimate coalescent tree shape. This method forms the basis for the phylogenetic parameter estimators discussed in this paper.

In Section 3, we show how to apply the simplest types of computations for measuring coalescent information in nucleotide sequence data, such as the level of polymorphism using the number of segregating sites, number of alleles, or the pairwise distance between sequences. Two recently introduced measures of coalescent information in a sample developed by Fu (1994b, 1995) are also discussed: the *size* and *type* of branch of a tree. These two summary statistics allow greater resolution of the pattern of polymorphism at the nucleotide level. All of these measures are based on various kinds of summary statistics of a sample that are time-scaled by mutation rate. These methods play an important part in applying methods of parameter estimation, statistical tests, and coalescent time estimation. Moreover, since they are among the fastest methods of statistical analysis, they may prove very useful in analyzing large-scale data sets.

Section 4 forms the second half of this paper: how to estimate population parameters using the summary statistics introduced in Section 3. Maximum-likelihood methods developed by Griffiths and Tavaré (1994) as well as Kuhner, Yamato and Felsenstein (1995) can be used to estimate not only ancestral parameters but also tree topology. In fact, these coalescent-based tree-building methods can be used to estimate phylogeny for intraspecific sequence data. These methods are covered in the chapter in this book by Beerli and his colleagues. However, the dual nature of these algorithms creates somewhat of a disadvantage in terms of speed of computation and biases that may be inherent in the process of tree reconstruction itself (Felsenstein, 1992b; Felsenstein *et al.*, 1999; Kuhner *et al.*, 1995). This may prove a problem when attempting to analyze large-scale data sets as well as apply computationally intensive statistical approaches such as the parametric bootstrap. Taking an alternative statistical approach to the tree-based parameter estimation problem based upon the method of least squares (LS), Fu (1994a; 1994b) developed a very fast recursive method of estimating population parameters that he called the UPBLUE method. In Section 4.4 we focus on how to use Fu's (1994a; 1994b) UPBLUE method. This method of estimation is useful because it takes full advantage of the information in the distance matrix. Also, this method places the coalescent part of the estimation algorithm on top of a previously derived phylogenetic tree. Separating these two processes allows a considerable increase in speed of computation and also may allow pinpointing sources of biases of estimation due to tree reconstruction errors (Fu, 1994a; 1994b; Deng and Fu, 1996).

In Section 4.5 of this chapter, we discuss recent extensions of LS methods to more complex population models. Utilizing all of the concepts developed in Sections 2-3 we show how a general theory of estimation can be constructed for ancestral population parameters. Several programs now exist that allow using genealogical summary statistics to LS estimate population parameters. These appear on the World Wide Web as a free package called EVE (Vasco, 1999). We show in Section 4 that this suite of programs allows analyzing sequence data so that efficient computation of statistical tests, estimation of ancestral population parameters, analyses of estimation bias, and hypothesis testing can be rapidly accomplished for even large data sets. For several cases we demonstrate the methods on real or simulated data sets. In others, we point out how the theoretical methods may be applied in the near future as the EVE package of programs is further developed. Section 4.6 is briefly devoted to examining the relationship between summary statistics estimators and phylogenetic estimators using a unified LS approach. Lastly, in Section 5, we argue the general merits of using summary statistics estimators. This includes analyzing samples that may have arisen as a result of the evolutionary forces of mutation, recombination, migration, and selection.

2. ESSENTIAL CONCEPTS FROM COALESCENT THEORY

As described by Hudson (1990) in his review of the coalescent, one of the most useful aspects of coalescent theory is that one can separate the genealogical process from the neutral mutation process. This division allows mathematically formulating statistical properties of coalescing genealogical and mutation processes separately from each other and then integrating them back together again in a consistent manner. In this chapter we will also take advantage of this property. First, we discuss the properties of coalescent trees in constant and varying environments. Second, we show, using these results, how a general model of the neutral mutation process can be developed. In the later part of this chapter, we will show how to apply this theory to construct inferences for nucleotide sequence data.

2.1 The Coalescent in a Constant Environment

In the 1930s, both R.A. Fisher (1930) and Sewall Wright (1931) developed a model that allows a mathematical description of the properties of binomial sampling in small populations over discrete generations. This model has become known as the Wright-Fisher model and is used widely in population genetics and coalescent theory. We now describe some of the basic properties of this model and use it to show how coalescent times arise in the neutral evolution of nucleotide sequences. Figure la shows a coalescent tree for a sample of n sequences from a finite population. The time (t_n) required for n sequences to coalesce to n - 1 sequences will be referred to as the nth coalescent time. The distribution of coalescent times for a given population genetic model will play a fundamental role in the theoretical development of this review. For this reason, we give a brief derivation that lends itself to immediate generalization to the case of coalescents in varying environments. Our review here follows Li and Fu (1999). Other reviews that stress the important effect of variable environments appear in Hudson (1990) as well as Donnelly and Tavaré (1995). We designate the population from which the sample was taken as generation 0 and look backward in time so that generation i represents the one that was i generations earlier than generation 0. For a finite population, there is a non-zero probability $q_n(i)$ that two of the n sequences at generation i came from one ancestral sequence at generation $i + 1$. The probability Q that the n sequences at generation i coalesce to n - 1 sequences at generation $i + t$ is therefore

$$Q(t_n = t) = [1 - q_n(i)][1 - q_n(i+1)]\cdots[1 - q_n(i+t-1)]q_n(t), \tag{1}$$

which is the distribution of coalescent time t_n. For the kth coalescent time, we have

$$Q(t_k = t \mid s_k) = [1 - q_k(s_k)]\cdots[1 - q_k(s_k + t - 1)]q_k(s_k + t), \tag{2}$$

where $s_k = t_n + \ldots + t_{k+1}$ with $s_n = 0$. The reason why t_k is dependent on s_k is that the period of t_k starts only when the n sequences coalesce to k ancestral sequences.

We assume that N sequences are evolving each generation. In a population in which a given sequence is selectively neutral, all N parental sequences are equally likely to have been a parent. Since sampling the sequences is done with replacement, the probability that a sequence is derived from a common parental sequence in the previous generation is $1/N$. Thus, if i represents a given generation then the probability of sampling the same sequence at generation i is

$$q_2(i) = \frac{1}{N} \tag{3}$$

In general, the probability that a random sample of k sequences came from k different parental sequences of the previous generation is

$$1 - q_k(i) = \prod_{j=1}^{k-1}\left(1 - \frac{j}{N}\right) \tag{4}$$

$$\approx 1 - \frac{k(k-1)}{2N}, \tag{5}$$

assuming k much smaller than N.

One can also ask, what is the probability that k distinct sampled sequences have exactly k distinct ancestors one generation earlier? We find

$$\begin{aligned} Q(t_k = t) &= (1 - q_k(t))^t q_k(t) \\ &\approx \frac{k(k-1)}{2N} e^{-\frac{k(k-1)}{2N}t}, \end{aligned} \tag{6}$$

which is an exponential distribution. Thus, equation (6) shows that the span of time back until a common ancestor occurs is geometrically distributed, and that this distribution can be approximated by an exponential distribution. This stochastic property gives rise to the "coalescing" of lineages. If one looks at the recent history of a sample of sequences, even as recently as a single generation (1-2 days for HIV), then one should observe between the "present" or the time when the sample was taken, and at generation $t + 1$, a single pair of lineages coalescing at the most recent common ancestor of two of the sample sequences. Each of the possible pairs of lineages coalesces with probability given by (6). With each generation this process keeps recurring until only a single sequence, the most recent common ancestor

(MRCA), is left.

Since equation (6) can be approximated by an exponential distribution the statistics of the coalescent time distribution for the t_k coalescent time are approximately determined by:

$$\mathrm{E}(t_k) = \frac{2N}{k(k-1)}, \tag{7}$$

$$\mathrm{Var}(t_k) = E^2(t_k). \tag{8}$$

Note that the average length of the coalescent time decreases with increasing k. This is because a larger k means more pairs of sequences, which means that there exists a larger chance that one of the pairs of sequences coalesces in one generation, resulting on average a shorter coalescent time.

We now consider the process by which neutral mutations accumulate along lineages of a genealogy. While the statistical properties of genealogies depend upon selection and population size, neutral mutations do not have an effect on how the topology of a genealogy evolves. Thus, we can study the mutational process without reference to a specific genealogical model. The model of mutation that we use is due to Kimura (1983) and is called an infinite sites constant rate mutation model because mutations accumulate on the branches of a genealogy in a clock-like fashion. Assume that the number of mutations that appear in a given time is a Poisson variable. Let μ be the mutation rate per sequence per generation. If a sample of sequences is examined at two separate time points along a lineage l_i from a completely homozygous population (one with no genetic variation between sequences on that branch), say time 0 and some time T in the future, then the number of mutations that will have occurred for a sampled sequence at T on that branch follows the Poisson distribution with mean rate μT (where μ is the mutation rate).

2.1.1 Superposition of the Genealogical and Mutational Processes

The coalescent and mutation processes can be considered as two independent simultaneous stochastic processes that together create the observed pattern of genetic variation in a sample of sequences as time progresses. Assume a constant environment model, then whether a mutation of a coalescent event takes place can be considered as two competing random evolutionary events. For example, the time at which a given evolutionary event (mutation or coalescence of a lineage) takes place can be thought of as being determined by two noisy clocks. At each generation the probability that either a coalescent or mutation clock goes off for a sample of k sequences is $k(k-1)/(2N)+k\mu = \frac{1}{2N}[k(k-1)+k\theta]$, where $\theta = 2N\mu$. Thus, the probability that a coalescent clock goes off first is,

$$\frac{k-1}{k-1+\theta} \tag{9}$$

while the probability that a mutation clock goes off first is,

$$\frac{\theta}{k-1+\theta} \tag{10}$$

By superimposing coalescent and mutation events constructed from these two noisy clocks we can simulate the molecular evolution of a sample of sequences. Working our way backwards in time we wait for a clock to go off and then implement the probability of an evolutionary event as an instruction in a computer program using a random number generator. We iterate this process backwards until we reach the MRCA. In this way we can rapidly simulate the evolution of coalescent trees such as the one shown in Figure 1. We now illustrate a simple method of simulating the coalescent in a constant environment.

(A)

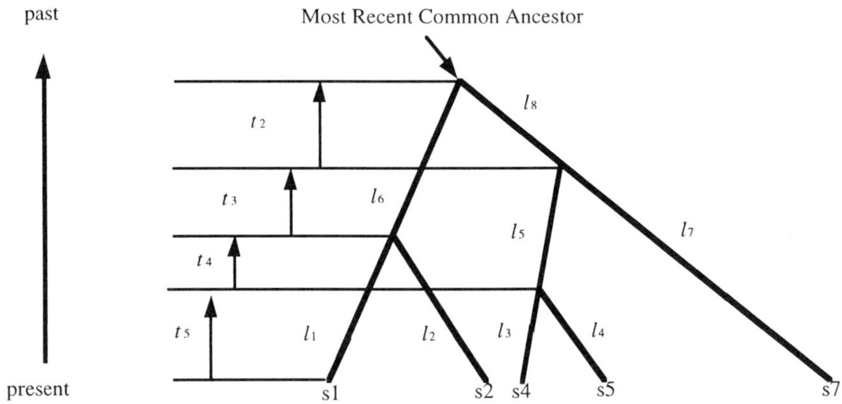

(B)

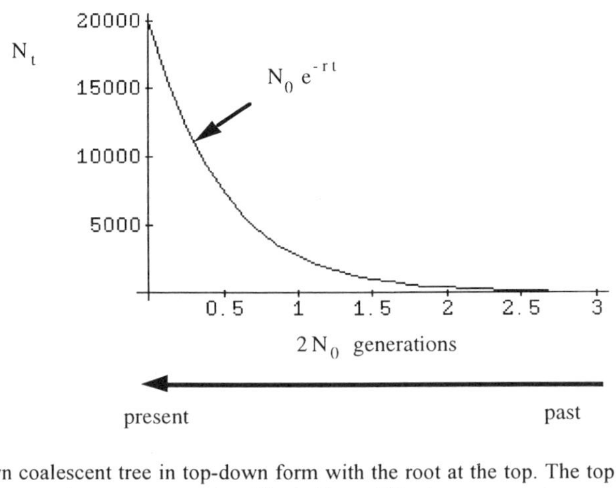

Figure 1 A. Known coalescent tree in top-down form with the root at the top. The top of the tree represents the most recent common ancestor. The bottom of the tree represents a sample of five sequences observed at the present. The symbols s1, s2, s4, s5 and s7 represent nucleotide sequences for a sample such the one discussed in Section 3. Known branch lengths and coalescent times are shown. B. Model of exponential growth backward in time. Time scales as two times the effective population size number of generations.

2.1.2 Simulating the Coalescent for a Sample of Nucleotide Sequences

A sample of nucleotide sequences can be created by simulating a random gene tree using Hudson's (1990; 1993) algorithm. It consists of three parts: create a gene tree topology, a set of branch lengths, and mutations. One first generates a random tree for the genealogy (using for example, the *maketree* C subroutine of Hudson (1990) for a sample n sequences). The n ancestral lineages are simulated backward in time, first coalescing to n-1 lineages and so on, until the n lineages are joined together to a common ancestor. In this process, two of the n individuals (represented as nodes in a C structure) in the sample are chosen at random to merge. These form the first two nodes of the genealogy. A new node is chosen as the ancestral node and this process is repeated on the remaining n-1sequences. The process stops when a single individual remains (the MRCA of the sample). The end result of the simulation is a bifurcating tree with tips representing the n sequences of the sample (see Figure 1a). Because the competing stochastic processes driving the two noisy clocks are independent of each other, we can simulate each set of coalescent and mutation events separately for the n sequences, and superimpose them on top of the random topology. Each kth coalescent time, t_k , occurs with probability determined by the exponential distribution. The number of mutations on a branch is determined as a random Poisson variable with mean rate θ T. The number of mutations that occur on a lineage is determined by the constant neutral mutation rate assumption, so that the number of mutations occurring on a lineage of length T is Poisson distributed with mean rate $2N\mu T = \theta T$. In this way we can rapidly simulate the evolution of coalescent trees such as the one shown in Figure la. By splitting the simulation of the genealogical processes from the mutation process, very fast and efficient computer codes can be constructed using coalescent statistics. Hence, many tens of thousands of simulated genealogies for a data set of a hundred sequences can be computed within seconds on a desktop computer.

2.2 The Coalescent in a Variable Environment

2.2.1 Models of Varying Environments

When population size is not constant, the mathematics of effective population size becomes more complicated than that developed using the Wright-Fisher model above. Indeed using standard prospective population genetic approaches there appears virtually no work on studying the concept of population effective size under size change population models. Using the retrospective coalescent approach however, much progress has recently been made in developing size change models (Tajima, 1989a; Slatkin and Hudson, 1991; Fu, 1997). Recently Kuhner, Yamato and Felsenstein (1998) and Vasco and Fu (submitted) developed methods of estimating effective population size in varying environments. In this section we concentrate on the method of Vasco and Fu. The maximum likelihood method of Kuhner and her coworkers (1998) is examined in the chapter in this book by Beerli and colleagues.

Let N_t be the effective population size at generation t. From (1) and (2) it follows that (Li and Fu, 1999)

$$q(t_k(\mathbf{p}) = t \mid s_k) = \frac{k(k-1)}{2N_{s_k+t+1}} \prod_{i=s_k+1}^{s_k+t} \left(1 - \frac{k(k-1)}{2N_i}\right) \tag{11}$$

$$\approx \frac{k(k-1)}{2N_{s_k+t+1}} \exp\left[-\frac{k(k-1)}{2} \sum_{i=s_k+1}^{s_k+t} \frac{1}{N_i}\right] \tag{12}$$

where **p** is the vector of population genetic parameters to be estimated (see below). Let $v(t) = N_0/N_t$ and scale the time so that one unit corresponds to $2N_0$ generations. Then a continuous approximation of the above equation results in the density function of t_k as

$$f(t_k(\mathbf{p}) = t \mid s_k) \approx \frac{k(k-1)}{2} v(s_k + t) \exp\left[-\frac{k(k-1)}{2} \int_{s_k}^{s_k+t} v(s)ds\right], \tag{13}$$

which was derived by Griffiths and Tavaré (1994).

2.2.1.1 *Exponential Growth*

Exponential growth is usually defined $N_t = N_0 e^{rt}$, where r is the growth rate (or decline when $r < 0$), t is the time since the initial generation, and N_0 is the initial effective population size, i.e., the size at the time of sampling. In using a coalescent approach, it is useful to reformulate the exponential growth equation going backwards in time as $N_t = N_0 e^{-rt}$. Thus, when we look backwards in time, the exponential growth of the population ($r > 0$) becomes an exponential decline in the population's size (see Figure 1b). One unit of time corresponds to $2N_0$ generations. Substituting $v(t) = N_0/N_t$ in (13) gives the density function of the kth coalescent time under the exponential growth model.

2.2.1.2 *Logistic Growth*

Let $N(T)$ be the effective population size of a logistically growing population that was sampled at time T. We can then determine the effect of sampling at different times on the pattern of sequence polymorphism using the model:

$$N(T_s - T) = N_{\min} + \frac{N_{\max} - N_{\min}}{1 + e^{-r(T_s - T - c)}} \tag{14}$$

where the time T_s-T is counted backwards starting at the sampling time T_s. The parameters N_{min} and N_{max} are the minimum and maximum effective population sizes, while r and c are both nonnegative parameters, r determines the speed of growth, while c is the reflection point of the growth curve. One unit of time corresponds to $2N_{max}$ generations. Setting $N(T_s$-$t)$ to $T = 0$ gives the population size at the time of sampling. For this model we can define the function

$$v(t) = N(T_s - T) / N_{\max} \tag{15}$$

Substituting $v(t)$ in (13) gives the density function of the kth coalescent time under the logistic growth model.

2.2.2 Expected Branch Lengths of Coalescent Trees

Consider the coalescent tree shown in Figure la. Let us assume that the number of mutations fix the topology, and determine the branching order independently of the mechanism of evolutionary change. Also, assume each of the five sequences can be traced back in time first to n-1 ancestral sequences, next to n-2, sequences and so on, until a single ancestral sequence remains (the MRCA). In Figure la, the quantity t_n, represents the time in $2N$ generations required for a coalescent event to have occurred from n to n-1 sequences. Let the coalescent tree in Figure la represent the known topology of a coalescent tree whose branch lengths are to be estimated. We will also assume that a coalescent model, with a specified $v(t)$ function (such as the exponential growth model shown in Figure 1b) has produced the tree dependent upon the parameter vector $\mathbf{p} = [p_1, \ldots, p_s]$. The $\mathbf{p}$ vector can be formed from any of the models discussed in Section 2.2.1 as well as many other coalescent models that satisfy equations (11)–(13). Examples of coalescent trees with typical topologies evolving in varying environments are shown in Figure 2a. Hence, the branch lengths of the coalescent tree must be approximated under the given model (if the true coalescent tree topology can be reasonably approximated). A schematic of how this can be accomplished is shown in Figure 2b. In order to compute the expected branch lengths we simulate the coalescent time distribution (13) for a sample of five DNA sequences many thousands of times and average the results. Thus, for the tree with the topology shown in Figure la we have an "average" coalescent tree shown in Figure 2b. The expectations of the branch lengths of the average tree are given by:

$$E(l_1(\mathbf{p})) = E(l_2(\mathbf{p})) = \bar{t}_4(\mathbf{p}) + \bar{t}_5(\mathbf{p}) \tag{16}$$

$$E(l_3(\mathbf{p})) = E(l_4(\mathbf{p})) = \bar{t}_5(\mathbf{p}) \tag{17}$$

$$E(l_5(\mathbf{p})) = \bar{t}_3(\mathbf{p}) + \bar{t}_4(\mathbf{p}) \tag{18}$$

$$E(l_6(\mathbf{p})) = \bar{t}_2(\mathbf{p}) + \bar{t}_3(\mathbf{p}) \tag{19}$$

$$E(l_7(\mathbf{p})) = \bar{t}_3(\mathbf{p}) + \bar{t}_4(\mathbf{p}) + \bar{t}_5(\mathbf{p}) \tag{20}$$

$$E(l_8(\mathbf{p})) = \bar{t}_2(\mathbf{p}) \tag{21}$$

Each average coalescent time $\bar{t}_k$ is computed using,

$$\bar{t}_k(\mathbf{p}) = \frac{1}{G}\sum_{j=1}^{G} t_k^j(\mathbf{p}). \tag{22}$$

The quantity G represents the number of genealogies that are simulated to obtain the average k^{th} coalescent time. Equation (22) can be used to study the statistical properties of genealogies under several different kinds of models involving population and selective change (Figure 2a). In general it is not difficult to show that for the branch lengths of any known coalescent tree one has

$$l_i(\mathbf{p}) = \sum_{k=2}^{n} s_{ik} t_k(\mathbf{p}) \tag{23}$$

The scalar s_{ik} represents a set of index variables for each branch that bookkeeps the number of times the coalescent time contributes to the length of the ith branch. Thus, for branch i, one can define n-1 s_{ik} index variables ($k = 2,\ldots,n$) such that $s_{ik} = 1$ if the branch has segment of length $t_k(\mathbf{p})$ between the kth and $(k\text{-}1)^{\text{th}}$ coalescence and $s_{ik} = 0$ otherwise. Thus, the branch lengths over the entire topology of a tree for a sample of n genes can be quantitatively characterized in terms of a set of $2\,(n\text{-}1)^2$ variables and corresponding coalescent times. For example, the tree shown in Figure la has $2(5\text{-}1)^2 = 32$ index variables. Detailed examples of how to use this bookkeeping device appear in Fu (1994a; 1994b) as well as Deng and Fu (1996).

Vasco and Fu (submitted) show that substitution of equation (22) into (23), gives the very general relationship

$$\bar{l}_i(\mathbf{p}) = \sum_{k=2}^{n} s_{ik}\bar{t}_k(\mathbf{p}) \tag{24}$$

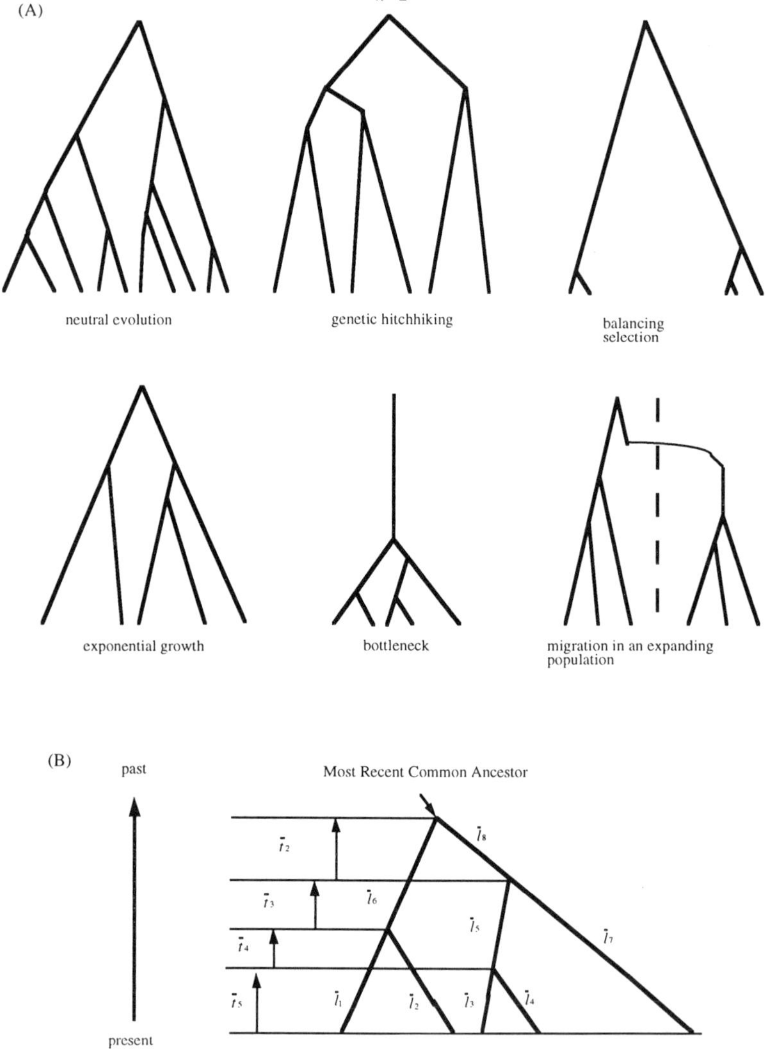

Figure 2 A. Typical phylogenies of coalescent trees observed under a given process of evolution in a varying environment. Migration, recombination and selection can all interact with demographic change over phylogenetic time scales to produce novel patterns in tree shape. One example is shown here with migration in an expanding population. B. Schematic of how Monte Carlo coalescent simulations can be used to approximate the branch lengths of the known or reconstructed tree such as that shown in Figure 1a.

We will show below that equation (24) allows developing efficient computational methods for calculating the expected branch lengths of a coalescent tree, that can be compared to empirically observed values obtained from sequence data. This forms an essential part of the theory of ancestral parameter estimation developed in this paper.

2.2.2.1 Constant Population Size Case

For the constant population case it is possible to go beyond deriving a closed-form expression for the expected branch lengths of a tree (23). From (7) we have the exact result for the average coalescent time. Hence for the tree shown in Figure la, assuming now that the branch lengths are constants, rather than functions of model parameters, equation (23) takes the simpler form,

$$\bar{l}_i = \sum_{k=2}^{n=5} \frac{2Ns_{ik}}{k(k-1)}. \tag{25}$$

2.2.2.2 Mutations on the Branches of a Genealogy

Earlier we saw that the number of mutations that occur at T on a lineage follows the Poisson distribution with mean μT (where μ is the mutation rate). This is true even if the coalescent process that created the lineage was undergoing a change in population size or a selective event. If the lineage is a function of some growth or selection parameter, $\mathbf{p}$, in the notation of the previous section, then for a branch of length $l_i(\mathbf{p})$ there exists a constant number of mutations μT as in the constant environment case, however this constant rate process is usually determined by the effective population size assumed at the time of sampling. For example, for the case of an exponentially growing population, the constant rate Poisson process occurs with mean rate μTN_0, where N_0 is the effective population size at the time of sampling. Hence, the only effect on the constancy of rate of mutation is determined by the endpoint or sampling time of a coalescent tree, and the process itself remains Poisson. This invariance of the mutation process is a very powerful way to model the evolutionary genetics of mutations and coalescent structure in populations evolving in variable environments. To see this, consider the following general model of neutral mutational change

$$m_i = x_i(\mathbf{p})\theta + \varepsilon_i(\mathbf{p}), \tag{26}$$

where $x_i\,(\mathbf{p})$ is some nonlinear function of the ancestral population parameter $\mathbf{p}$.

We now show how the coalescence theory can be used to compute a set of nonlinear regression equations that determine the statistical properties of the number of segregating sites in terms of easily computed expectations, variances and covariances of the branch lengths of a phylogenetic tree. Assume for the moment that we know the exact branch lengths of the coalescent tree. Let $l_i(\mathbf{p})$ be the scaled time lengths of branch i (with one unit of time equal to $2N$ generations) and m_i be the number of mutations on branch i. Further assume for each i, the m_i follows a Poisson distribution with parameter $\theta l_i\,(\mathbf{p})$, conditional on $l_i(\mathbf{p})$. Then, it can be shown that $x_i = \bar{l}_i(\mathbf{p})$ and that the theoretically expected number of mutations on the ith branch is given by

$$\mathrm{E}(m_i) = \bar{l}_i(\mathbf{p})\theta \tag{27}$$

Substituting (24) into (27) gives

$$\mathrm{E}(m_i) = \theta \sum_{k=2}^{n} s_{ik} \bar{t}_k(\mathbf{p}) \tag{28}$$

The equation for the variance of the m_i is

$$\mathrm{Var}(m_i) = x_i(\mathbf{p})\theta + \beta_{ii}(\mathbf{p})\theta^2, \tag{29}$$

where $x_i(\mathbf{p})$ is defined as before and $\beta_{ii}(\mathbf{p})$ is the variance of the ith branch lengths. For each sample the covariances of mutations along the ith and j^{th} branches of a phylogenetic tree can also be computed.

$$\begin{aligned}
\mathrm{Cov}\ (m_i, m_j) &= \mathrm{E}(m_i m_j) - \mathrm{E}(m_i)\mathrm{E}(m_j) \\
&= \mathrm{E}\left[\mu^2 l_i(\mathbf{p}) l_j(\mathbf{p})\right] - \mathrm{E}(m_i)\mathrm{E}(m_j) \\
&= \mu^2\left[\mathrm{E}\left(\sum s_{ik}\bar{t}_k(\mathbf{p}) \sum s_{jk}\bar{t}_k(\mathbf{p})\right)\right] - x_i(\mathbf{p})x_j(\mathbf{p})\theta \\
&= \mu^2\left[\sum s_{ik}\bar{t}_k(\mathbf{p}) \sum s_{jk}\bar{t}_k(\mathbf{p}) + \sum s_{ik}s_{jk}\bar{t}_k^2(\mathbf{p})\right] - x_i(\mathbf{p})x_j(\mathbf{p})\theta \\
&= \beta_{ij}(\mathbf{p})\theta^2
\end{aligned} \tag{30}$$

As was the case for computing the average branch lengths, one can derive analytic expressions (Fu, 1994a) for the average number of mutations along the ith branch of a coalescent tree in a constant environment,

$$\mathrm{E}(m_i) = \theta \sum_{k=2}^{n=5} \frac{2Ns_{ik}}{k(k-1)} \tag{31}$$

Fu (1994a) also derived exact results for the variances (29) and covariances (30) in the constant population case.

It is important to note that the coefficients of the nonlinear regression equations (28-30): $x_i(\mathbf{p})$, $\beta_{ii}(\mathbf{p})$, and $\beta_{ij}(\mathbf{p})$, are fixed *functions* (dependent only upon the vector $\mathbf{p}$) once the topology of the phylogenetic tree is determined (Vasco and Fu, submitted). While for the constant population size case the coefficients become fixed *constants* (Fu, 1994a; 1994b). Below we show how the nonlinear regression equations allow determining the least squares fit of the observed number of mutations along a branch of a phylogenetic tree to theoretical expectations of the branch lengths computed from a specified coalescence model. Besides the observed number of mutations on the branch of a phylogeny, there are two other kinds of summary statistics that allow quantifying the amount of polymorphism in a sample of sequences. We now describe these alternative phylogenetic information measures. We then show that all of the theory developed in this section applies to these summary statistics as well.

3. SUMMARY STATISTICS AND THEIR PROPERTIES

The computation of summary statistics can be used to quantify the amount of polymorphism in a sample. The coalescent theory developed in the last section shows that summary statistics describing DNA polymorphism of a sample can be used to build a very general analysis of coalescing sequences. Some of the earliest applications of the coalescent showed that a complete specification of the simultaneous coalescent and mutation processes allows the pattern of polymorphism for a sample of sequences to be qualitatively and quantitatively analyzed (Hudson, 1993). In the first part of this section, we introduce two of the most commonly used summary statistics for analyzing a sample of DNA sequences. The first is the number of mutations (K) and the second is the mean number of pairwise nucleotide differences (Π) between each sequence in the sample. After this we introduce some newer, less widely known summary statistics recently developed by Fu (1994b; 1995). Fu (1994b; 1995; 1997) has found that the statistics K and Π convey only a small amount of the information that can be computed for a sample. Hence, an alternate approach is to develop statistical methods based upon the complete nucleotide sequence of a set of genes. By taking advantage of the infinite sites property that segregation at any site starts as a result of a unique mutational event so that at most two nucleotides segregate at a site, Fu (1994b) showed that the statistics of a set of mutations of a sample could be computed on a much finer scale. He developed two classes of summary statistics based upon classifying frequency of mutations by category. In the sections following this one, we shall use all of the summary statistics of this and the previous sections to show how population parameters can be rapidly estimated from sequence data.

Assume that one has sequenced a population of individuals and wishes to apply the summary statistics we are developing in this chapter. Several questions would be posed by such an investigator: Are the data compatible with the infinite sites model? If so, how does one go about applying coalescent methods to the data set? In this section, we will attempt to answer these questions. For concreteness consider the following set of seven hypothetical sequences:

s1 A T C A A **A** G C A T T **G** C A **A** C
s2 A T **G** A A T G C A T T C C A T C
s3 A T C A A **A** G C A T T **G** C A **A** C
s4 A T C A A **A** G C A T T C C A T C
s5 A T C A A **A** G C A **A** T **G** C **T** T C
s6 A T C A A **A** G C A T T **G** C A **A** C
s7 A T C A A T G C A **A** T **G** C **T** T C

We assume at this point that we have already obtained aligned nucleotide sequences. Now we can compute summary statistics that will allow us to construct inferences about the past evolutionary history of these sequences.

3.1 The Number of Alleles in a Sample

The total number of alleles or unique sequences in this sample is 5 since s1, s3 and s6 are identical. In order to approximate the infinite sites model we only use the number of unique haplotypes in a sample

Frequency	*Pattern of Sequence Polymorphism*
3 s1	...0...1...0...1...0...1...
1 s2	...1...0...0...0...0...0...
1 s4	...0...1...1...0...0...0...
1 s5	...0...1...0...1...1...0...
1 s7	...0...0...1...1...1...0...

where 0 and 1 represent the ancestral and mutant nucleotides, respectively, and dots represent the intervening sequence segments between segregating sites. Thus, we eliminate sequences s3 and s6 from the sample when reconstructing the genealogy of the sample in any coalescent analysis. However, note that the frequency of each haplotype is recorded at the left. Frequency information will be used in some of the summary statistics.

3.2 The Number of Segregating Sites in a Sample

One of the most commonly used summary statistics is the expected number of polymorphic or segregating sites in a sample. The number of segregating sites *(K)* of a sample is the number of sites that are occupied by at least two different nucleotides. Thus, a segregating site is a site that shows variation among the sequences in a sample. In the sequences above there are six polymorphic sites, giving $K = 6$. The theoretical expectation of K can be computed very simply under the assumption of the infinite sites model. Let K_i be the number of mutations during the period t_i, so that $K = K_{2i}+\ldots+K_n$. In assuming the infinite sites model, we can be sure that each observed mutation in a sample is a segregating site. Since the number of segregating sites follows the Poisson distribution with mean θT, it is straight forward to show that $E(K \mid t_2,\ldots,t_n) = \mu(2t_2 + \cdots + nt_n)$ (Hudson, 1990; Li and Fu, 1999).

It follows simply that the expectation of K is

$$\mathrm{E}(K) = \mu\mathrm{E}\left(\sum_{k=2}^{n} kt_k\right) \tag{32}$$

$$= a_n\theta, \tag{33}$$

where

$$a_n = 1 + \frac{1}{2} + \cdots + \frac{1}{n-1} \tag{34}$$

The variance can be readily computed and shown to be (Watterson, 1975; Hudson, 1990)

$$\mathrm{Var}(K) = \mathrm{E}(K^2) - \mathrm{E}^2(K) = a_n\theta + b_n\theta^2 \tag{35}$$

where

$$b_n = 1 + \frac{1}{4} + \cdots + \frac{1}{(n-1)^2}. \tag{36}$$

3.3 Distance and the Mean Number of Nucleotide Differences between Two Sequences

A very useful summary statistic, in addition to the number of segregating sites, is the number of nucleotide differences between two sequences. Define Π as the mean number of nucleotide differences between two sequences and Π_{ij} as the number of nucleotide differences between sequences i and j. Then Π is defined as (Tajima, 1983),

$$\Pi = \frac{2}{n(n-1)} \sum_{i<j} \Pi_{ij} \tag{37}$$

One can alternately estimate Π by using

$$\Pi = \frac{n}{n-1} \sum_{i,j} \phi_i \phi_j \Pi_{ij} \tag{38}$$

where ϕ_i and ϕ_j are the frequencies of the ith and j^{th} alleles in the sample. The factor $n/(n-1)$ is a correction factor for the sampling bias.

The distance matrix for the sample of 5 haplotypes is shown in Table 1. Substituting $n = 5$ and the elements of the distance matrix into Π_{ij}, summing over all i,j when $i < j$ gives $\Pi = (.1)(30) = 3.0$.

Table 1 Pairwise distance matrix for 5 polymorphic sequences

	s1	s2	s4	s5	s7
s1	0				
s2	3	0			
s4	1	2	0		
s5	2	5	3	0	
s7	3	6	4	1	0

3.4 Classifying Frequency of Mutations by Category

Fu (1994b; 1995) showed that the frequencies of mutations in a genealogy can be partitioned into different categories. The genealogy of a sample of a sequences consists of $2(n\text{-}1)$ branches and each branch has at least one sequence in the sample as its descendant. Define the number of sequences in a sample that are descendants of a branch as the *size* of that branch. That is, a mutation that is inherited by i descendent sequences is said to be of *size i.* Just as there exist $2(n\text{-}1)$ branches of a coalescent tree, there exist n-1 different sizes of mutations for a tree. It is easy to see that a mutation of size 1 can only occur in an external branch, i.e., a branch that directly connects to an external node (sequence). For this reason, a mutation of size 1 is often referred to as an *external mutation* (Li, 1997, p. 244). Let ξ_i be the number of mutations of size i. Fu and Li (1993) showed that $E(\xi_1) = \theta = 2N\mu$, so that the expected number of external mutations does not depend on the sample size. Fu (1995) showed that

$$E(\xi_i) = \frac{1}{i}\theta. \tag{39}$$

The variance and covariance between ξ_i and ξ_j are also given by Fu (1995). Below we will find it useful to define the state vector ξ,

$$\xi = (\xi_1, \cdots, \xi_{n-1})^T \tag{40}$$

where T represents the transpose of the vector. This vector that we are considering is a primary source of information for a large class of θ estimation models. If we assume the infinite sites model and that an outgroup sequence is available, then we can infer ξ directly from the sample of sequences. Otherwise, ξ must be inferred using a genealogy obtained from a method of tree reconstruction. Figure 3 shows the reconstructed genealogy for the sample of five sequences we are analyzing. There exists a total of 7 mutations. Four of these mutations are of size 1 and three of these mutations are of size 2.

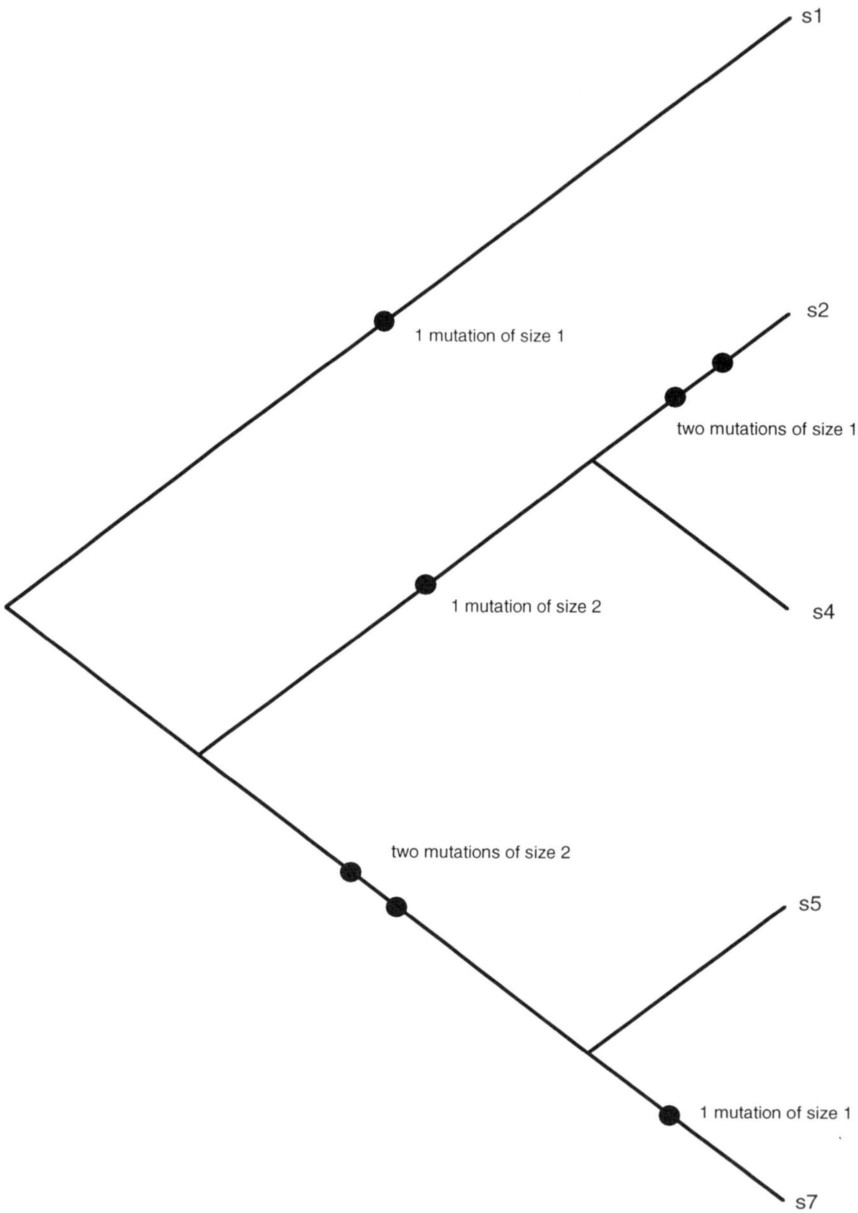

Figure 3. Reconstructed genealogy of five sequence example developed in text. There exist a total of seven mutations. Four of these mutations are of size one and three are of size two.

A second vector of information that will prove useful, and also determines a large class of θ estimation models is defined:

$$\eta = (\eta_1, \cdots, \eta_{[n/2]})^T \tag{41}$$

where $[n/2]$ denotes the largest integer contained in $n/2$, and η_i (i = 1, ..., $n/2$) is the number of mutations of type i. The ith element of η is defined as

$$\eta_i = \frac{\xi_i + \xi_{n-1}}{1 + \delta_{i,n-1}} \tag{42}$$

where $\delta_{i,n-1}$ is the Kronecker delta function:

$$\delta_{i,n-i} = \begin{cases} 1, & \text{if } i = n - i \\ 0, & \text{otherwise} \end{cases} \tag{43}$$

Under the infinite sites model, η_i is the number of segregating sites at which the frequencies of the two segregating nucleotides are i and n - i (i < n - i). This type of segregating site is called a *type i* or i-segregating site. As shown in Figure 3, this summary statistic can be computed directly from a sample without the help of an outgroup sequence. For the five sequences we see there are four mutations of type 1 and three mutations of type 2.

The expectation of η_i is

$$\mathrm{E}(\eta_i) = \begin{cases} \dfrac{\theta}{i} + \dfrac{\theta}{n-i}, & \text{if } i \neq n - i \\ \dfrac{\theta}{i}, & \text{if } i = n - i \end{cases} \tag{44}$$

Note that the estimate of K derived from the parsimony tree shown in Figure 3 is not equal to the actual amount of polymorphism in the n = 5 sample that, as was shown above, had six mutations. Therefore, it is possible when using genealogical reconstructions for inferring K, to under or overestimate the total number of mutations of a sample. Alternative methods of tree reconstruction give different estimates of this total number. For example, the UPGMA method for reconstructing phylogeny gives the correct value of K = 6. Bias or error in tree reconstruction is an important source of error. However, we argue below, that recent theoretical work in population genetics shows that estimation of the number of mutations on a given branch or size of a branch are the primary determinants of accurate ancestral parameter estimation. Hence, summary statistics based upon nucleotide level polymorphism, such as the size of branch, are critical information when using coalescent methods.

4. ESTIMATION OF POPULATION PARAMETERS

It is known from extensive simulation studies that the processes of population growth, natural selection, and geographic variation produce characteristic shapes of coalescent trees (Hudson, 1990; Fu, 1995; 1997; Simonsen *et al.*, 1995). Some of these patterns are shown in Figure 2a. In reality, of course, we have no knowledge of the complex stochastic processes that created a sample of sequences, and hence we must develop computational methods to *infer* time underlying parameters **p**. For example, we may want to estimate the mutation rate, growth rate, or selection coefficients. Let **p** be a vector of population parameters that we wish to estimate. For example, the vector **p** = $[r, \theta]$ might represent the parameters, population growth rate (r) and θ that we wish to simultaneously estimate using sequence data obtained from a population suspected of having experienced a history of population expansion. One of the major goals of statistically analyzing a stochastic process, such as the coalescent in a varying environment, is to take the resulting sequence data and reduce them to statistics and estimators for the underlying parameters of the process (**p**). The underlying machinery of parameter estimation throughout this chapter will lie in computing the branch lengths of a coalescent tree and different classes of mutations on the tree.

4.1 Concept of Inbreeding Effective Population Size

In this section we show how to estimate two fundamental parameters of theoretical and experimental population genetics: the effective population size and genetic diversity (θ). The *census size* of a population is the number of individuals assayed, as for example, in the estimation of an HIV patient's viral load. The *effective population size* (N) is the size of an ideal population that has the same amount of genetic randomness as the actual population. To understand why this is so, consider first the case of a deterministic population. In such a population, if we had an exact knowledge of the gene frequency, selection coefficients and number of individuals, we could specify with certainty, one specific value of the gene frequency. In a stochastic population we can only predict the *probability* that a specified value of the gene frequency is one of several values. We must assume that the population can be in many possible states. Mathematical and computational theory from population genetics allows predicting the probability that the population of alleles exists in a given state at a given time. The Wright-Fisher model is a genetic model in which each individual is considered to be a random sample of genes from the gene pool of the previous generation. It is a simple binomial model of the amount of genetic randomness in a population of alleles created due to sampling. Sampling error introduces noise into estimation and this noise is propagated through the population generation by generation. This form of noise is often called *genetic drift* in evolutionary theory. The concept of effective population size allows rigorous measurement of the effect of genetic drift in a population. To show why this is the case, consider the following simplified example.

Let p_r be the probability that two randomly chosen individuals come from the same parent (in the previous generation). Then we have,

$$p_r = \frac{1}{N}. \tag{45}$$

The effective population size can be obtained by inverting this probability,

$$N = \frac{1}{p_r}. \tag{46}$$

Although only two generations are needed to estimate effective population size it is often useful to define effective population size over several generations. Thus, one can define within a host population of HIV, say, a *short-term effective population size* over days or weeks. Or one can develop a *long-term effective population size* over months or years. For transmissions between individual hosts the time scale again could be varied according to the frequency of transmissions. The advantages of these applications is that the one might expect the short-term effective population size to closely track the actual population dynamics or at least fluctuations in viral load, while the long-term definition is more useful in gaining an understanding of the dynamics of genetic diversity. For example, in averaging over many generations, one can show that a small population at some point in the evolution of the virus can have a large influence in determining the outcome of an evolutionary event.

4.1.1 The Wright-Fisher Model and Effective Population Size

In this section we develop a slightly more mathematical basis of effective population size concept. Assume a haploid population of size *N*. We want to describe a population in terms of the variation in the number of descendant sequences contributed by a parental sequence to the next generation. We can consider any member of the population to represent the parental sequence. Since it is assumed that all sequences are neutral, we will call the parental sequence *A* and all the other (N-1) sequences *a*. Then, the probability that the *A* sequence gives rise to *j* offspring is equal to the probability that a parental population with frequencies $1/N$ of *A* sequences and $1-1/N$ of *a* sequences gives rise to an offspring population with *j A* sequences. The Wright-Fisher model computes this assuming it to be a binomial probability:

$$P_{ij} = \binom{N}{j}\left(\frac{1}{N}\right)^{j}\left(1-\frac{1}{N}\right)^{n-j} \tag{47}$$

The generalization to the case of *i* parental sequences is immediate. For this case, define the transition probability P_{ij} of a population with *i* parental sequences to an offspring population with *j* sequences at time $t + 1$ to be given by

$$P_{ij} = \binom{N}{j}\left(\frac{i}{N}\right)^{i}\left(1-\frac{i}{N}\right)^{n-j} \tag{48}$$

The Wright-Fisher model has been extensively studied by Ewens (1972; 1979) and Feller (1968) and these references serve as a useful starting point for understanding the population genetic basis of the coalescent approach. For our purposes we wish to note two important definitions that follow from this model. First, using standard mathematical methods in population genetics, one can compute three quantities for the transition matrix called its eigenvalues. One of these eigenvalues is equal to

$$\lambda_{\max} = 1 - \frac{1}{N} \tag{49}$$

This allows defining the population size *N* in terms of λ_{max},

$$N = \frac{1}{1-\lambda_{\max}} \tag{50}$$

and is called the *eigenvalue effective population size.* A second important definition follows from this model, if we ask: given that two genes are taken at random in generation $t + 1$, what is the probability that they have the same parental sequence? This turns out to be the same probability computed in the previous section: $p_r = 1/N$. And now we see, as in the ease of deriving the eigenvalue effective size, we can invert p_r to obtain what is called the *inbreeding effective population size.* This is the same definition of effective population we presented in the last section using an intuitive derivation.

The inbreeding effective size is the definition of effective population size used throughout this paper, as well as in much work in coalescent theory. The relationship of the inbreeding effective population size and the Wright-Fisher model to the approximation of the coalescent times of a genealogical tree is shown in equation (7) above. We thus see that the phylogenetic information contained in the tree can significantly contribute to the estimation of the effective population size from sequence data. Environmental factors that can dramatically affect the branch lengths of a coalescent tree such as selection and population growth will also affect estimation of effective population size.

4.1.2 Non-Phylogenetic Versus Phylogenetic Estimators

Recently several methods of estimating effective population size and genetic diversity have been developed (Watterson, 1975; Tajima, 1983; Fu, 1994a; 1994b; Kuhner *et al.*, 1995; Griffiths and Tavaré, 1994). In general, we can divide these methods into those that efficiently utilize the information contained in a genealogy and those that do not (Fu and Li, 1993; Felsenstein, 1992a). Also, we will focus on methods that use the major concepts of coalescent theory developed in the previous sections of this chapter, i.e., those methods that utilize summary statistics of a sample. These summary statistics include all of those covered thus far in this chapter: statistics of the tree branch lengths, segregating sites, and distance information of a sample. For alternative methods of effective population size estimation, based upon maximum likelihood approaches, see the chapter by Beerli and his colleagues.

4.2 Watterson's Estimator

Using the number of polymorphic sites (or segregating sites) in a sample computed using equation (33) Watterson (1975) derived the following estimate of genetic diversity in a sample, estimate of θ,

$$\theta_\omega = \frac{K}{a_n}. \tag{51}$$

For the set of seven sequences presented above, we have $K = 6$ haplotypes (unique sequences in the sample) so that $\theta_\omega = (6)/(2.083) = 2.88$. If it is assumed that there is no recombination the variance of θ_ω is given by

$$\mathrm{Var}(\theta_\omega) = \frac{\mathrm{Var}(K)}{a_n^2} \tag{52}$$

Thus, the variance can be derived using the estimate of the variance of K in equation (35). For the example we obtain a variance of θ_ω is equal to 4.58. Because this estimator does not efficiently use phylogenetic information, it has a high variance (Fu and Li, 1993). If we know the mutation rate, μ, N can also be estimated by

$$N = \frac{K}{2\mu a_n}. \tag{53}$$

Taking $\mu = .05$ per locus per generation as an estimate of mutation rate in HIV, then for the sample of 5 haplotypes of length 16 nucleotides, the estimate of effective population size is $N = (2.88)/(.1) = 28.8$.

4.3 Tajima's Estimator

Watterson (1975) showed using (37) that

$$\mathrm{E}(\Pi) = \theta_T = 2N\mu \tag{54}$$

so that by estimating the average number of nucleotide differences between two sequences in a sample we have also computed an estimate of θ. Thus, $\theta_T = \Pi = 3.0$. The variance of θ_T was derived by Tajima (1983) and is given by

$$\mathrm{Var}(\Pi) = \frac{n+1}{3(n-1)}\theta + \frac{2(n^2+n+3)}{9n(n-1)}\theta^2 \tag{55}$$

For the example we obtain a variance of θ_T equal to 4.80. The effective population size can be easily estimated using Tajimas's estimate of Π,

$$N = \frac{\Pi}{2\mu} \tag{56}$$

and gives a value of 30 that is close to the Watterson's estimate.

4.4 Fu's Estimators

Because of the low efficiency of the Watterson and Tajima estimators, and analytical work by Fu and Li (1993) showing that better estimators of θ exist if one utilizes the information given the total number of mutations K in a sample more efficiently, Fu (1994a; 1994b) developed a new set of θ estimators.

A fundamental property of the Watterson and Tajima estimators of θ is that they can be written in a form that shows that their coefficients are predetermined constants (Fu, 1994b). For example, the constants a_n and b_n in the Watterson estimate are fixed by the number of sequences in the sample n. Fu (1994b) demonstrated using elementary examples that, in general, such linear estimators do not perform as well as those in which the coefficients are functions of more detailed information in the coalescent process. Estimators that incorporate this information with alternative coefficients are more complex because of the way information in the genealogy, such as summary statistics based on the estimated number of mutations on each branch of a tree (**m**), or the size (ξ) and type of mutations (η) must be incorporated into the computation of the estimate. In this section, three of these kinds of estimators are examined. All of these methods of estimation of θ are based on recursive least-squares methods. In the following sections, we will show how all of these methods can be easily generalized to include factors that affect the coalescent tree structure such as population growth, selection, and mutation models.

4.4.1 The BLUE of θ

Fu and Li (1993) showed that better estimators of θ could be found. These depend upon how one partitioned the total number of mutations in a genealogy and then used this information to construct the estimator. Fu (1994a) partitioned the number of mutations according the branch on which the mutations occurred. Define a linear model of the observed number of mutations on each branch of a coalescent tree to be given by:

$$\mathbf{m} = \mathbf{X}\theta + \varepsilon \tag{57}$$

where $\mathbf{m} = (m_1,\ldots,m_{2(n-1)})^T$, with the m_i being the number of mutations on the ith branch of the tree [see equations (26) - (31)] and the vector $\mathbf{X} = (x_1, \ldots, x_{2(n-1)})^T$ with each component x_i being equal to the ith average branch length and for which the constant population case are given by equation (25). These vectors allow defining the vector error term $\varepsilon = \mathbf{m} - \mathbf{X}\theta$. The variance of the error term is given by

$$\mathrm{Var}(\varepsilon) = \mathbf{d}\theta + \beta\theta^2, \tag{58}$$

with

$$\mathbf{d} = \text{diag}(x_1, \ldots, x_{2(n-1)}), \tag{59}$$

$$\beta = \{\beta_{ij} : i, j = 1, \ldots, 2(n-1)\}. \tag{60}$$

From (57) it is straightforward to define a linear statistical model of the errors (residuals) using the observed number of mutations along all branches of the coalescent tree, equation (31):

$$\varepsilon^T \varepsilon = \sum_{i=1}^{2n-2} \left(m_i - \sum_{k=2}^{n} x_{ik}\theta \right)^2 = \sum_{i=1}^{2n-2} (m_i - \bar{l}_i)^2 \tag{61}$$

where

$$x_{ik} = \frac{s_{ik}}{k(k-1)} \tag{62}$$

$$x_i = \bar{l}_i = \sum_{k=2}^{n} x_{ik} \tag{63}$$

By minimizing the theoretically expected variation in the branch lengths of a coalescent tree with respect to the accumulation of the observed number of mutations along a branch of that tree, one obtains an estimator of θ upon solution of the nonlinear program: minimize $\varepsilon^T\varepsilon$ with respect to θ.

Fu's (1994a) method of obtaining a nonlinear minimum variance estimator (that he misleadingly calls a BLUE-Best Linear Unbiased Estimator) of θ for a known genealogy is in essence the solution of this nonlinear (quadratic in ε) program under the linear model (57). He developed a computational method for estimating the BLUE of θ for a given genealogy based upon the following,

Estimator 1 (Fu, 1994a): *Let m_i be the number of mutations occurring on branch i of a coalescent tree. Then it follows that,*

$$\text{E}(m_i) = x_i\theta, \tag{64}$$

$$\text{Var}(m_i) = x_i\theta + \beta_{ii}\theta^2, \tag{65}$$

$$\text{Cov}(m_i, m_j) = \beta_{ij}\theta^2, \tag{66}$$

where x_i, β_{ij} are constants that are determined by the topology of the genealogy for the sample. Both the x_i and β_{ij} constants can be computed using analytic solutions of the coalescent expectations of the branch lengths of a genealogy (25). Then an optimal nonlinear estimator of θ is given by the solution of the unconstrained quadratic program for the linear mutation model (57) in appropriate norm. A numerical solution to this quadratic program is given by the recursion:

$$\theta_{k+1} = \left[\frac{\mathbf{X}^T(\mathbf{d} + \theta_k\beta)^{-1}}{\mathbf{X}^T(\mathbf{d} + \theta_k\beta)^{-1}\mathbf{X}} \right] \mathbf{m}, \tag{67}$$

taking θ_0 as an arbitrary non-negative number.

Note that although these become constants once they are fixed by a gene tree, they explicitly incorporate information about the coalescent process. Both the x_i and β_{ij} constants can be instantly computed using the analytic forms of the expected branch lengths. By taking advantage of the theory of linear models and the recursive structure of the mean and variance of **m** (Searle, 1971), Fu (1994a) was able to develop an efficient algorithm to compute a nonlinear minimum variance estimator of θ. The "appropriate norm" referred to in Estimator 1 is derived in Vasco (submitted).

4.4.2 *The BLUEs for η and ξ*

Fu's (1994a) BLUE of θ shows that summary statistics of a genealogy can be very useful in obtaining fast and accurate estimators for population parameters. However, this estimator was based on using the summary statistics determined from the *average branch lengths* of a coalescent tree and the *estimated number of mutations from a reconstructed genealogy.* Two alternative BLUE estimators were also developed by Fu (1994b). These are based upon two other summary statistics we have mentioned above in this review: the size (η) and type (ξ) of branches of a coalescent tree.

Recall that a branch is of *size i* if exactly i sequences in the sample are descendants of the branch. The quantity ξ_i is the number of mutations of size i. Results from Fu (1994b; 1995) show that the total number of K mutations of a tree for sample of n sequences is equal to

$$\begin{aligned} \mathrm{E}(K \mid \xi_1,\ldots,\xi_{n-1}) &= \mathrm{E}(\xi_1)+\ldots+\mathrm{E}(\xi_{n-1}) \\ &= \theta(1+1/2+\ldots+1/(n-1)) \\ &= \mu(2t_2+\ldots+nt_n) \\ &= \mathrm{E}(K \mid t_2,\ldots,t_n) \end{aligned} \tag{68}$$

Hence, one can partition the number of mutations in a genealogy according to the number of mutations of a branch of length *i,* or the number of mutations of a branch of size *i,* or the number of segregating sites of type *i.*

Fu (1994b) showed that this interesting result implies the following general model of neutral sequence evolution. Consider a sample of n sequences. The time interval between the present, that is the time when the sample is collected, and the time to the MRCA, can be divided into a number of epochs in which population genetic events influencing the phylogenetic history of the sample have occurred (see Figure la). These population genetic events can be coalescences, recombinations, or migration between subpopulations. Assuming the neutral Wright-Fisher model, each change in any of the information states of a genealogy ($\mathbf{m}$, ξ, η) will always be determined by a population genetic event. Also, note that only coalescent and mutation events can occur in the neutral Wright-Fisher model, so that for the moment we ignore the effect of recombination and migration. From our arguments in Section 2 and Figure la, we can see that the time length of any branch in a tree must be of the form:

$$t_i+\cdots+t_j \tag{69}$$

where i is the number of the coalescent events after which the branch starts and j is the number of the coalescent events at which the branch ends. We can now compute all branches in a tree of size k using

$$l_k = s_{k1}t_1+\cdots+s_{k\rho}t_\rho \tag{70}$$

where s_{ki} is an index variable that represents the number of times that t_i appears in the time lengths of branches of size k, and ρ is the total number of events.

We see immediately that the average time length of all branches of size i is equal to

$$\bar{l}_i = \sum s_{ki}\bar{t}_{i.} \tag{71}$$

Hence all of the coalescent methods discussed thus far for computing estimates of the **m** vector can be used to compute estimates of the means of the ξ and η vectors.

In fact, all of the state changes of the coalescent can be expressed in a single linear model

$$y_i = x_i\theta + \varepsilon_i \tag{72}$$

where i represents the state of the coalescent process and $y_i = m_i, \eta_i, \xi_i$. Note that this linear model is a special case of the non-linear mutation model discussed above. For the case where the component of the information state of the genealogy is equal to $y_i = m_i$, and the coalescent expectation of the branch length is $x_i = \bar{l}_i$ we immediately obtain equation (57) of the BLUE estimator. Fu (1994b) also shows how to analytically estimate the variance-covariance matrices of ξ and η for the constant environment case. These results allowed him to use the correspondences between a given information state (y_i) of a genealogy and its coalescent expectation (x_i) to build a set of nonlinear minimum variance estimators. We now briefly discuss some of the properties of coalescents that allowed him to do this.

Because the number of segregating sites, K, is equal to $\xi_1 + \ldots + \xi_{n-1}$ and $\eta_1 + \ldots + \eta_{[n/2]}$, Fu (1994b) was able to show that Watterson's estimator can be expressed using the information state vectors η and ξ,

$$\theta_\omega = \sum_{i=1}^{n-1}\left(\frac{1}{a_n}\right)\xi_i = \sum_{i=1}^{[n/2]}\left(\frac{1}{a_n}\right)\eta_i \tag{73}$$

Similarly, because a mutation of size i can be counted in $i(n - i)$ pairwise comparisons he was also able to show that Tajima's estimator can be written in the form,

$$\theta_T = \sum_{i=1}^{n-1}\left(\frac{2i(n-i)}{n(n-1)}\right)\xi_i = \sum_{i=1}^{[n/2]}\left(\frac{2i(n-i)}{n(n-1)}\right)\eta_i \tag{74}$$

Again, it should be noted that the coefficients are predetermined coefficients and hence we do not expect them to exhibit the properties of an optimal estimator. However, using equation (72), it is obvious that least squares estimators can be derived for these mutation frequency classes:

$$\varepsilon_i = y_i - x_i\theta \tag{75}$$

Squaring both sides, adding over all states of the coalescent, and minimizing this function with respect to θ, allows computing estimators that minimize the distance between the expected number of mutations in a given state of the coalescent the ith size and type of a mutation. Hence, Fu (1994b) was able to use the same framework in Estimator 1 for the information state vectors, ξ and η, whose means are linear functions of θ and whose variances and covariances are quadratic functions of 9. This gives two more BLUE models for estimating θ. Because of the linear structure of these estimators in θ all three can be expressed in terms of a single unified framework:

Estimator 2 (Fu, 1994a; 1994b): *Let* **Y** *represent the information state* vector *of the coalescent in the variables* ***m***, *ξ, or η. Then it follows that,*

$$\mathrm{E}(y_i) = x_i\theta \tag{76}$$

$$\mathrm{Var}(y_i) = x_i\theta + \Sigma_{ii}\,\theta^2 \tag{77}$$

$$\mathrm{Cov}(y_i, y_j) = \Sigma_{ij}\,\theta^2 \tag{78}$$

where x_i, Σ_{ij} are constants that are determined by the topology of the genealogy for the sample and the variance-covariance matrix for the given information state vector. An optimal estimator of θ is given by the solution of the unconstrained quadratic program for the linear mutation model (72) in appropriate norm. A numerical solution to this quadratic program is given by the recursion:

$$\theta_{k+1} = \left[\frac{\mathbf{X}^T(\mathbf{d}+\theta_k\,\Sigma)^{-1}}{\mathbf{X}^T(\mathbf{d}+\theta_k\,\Sigma)^{-1}\mathbf{X}}\right]\mathbf{Y}, \tag{79}$$

taking θ_0 as an arbitrary non-negative number.

4.4.3 Why the BLUEs Work so Well

The BLUEs produce nearly unbiased estimates of θ with nearly minimum possible variance: why? Because the estimation depends upon determining the number of mutations in state i of the coalescent. This state of the coalescent process can be characterized in terms of the mutations on a branch (m_i), the size of a branch (ξ_i) or the type of a branch (η_i). The topological structure of the genealogy in and of itself has nothing to do with the estimation of θ. The primary effect of tree reconstruction is the estimation of the number of mutations in the ith state of a coalescent process. Thus, summary statistics of the coalescent process describing the ith state change (m_i, ξ_i, η_i) of the coalescent tree, form the primary sources of information required to reconstruct the number of segregating sites on each branch. While errors in tree reconstruction can occur, they do not significantly affect the estimates obtained. This is probably why extensive simulation studies (Fu, 1994a; 1994b; Deng and Fu, 1996) show that using a UPGMA tree does not appear to greatly affect the quality of estimation of θ.

Felsenstein (1992a) conjectured that, for the constant population case, the topology of a coalescent tree essentially corresponds to a random topology, and, further, that the stochastic (coalescent time generating) process creating the lengths of the tree should be relatively independent of the topology. So, if his conjecture is correct, we do not necessarily expect the topology in itself to greatly affect estimation. However, this conjecture has yet to be explored.

4.4.4. Application to Nucleotide Sequence Data

As an example, we consider a sample of HIV sequences taken from Holmes and his coworkers (1992) shown in an Appendix to this chapter. To use the BLUE to estimate θ for this sample one needs to compute the observed number of mutations separating each pair of sequences. These can be estimated using Fitch parsimony. For all computations, except those requiring Fu's BLUE estimation program, the test version PAUP*4.0bl (PPC) was used on a Macintosh PowerPC. A Fitch parsimony tree for this set of sequences (one of seventeen such trees) is shown in Figure 4. Using the "Describe Trees" option in the PAUP "Trees" menu, a list of branch lengths and linkages for the tree was computed. These data are taken and used to construct a text file consisting of three columns (see Table 2). Column one represents the first set of nodes shown in Figure 5. Column two represents the second set

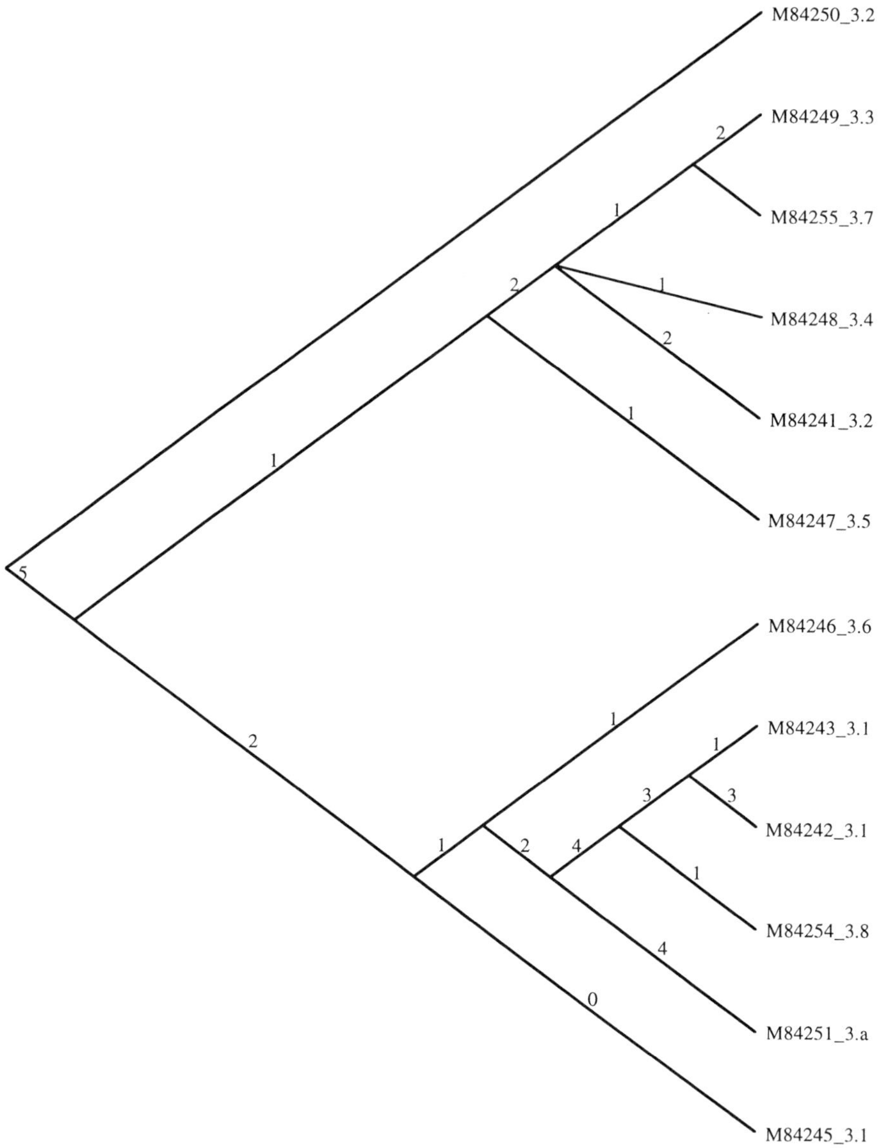

Figure 4 Reconstructed genealogy of sampled sequences shown in the Appendix to this chapter. Maximum parsimony was used to obtain the tree. Numbers on each branch represent the estimated number of mutations on that branch.

of nodes labeled as "Connected to node." Column three represents the data labeled "Assigned branch length." The data shown in Table 2 represent the input file to the UPBLUE estimation program and for which it computes a distance matrix. The output for this is shown in Table 3. Using the distance matrix an UPGMA tree is constructed. The UPGMA tree is then used to estimate the observed number of mutations along each ith branch. These data are stored in the vector **m** and are input into the recursion equation (67) that is then used to estimate θ. Since there is a slight downward bias created by tree reconstruction in the estimate of θ, a bias correction must be applied using the following regression equation:

$$\hat{\theta} = \left(0.0335\sqrt{n-2} + 0.997\sqrt{\hat{\theta}_U}\right)^2 \tag{80}$$

where $\hat{\theta}_U$ is the UPBLUE estimate of θ, based on the genealogy reconstructed by UPGMA. Fu (1994a) called this estimator UPBLUE and found that UPBLUE is nearly unbiased and has a variance close to the theoretical minimum variance V_{min} (that is, the smallest possible variance assuming that the coalescent tree is known exactly). Estimates of θ and their variances for the Holmes et al. (1992) data using the Watterson, Tajima, and BLUE estimators are shown in Table 4. It should be noted that there appears to be a hierarchical pattern in the means and variances of these estimators. As we discuss below, this is a signature that may imply rapid population expansion in the effective population size (Vasco and Fu, submitted). Recall that all of the estimators discussed thus far assume the existence of a constant effective population size. In several cases where we suspect the effective size is

Table 2 Raw phylogenetic data that are input into the UPBLUE and EVE programs.

node[a]	connected nodes[b]	assigned branch length
1	21	5
15	21	1
14	15	2
13	14	1
2	13	2
5	13	0
3	14	1
4	14	2
6	15	1
20	21	2
19	20	1
7	19	1
18	19	2
17	18	4
16	17	3
9	16	1
10	16	3
12	17	1
11	18	4
8	20	0

[a] the tree nodes shown in Figure 6 for the Holmes et al. (1992) data
[b] set of tree nodes labeled as "Connected to node" in Figure 5

constant, we have found that the estimates of θ computed using these three estimators are nearly the same (i.e. that the estimators are nearly unbiased) and the BLUE always has the smallest variance.

Many assumptions were made by Fu (1994a) in the derivation of the BLUE of θ; constant population size, the genealogy of the sample is assumed to be correctly reconstructed, and bias in estimation due to tree reconstruction can be corrected. Intensive simulation studies show that the estimator and its variance is nearly unbiased and very robust under changing assumptions such as introducing heterogeneity of mutation rate across sites (Deng and Fu, 1996). Moreover, this estimator appears to give nearly the theoretical minimum variance of any possible θ estimator. In practical applications to several HIV sequence data sets, we have found that the maximum likelihood estimator of Kuhner, Yamato and Felsenstein (1995) gives nearly the same estimate as Fu's (1994a) distance based BLUE estimator. However, it is likely that differences in properties of these two estimators will be seen when very large sample sizes and data from multiple independent loci are included in estimation. Finally, as we shall show in the next few sections, the

```
Tree number 1 (rooted using default outgroup)

Branch lengths and linkages for tree #1 (unrooted)

       Assigned Minimum Maximum
    Connected branch possible possible
 Node    to node   length length   length
-------------------------------------------------------------
M84250_3.2 (1)    21    5  5    5
           15     21    1  1    1
           14     15    2  2    2
           13     14    1  1    1
M84249_3.3 (2)    13    2  2    2
M84255_3.7 (5)    13    0  0    0
M84248_3.4 (3)    14    1  1    1
M84241_3.2 (4)    14    2  2    2
M84247_3.5 (6)    15    1  1    1
           20     21    2  1    2
           19     20    1  1    1
M84246_3.6 (7)    19    1  1    2
           18     19    2  1    2
           17     18    4  3    4
           16     17    3  3    3
M84243_3.1 (9)    16    1  1    2
M84242_3.1 (10)   16    3  2    3
M84254_3.8 (12)   17    1  1    2
M84251_3.a (11)   18    4  4    4
M84245_3.1 (8)    20    0  0    1

Tree length = 37
Consistency index (CI) = 0.6486
Homoplasy index (HI) = 0.3514
CI excluding uninformative characters = 0.4800
HI excluding uninformative characters = 0.5200
Retention index (RI) = 0.6286
Rescaled consistency index (RC) = 0.4077
```

Figure 5 Branch lengths and nodes of the reconstructed genealogy shown in Figure 4. These were obtained using the "Describe Trees" option in the Paup "Trees" menu.

straight forward application of Least Squares concepts allows computationally efficient use of alternative coalescent models of population growth, selection, geographic structure, and mutation for the estimation of population parameters. This opens up a rich space of parameters available for estimation using summary statistics of large-scale DNA sequence data sets.

4.5 EVE Estimators

In this section, we discuss the development of estimators of population parameters that do not assume the coalescent evolves in a constant environment. Population growth patterns such as exponential or logistic growth can greatly affect the shape of the coalescent tree (Figure 2a). For the constant population case there exist analytical formulas for the mean and variance of the coalescent time distribution (8) that allow determining the mean, variance and covariances of the branch lengths. We saw in the last section that these analytic expressions can be used to derive Fu's BLUE estimators. In Section 2.2 we saw that such expressions do not exist for the case of the coalescent in a variable environment. However, by estimating the coalescent times of a gene tree using summary statistics over a set of Monte Carlo generated genealogies, a set of branch lengths can be inferred if the topology of a tree can be reconstructed. These can then be used to approximate the coalescent expectations. Once this is accomplished, we show that it is possible to develop efficient Least Squares *Estimators for Variable Environments* (EVE estimators, Vasco and Fu, submitted). A set of programs that use genealogical summary statistics to estimate population parameters, based upon the phylogenetic estimators described in this article, appears on the world wide web as a free package called EVE (*http://bioag.byu.edu/zoology/crandall_lab/Vasco/eve.htm*; Vasco, 1999).

In this section, we first develop three estimators of the coalescent and show how each can be used to approximate the shape of a tree when the coalescent evolves in a varying environment. Secondly, utilizing the information on tree shape we show how EVE estimators can be derived for a variety of alternative coalescent models. However, for simplicity we use only two coalescent models. One representing exponential population growth and one representing logistic population size change. We show elsewhere that this Least Squares estimation

Table 3 Pairwise distance marix computed from data shown in Table 2 and Figure 4.

```
 2   --The input type, internal use for upblue
12   -- # of alleles. Occurrences of alleles are ( change if not true)
 1    1    1    1    1    1    1    1    1    1    1    1    1
 1    0
 2    6    0
 3    4    4    0
 4    5    5    3    0
 5    4    2    2    3    0
 6    2    6    4    5    4    0
 7    4   10    8    9    8    6    0
 8    2    8    6    7    6    4    2    0
 9   13   19   17   18   17   15   11   11    0
10   15   21   19   20   19    1   13   13    4    0
11    9   15   13   14   13   11    7    7   12   14    0
12   10   16   14   15   14   12    8    8    5    7    9    0
```

Table 4 Output of UPBLUE for Holmes *et al.* (1992) data. Raw input files are shown in Tables 2 and 3.

```
There are    12 alleles and    12 sequences in your data
 Starting the iterations of BLUE ...

      cycle=  1 Current estimate of theta =     17.7640
      cycle=  2 Current estimate of theta =     18.4030
      cycle=  3 Current estimate of theta =     18.4584
      cycle=  4 Current estimate of theta =     18.4630
      cycle=  5 Current estimate of theta =     18.4634

 Estimate of theta before bias correction =     18.4634


EVE estimate of theta    =  53.73
UPBLUE estimate of theta  =   19.309, Variance=  40.504
Sequence  diversity (pi)  =    9.606, Variance = 28.52
Watterson's estimate      =   12.047, Variance = 28.81

See the discussion of Table 3 (Fu, 1994a)
for a comparison of the variances of various estimators.


EVE estimate of population growth rate r = -16.00
```

method is general enough to apply to many different models of the coalescent in variable environments, including those describing balancing selection, background selection, and population genetic events such as the estimation of recombination and migration rates.

4.5.1 *Least Squares Parameter Estimation for a Nonlinear Mutation Model*

We now show how to construct an EVE estimator that simultaneously estimates any model parameters. In this chapter we use the method of Vasco and Fu (submitted) that is based on estimating the **m** information state vector of a genealogy. Else where, results from EVE estimators using estimated ξ and η state vectors will be presented. However, for completeness, at the end of this section we state the general estimation problem as we did for the case of the BLUE estimators. Assume the nonlinear mutation model,

$$\mathbf{m} = \mathbf{X}(\mathbf{p})\theta + \varepsilon(\mathbf{p}) \tag{81}$$

where $\mathbf{m} = (m_1,\ldots,m_{2(n-1)})^{\mathrm{T}}$ and $\mathbf{X}(\mathbf{p}) = (x_1(\mathbf{p}),\ldots,x_{2(n-1)}(\mathbf{p}))^{\mathrm{T}}$ and the error term is $\varepsilon = \mathbf{m} - \mathbf{X}(\mathbf{p})\theta$. The variance of the error term is given by

$$\mathrm{Var}(\varepsilon) = \mathbf{d}(\mathbf{p})\theta + \beta(\mathbf{p})\theta^2 \tag{82}$$

with

$$\mathbf{d}(\mathbf{p}) = \mathrm{diag}(x_1(\mathbf{p}),\ldots,x_{2(n-1)}(\mathbf{p})) \tag{83}$$

$$\beta(\mathbf{p}) = \{\beta_{ij}(\mathbf{p}) : i,j = 1,\ldots,2(n-1)\}. \tag{84}$$

Using the nonlinear statistical model of the errors for the observed number of mutations in a coalescent tree given by equation (81), define the scalar function:

$$V(\mathbf{p}) = \sum_{i=1}^{2n\text{-}2}\left(m_i - \sum_{k=2}^{n} x_{ik}\theta\right)^2 = \sum_{i=1}^{2n-2}\left(m_i - \bar{l}_i(\mathbf{p})\right)^2 \tag{85}$$

where

$$x_{ik} = s_{ik}\bar{t}_k(\mathbf{p}) \tag{86}$$

$$x_i(\mathbf{p}) = \bar{l}_i(\mathbf{p}) = \sum_{k=2}^{n} x_{ik} \tag{87}$$

$$V(\mathbf{p}) = \varepsilon^T \varepsilon. \tag{88}$$

We then have

Estimator 3 (Vasco and Fu, submitted): *Assume a coalescent model with parameters* **p** *and density function* $v(t)$. *For the model let* m_i *be the number of mutations occurring on branch i of a coalescent tree. Then it follows that,*

$$\mathrm{E}(m_i) = x_i(\mathbf{p})\theta, \tag{89}$$

$$\mathrm{Var}(m_i) = x_i(\mathbf{p})\theta + \beta_{ii}(\mathbf{p})\theta^2 \tag{90}$$

where $x_i(\mathbf{p})$, $\beta_{ii}(\mathbf{p})$ *are fixed functions that are determined by coalescent simulation of the genealogy of the sample. Minimize the theoretically expected variation in the branch lengths of a coalescent tree with respect to the accumulation of the observed number of mutations along all branches of the gene tree. There exists a set of estimators for* **p** *upon solution of the nonlinear program with respect to the nonlinear mutation model (81):*

minimize $V(\mathbf{p})$ *with respect to* **p**.

In the next two sections we show how to solve this nonlinear program for the estimators **p** using a combination of coalescent and standard computational methods. The first stage involves computing the expectations of the branch lengths of the coalescent tree. We then use these expectations to construct the $V(\mathbf{p})$ function and discuss pseudocode for implementation of a nested grid search for finding the minimum of this function for all **p**. Further details describing the full computational implementation of this statistical theory may be found in the papers by Vasco and Fu (submitted). Programs that compute estimates for the models described in this chapter may be accessed from links on the home web pages of the Fu and Crandall labs (*http://hgc.sph.uth.tmc.edu/fu* and *http://bioag.byu.edu./zoology/crandall_lab/Vasco/eve.htm*, respectively).

4.5.2 Computation of the Expected Branch Lengths.

We now describe a very fast method to approximate the expected branch lengths of a coalescent tree for the models described in the last three sections. The method is quite general and can be readily adapted for other models. Several patterns of shapes for coalescent trees are shown in Figure 2a. In every case the topology of the tree is fixed for the five sequences but the branch lengths have been dramatically altered. For example, in the case of balancing selection one expects to see a tree with the branches near the root elongated.

To compute the expected branch lengths of a coalescent tree use the following algorithm:

3) Construct the topology of a random gene tree for the sample using Hudson's (1990) *maketree* subroutine outlined above.
3) Simulate the distribution of coalescent times under a given model with parameters **p** and function $v(t)$. Compute the random value of $t_k(\mathbf{p})$ conditional on s_{k+1} in equation (13) by iteratively solving the integral,

$$\int_{s_k}^{s_k+t_k(\mathbf{p})} v(s)ds = \log(U) \tag{91}$$

where U is a random variable chosen from the uniform distribution over (0, 1). To do this we start with a sample of size n sequentially by solving the integral for $t_k(\mathbf{p})$ first, then $t_{n-1}(\mathbf{p})$ and so on until one obtains $t_2(\mathbf{p})$. Repeat this computation for G genealogies. This gives a set of coalescent times for each genealogy corresponding to each time interval in a genealogy. As an example, for the first coalescent time, $t_2(\mathbf{p})$, one has G values, labeled $t_2^1(\mathbf{p}),\ldots,t_2^G(\mathbf{p})$. Store all of these randomly generated coalescent times in a set of G vectors:

$$\begin{pmatrix} t_2^1(\mathbf{p}) \\ t_3^1(\mathbf{p}) \\ \vdots \\ t_k^1(\mathbf{p}) \end{pmatrix}, \begin{pmatrix} t_2^2(\mathbf{p}) \\ t_3^2(\mathbf{p}) \\ \vdots \\ t_k^2(\mathbf{p}) \end{pmatrix}, \ldots, \begin{pmatrix} t_2^G(\mathbf{p}) \\ t_3^G(\mathbf{p}) \\ \vdots \\ t_k^G(\mathbf{p}) \end{pmatrix}$$

3) From the set of vectors at Step 2, compute the average coalescent times over all of the G genealogies, $\bar{t}_k(\mathbf{p}) = \frac{1}{G}\sum t_k^j(\mathbf{p})$. Store this set of average coalescent times in a vector of dimension n-1, $\begin{pmatrix} \bar{t}_2(\mathbf{p}) \\ \bar{t}_3(\mathbf{p}) \\ \vdots \\ \bar{t}_n(\mathbf{p}) \end{pmatrix}$.
3) Sum over the average coalescent time vector using the indexing matrix $[s_{ik}]$ to obtain an estimate of each ith average branch length of the tree, $\bar{l}_i = \sum s_{ik}\bar{t}_k(\mathbf{p})$.

Store these in a vector of dimension $2n$–2, $\begin{pmatrix} \bar{l}_1(\mathbf{p}) \\ \bar{l}_2(\mathbf{p}) \\ \vdots \\ \bar{l}_{2n-2}(\mathbf{p}) \end{pmatrix}$

4.5.3 A Grid Search Method of Solving the Nonlinear Program.

Because computational methods of solving for the vector of coalescent times are so fast, it is practical to solve the nonlinear programming problem, at least for the case of estimating two simultaneous parameters, using a brute force grid search. In nonlinear programming problems, it is sometimes useful to consider a variety of $V(\mathbf{p})$ state functions. For example, instead of using $\varepsilon^T\varepsilon$, one could use a "chi-square" type function,

$$V(\mathbf{p}) = \sum_{i=1}^{2n-2} \frac{(m_i - \bar{l}_i(\mathbf{p})\theta)^2}{\bar{l}_i(\mathbf{p})\theta} \tag{92}$$

Once a state function is chosen, it must be minimized with respect to $\mathbf{p}$. Since the actual coalescent tree is a function of the unknown parameters $\mathbf{p}$, in principal, it should be computed for *every possible value of* $\mathbf{p}$. Since this is not possible, a search routine must implemented over as large a portion of the parameter space as possible to find the global minimum. We have found that a fast method of doing this for two parameters is the nested grid search shown in Figure 6. Since, in coalescent theory, a parameter of major interest is θ, we designate this and a free parameter p as the two estimators being targeted in the search for the global minimum of the V function. The search begins at a starting value p^0, where the superscript represents the iterate of the search step, and each iterate is incremented by a preset amount Δp. Each line in Figure 6 represents a candidate coalescent tree (with branch lengths

Candidate p	Set of Average Coalescent Times for Candidate p		Set of Average Branch Lengths for Candidate p
p^0	$\{\bar{t}_2(p^0), \bar{t}_3(p^0), \ldots, \bar{t}_n(p^0)\}$	$\xrightarrow{\bar{l}_i = \sum_{k=2}^{n} s_{ik}\bar{t}_k}$	$\{\bar{l}_1(p^0), \bar{l}_2(p^0), \ldots, \bar{l}_{2n-2}(p^0)\}$
p^1	$\{\bar{t}_2(p^1), \bar{t}_3(p^1), \ldots, \bar{t}_n(p^1)\}$	$\xrightarrow{\bar{l}_i = \sum_{k=2}^{n} s_{ik}\bar{t}_k}$	$\{\bar{l}_1(p^1), \bar{l}_2(p^1), \ldots, \bar{l}_{2n-2}(p^1)\}$
	...	...	...
p^x	$\{\bar{t}_2(p^x), \bar{t}_3(p^x), \ldots, \bar{t}_n(p^x)\}$	$\xrightarrow{\bar{l}_i = \sum_{k=2}^{n} s_{ik}\bar{t}_k}$	$\{\bar{l}_1(p^x), \bar{l}_2(p^x), \ldots, \bar{l}_{2n-2}(p^x)\}$

Figure 6. Fast coalescent simulation method for estimating the known coalescent tree. Each line starts with a candidate parameter (p), the value of that is used to simulate the coalescent. For each p one then obtains a set of average of coalescent times (left of arrows) that Vasco and Fu showed can be used to compute a set of average branch lengths for the candidate p (right of arrows). The arrows themselves represent a transformation of the coalescent time distributions into a set of expected branch lengths. Vasco and Fu were able to show that this transformation exists because the jump chain of the coalescent is fixed under the assumption that the true topology of a coalescent tree can be approximated.

$l_i(p^i)$) that can be used to compute a nonlinear Least Squares fit to known (or estimated) branch lengths $l_i(p)$. For each candidate p a set of average branch lengths is computed using the tree shape estimation algorithm described in the previous section. Thus, we obtain a set of branch lengths over a grid of length x: $l_i(p^0)$, $l_i(p^1)$, ..., $l_i(p^x)$. This gives an estimate of the range in shape of the coalescent topology over the parameter space, $p^0,\dots, p^x$. Moreover, since estimation of θ does not play a role in the integration of the coalescent times, which involves only p, this set of branch lengths can be stored in a vector until the next step of the search routine is implemented. Once computed, this set of functions, the estimated **m** vector, and a range of θ values suffice to compute any value of the V function.

4.5.4 *An Optimal Estimator of* θ.

Before we describe how to apply EVE estimators to nucleotide sequence data we discuss how to use Estimator 3 to derive a minimum variance estimator of θ. Proof of the optimality properties of this phylogenetic estimator is given in Vasco (submitted). It is demonstrated there that the method gives rise to the most efficient estimator of θ for a given model, assuming that all other **p** are known for that model.

We now briefly discuss the major points required to compute the estimate. Generalizing (70) to the variable environment case it follows that the average length of all branches of size i is equal to $\bar{l}_i = \sum s_{ki}\bar{t}_i(\mathbf{p})$. Substituting $\bar{t}_i(\mathbf{p})$ into equation (68) allows computing the total expected number of mutations for a sample of size n. This result can be used to construct generalized Watterson and Tajima estimators for the coalescent in a varying environment (Vasco, Crandall and Fu, submitted). However, since we wish to utilize as much information in the branch lengths of a genealogy as possible, we take advantage of another result that follows from equation (68): that on average $\theta x_i(\mathbf{p})$ mutations occur during the coalescent time of length $\bar{t}_i(\mathbf{p})$, where $x_i(\mathbf{p})$ represents average branch lengths of size *or* type i. Or alternatively, on average $\theta x_i(\mathbf{p})$ mutations occur on branches of size i in the coalescent tree. This simple result allows all of the state changes of the coalescent to be expressed in a single nonlinear mutation model

$$y_i = x_i(\mathbf{p})\theta + \varepsilon_i \tag{93}$$

where

$$x_i(\mathbf{p}) = \begin{cases} \bar{l}_i(\mathbf{p}), & \text{if } y_i = m_i \\ \bar{l}_i(\mathbf{p}) \text{ of size } i, & \text{if } y_i = \xi_i \\ \dfrac{\bar{l}_i(\mathbf{p}) + \bar{l}_{n-1}(\mathbf{p})}{1+\delta_{i,n-1}} \text{ of size } i, & \text{if } y_i = \eta_i \end{cases} \tag{94}$$

and quantities "of size i" always represent average branch lengths of size i in (94). Using vector notation (93) can be expressed as the system

$$\mathbf{Y} = \mathbf{X}(\mathbf{p})\theta + \varepsilon(\mathbf{p}) \tag{95}$$

Fu (1994b, 1995) derives analytic forms for the variance-covariance matrix of ξ and η for the constant environment case. Generalization of the variance-covariance matrix for (93) to the variable environment case is straightforward, using the computational methods developed in Vasco and Fu (submitted) and discussed extensively in this review. The closed-form expressions take the general form of equations (30). Consider representing the variance-covariance matrix for the generalized case by $\Sigma(\mathbf{p})$, with elements $\Sigma_{ij}(\mathbf{p})$. Assume now that we have used (95) to compute a vector of EVE estimates using Estimator 3. Call them $\mathbf{p}^*$. Then, we have,

Estimator 4 (Vasco, submitted): Assume *a coalescent model with parameters* $\mathbf{p}^*$ *and density function* $v(t)$*. For the model let* y_i *be a component of the information state vectors* $\boldsymbol{m}$*,* ξ *or* η *for state i of a coalescent tree and* $x_i(\mathbf{p}^*)$ *the corresponding coalescent expectation as defined by (94). Then it follows that,*

$$E(y_i) = x_i(\mathbf{p}^*)\theta, \tag{96}$$

$$\mathrm{Var}(y_i) = x_i(\mathbf{p}^*)\theta + \Sigma_{ii}(\mathbf{p}^*)\theta^2, \tag{97}$$

where $x_i(\mathbf{p}^*)$*,* $\Sigma_{ii}(\mathbf{p}^*)$ *are fixed constants (for given* $\mathbf{p}^*$*) that are determined by coalescent simulation of the genealogy of the sample. Using the regression equations (linear in* θ*) define the vector error function,* $\varepsilon = \mathbf{Y} - \mathbf{X}(\mathbf{p}^*)\theta$ *and minimize the sum of the LS difference determined from the theoretical coalescent expectation corresponding to the* i^{th} *information state (branch length or size or type of a branch) of a coalescent tree with respect to the accumulation of the observed number of mutations in that state. There exists an optimal estimator of* θ *upon solution of the quadratic program for the linear model, assuming all other* $\mathbf{p}$ *are known in (94), in appropriate norm:*

minimize $V(\theta) = \varepsilon^T\varepsilon$ *with respect to* θ*.*

4.5.5 Application to Nucleotide Sequence Data

Application of EVE to sequence data can be accomplished by using the same methods used for inputting of data into UPBLUE. Thus, in computing simultaneous independent estimates of θ and exponential growth rate *(r)* for the Holmes *et al.* (1992) example, the input to EVE is the same as shown in Table 2. EVE also automatically computes the same distance matrix shown in Table 3 to construct a UPGMA tree. This is used to estimate the **m** vector. Hence, the same preparation (using PAUP or some other tree reconstruction package) that is used to create the input file to UPBLUE, allows the investigator to immediately use the EVE programs. Instead of using a recursion equation to compute the parameter estimates, however, a brute force grid search of the type we have discussed in this section is used. The EVE parameter estimates for the Holmes *et al.* (1992) data are shown in Table 4 using an exponential growth model. We see that the EVE estimate of θ = 53.73 is higher than any of the other theta estimators. The r = -16.0, shows that a rapid population expansion of the HIV is inferred for this patient (recall that with the coalescent we are looking backwards in time so that a negative r is positive in forward time). An optimal estimate of θ can be found using Estimator 4.

4.6 Relationship of Summary Statistics Estimators to Phylogenetic Estimators

In recent work (Vasco, Crandall and Fu, submitted; Vasco and Fu, submitted), we have developed a set of generalized Watterson and Tajima estimators that allow simultaneously estimating **p**. These estimators use precisely the same information encoded in the genealogy that the standard Watterson and Tajima estimators use: K, the total number of segregating sites in the sample and π the average overall pairwise differences between sequences, respectively. However, using theory developed in our recent work it has become clear that the coalescence process (or its representation as a phylogenetic tree), when characterized by the information state triplet $(\mathbf{m}, \xi, \eta)$, follows a simple Least Squares principle that can be used to develop powerful statistical methodologies. Define the state function $\Phi(\mathbf{p}, \Xi)$ where Ξ represents the information measure of the genealogy and can be either scalar or vector depending upon whether a summary statistic or a phylogenetic measure is used. If a summary statistic is used to measure the information, then Ξ is scalar and is either K or π. The LS function for the coalescence process then takes the form:

$$\Phi = (1/2)(\Xi - x(\mathbf{p})\theta)^2 \tag{98}$$

where $x(\mathbf{p})$ is the theoretical coalescent expectation and depends upon the summary statistic used. If the total number of segregating sites is used, then the theoretical expectation is simply the expected total length of the coalescent tree. If the average pairwise difference in the sample is used, then $x(\mathbf{p})$ is determined by a slightly more complex theoretical expectation: the sum over all expected branches of size i, weighted by the number of descendents of the branch and the total number of sequences in the sample. By minimizing Φ using the computational methods discussed in this paper, summary statistic estimators for **p** can be developed. If phylogenetic information is used, then Ξ is the vector $\boldsymbol{Y}$ in (95) and is either $\mathbf{m}$, ξ, or η. One then has the LS function described in Estimator 3.

Using Φ it is also possible to immediately derive simple closed-form estimators of θ for more complex population models. For this class of models we are minimizing the sum of the Least Squares error between the observed information state of the tree and the theoretical expectation from the coalescent theory:

$$\frac{d\Phi}{d\theta}\Big|_{\mathbf{p}=\mathbf{p}^*, \theta=\hat{\theta}} = 0 \tag{99}$$

Computational methods can then be used to determine properties of bias and consistency of estimation. Our recent work suggests that when all **p** are known for a given population model, except θ, then it is possible to obtain nearly unbiased estimation of θ for the model. A possible reason for this is that the effect of θ on generating the number of segregating sites on branches of a genealogy during coalescence is always *linear* regardless of whether the stochastic mutation process itself is driven at a linear rate (constant population size) or a nonlinear rate (expanding populations or variable changes in size). These results suggest that Estimator 4 may be useful for determining the optimal efficiency of summary statistics versus phylogenetic estimators for more complex population models, e.g., providing minimum variance estimation of θ.

5. IMPORTANCE OF SUMMARY STATISTICS IN ESTIMATING POPULATION PARAMETERS

HIV population change is an important phenomenon and occurs across a wide spectrum of host populations. HIV population change leaves a signature on the genetic structure of host viral populations by "freezing in" patterns of nucleotide polymorphism within its genealogy. With the advent of new sequencing technologies it is now possible to obtain large samples of nucleotide sequences from host viral populations. This allows reconstructing large population genealogies. To efficiently assess patterns of nucleotide polymorphism in populations with varying sizes it is important to have fast and accurate estimators of θ.

5.1 Past Work

The roots of coalescent theory can be found in work as early as the 1930's (Wright, 1931; Fisher, 1930) and 1940's (Malecot, 1941). However, it took the imagination and persistence of Kimura (1983). Ewens (1972; 1979), Watterson (1975) and others to propose the importance of neutral evolution of nucleotide sequences and develop the theory required to give birth to the coalescent theory (Kingman, 1982; Hudson, 1990). The first developments of neutrality theory and coalescents that showed the power of summary statistics methods appear in work by Watterson (1975), Hudson (1990; 1993) and colleagues (Kaplan *et al.* 1989), Strobeck (1987), Tajima (1983; 1989a; 1989b; 1989c; 1993; 1997), Griffiths (1989), Tavaré (1984) and Takahata (1991). Some of these works pointed to new methods utilizing summary statistics for hypothesis testing of neutrality, estimation of θ using sequence data, and investigation of the role of migration and recombination rates in producing observed patterns of neutral sequence variation. Other work developed by R.C. Griffiths and S. Tavaré (1994) and C. Strobeck (1983) investigated the more complex aspects integrating gene tree topology into coalescents.

The synthesis of this past work with phylogenetic tree building methods, such as the maximum likelihood methods of Felsenstein (1992a; 1992b), Griffiths and Tavaré (1994) and the Least Squares methods of Fu (1994a; 1994b), lead to some of the methods discussed in this review. We have concentrated on the Least Squares methods advocated by Fu (1994a; 1994b) because of the close relationship of his phylogenetic methods to summary statistics.

5.2 Prospectus

Despite the advances of past work, estimation of population genetic parameters from nucleotide sequence data is a field that is still in its infancy. Many published methods have not even been adequately benchmarked on simulated data let alone applied to actual data. Part of the reason for this is slowness in speed of computation. Developers hope this will ultimately be overcome by rapidly changing technology. However, our ability to collect large data sets is also increasing at a rapid pace. Given the extraordinary increase in tree space with increasing sample size, it is unlikely that increases in computational power, alone, will overcome the difficulties posed in phylogenetic data sets. Another basic problem with these methods is that analytic or even closed form mathematical expressions are not easily obtainable making direct connections between computational genetics methods and standard population genetics theory difficult to ascertain. This is particularly important if computational genetic methods are to be benchmarked with more mathematical developments of population genetics theory. In this paper and other work (Vasco and Fu, submitted; Vasco, Crandall and Fu, submitted) we have attempted to develop a computational theory that integrates statistical principles and coalescent theory and has direct connections with population genetics theory. We have demonstrated that complex phylogenetic estimation methods for estimation of parameters in varying environments (Kuhner *et al.* 1998; Vasco and Fu, submitted) have direct lineage to the basic work in coalescent estimation developed by Watterson (1975) and Tajima (1983). Although we have focused on the LS methods developed by Fu and colleagues, several recent summary statistics methods using coalescent likelihoods have appeared (Grassly *et al.*, 1999; Weiss and Von Haeseler, 1996; Weiss et al., 1997; Griffiths, 1999). Work by Wakely and Hey on speciation (1997), recombination (Hey and Wakely, 1997), and migration (Wakely, 1999) suggest that summary statistic coalescent methods can be broadly applied to fundamental problems in evolutionary theory.

In the near future, Least Squares phylogenetic estimation is likely to play an important role in analysis of large scale sequencing studies. Eventually more powerful computing technology involving parallelization of coalescent algorithms will allow application of coalescent likelihood methods to large scale studies. Thus, currently the jury is still out on which methods are most useful although it is likely that both methods will be useful in future applications. Certainly, the statistical properties maximum likelihood estimators are most desirable. However, it is also likely to be true that each statistical methodology has its own weaknesses. Future work may show that a close relationship exists between maximum likelihood and Least Squares phylogenetic methods. This may allow a hybrid theory to be develop that exploits the strengths of these two methods and minimizes their respective weaknesses.

Recently, Felsenstein and colleagues (1999) proposed what they called "an object oriented fantasy." In this fantasy world, one reduces the combinatorial complexity of writing code for Monte Carlo Markov Chains using object-oriented programming (C++, Java) methods that allow the program to self-assemble in response to a user's requirements. They propose creating such an environment. We would also like to develop such an approach based upon our Least Squares methodology. How would such an approach work? Consider the schematic shown in Figure 7. On the left, one has the evolutionary dynamics producing observed patterns of sequence evolution. Several forces are at work: including the effect of the variable environment, recombination, migration and selection to name a few. There are also complex mutation processes such as varying rates across sites. These processes acting together produce the complex phylogenies that must be mined for statistical inferences. Some of the inference problems we have discussed in this review are listed on the right of the figure: (a) LS estimation of parameters, (b) minimum variance estimation of parameters (probably θ although LS linearization methods could be used to explore other parameters of interest), (c) hypothesis testing, perhaps involving different regions of the phylogeny, and (d) exploring the role of the gene

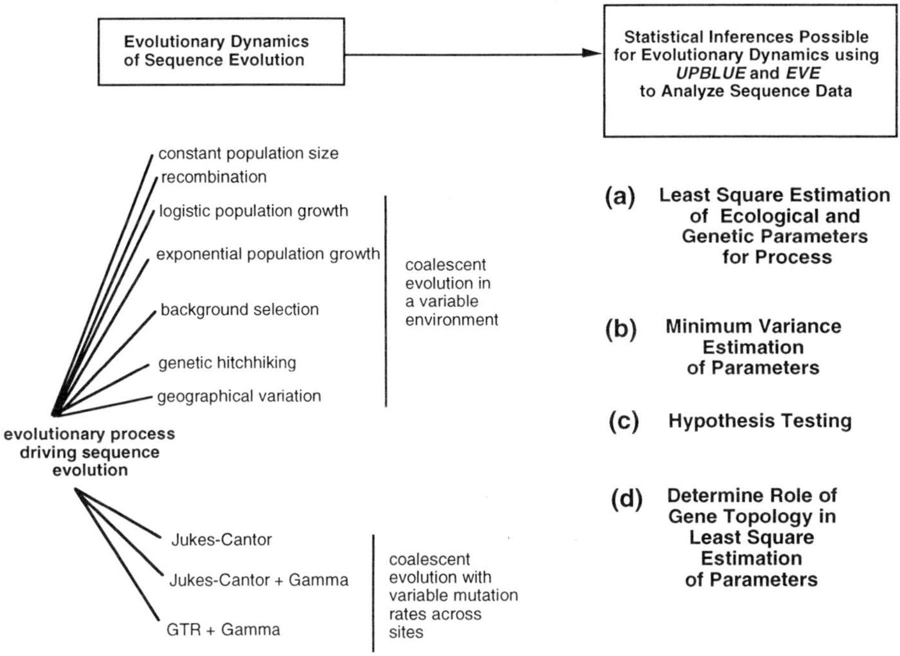

Figure 7 Relationship between evolutionary dynamic forces producing observed patterns of sequence evolution and statistical inference possible using UPBLUE and EVE in analyzing sequence data.

tree topology on (a-c). Using an object-oriented programming approach one could tailor the desired statistical analysis to a particular user or lab's needs.

It is not commonly appreciated that the summary statistics and phylogenetic methods we have developed in this review can be readily generalized to develop estimators for both recombination and migration rates in variable environments. Indeed, earlier work by Fu (1994b) showed how this can be done for the constant population case. We are currently extending these methods to the variable environment case. These LS estimators will be fast enough to explore wide regions of parameter space for complex phylogenetic models. Thus, excellent prospects exist for gaining an improved understanding of the major population genetic forces listed in Figure 7.

Finally, we point out an important conceptual advantage of the Least Squares approach over standard coalescent likelihood methods (Felsenstein et al., 1999). Using Least Square methods it is possible, in principle, to simultaneously estimate each component of the compound parameter $\theta = 2N\mu$, i.e. the pair (N, μ). Since coalescent likelihood methods must integrate over all potential genealogies, it is essentially impossible to independently estimate μ and N since they are confounded. Simultaneous estimation of products, such as N and μ is likely to be particularly important in the estimation of the coalescent time to the MRCA when information from the underlying distributions of the estimators is required to more efficiently implement Baysian methods (Fu and Li, 1997). Another application lies in the development of more accurate statistical tests of neutrality (Tajima, 1989b;

1993; 1997; Simonsen *et al.*, 1995; Braverman *et al.*, 1995; Li and Fu, 1994), where knowledge of the underlying distributions for the coalescent expectations is crucial. Again, the conceptual advantage of the Least Squares theory in this case, lies in its ability to develop novel theoretical expectations, based upon single or multiple branches of a coalescing genealogy.

Progress in understanding complex evolutionary processes will require coalescent statistical methods with enough flexibility to rapidly incorporate advances in standard phylogenetic reconstruction methods (maximum parsimony, maximum likelihood, and distance). Coalescent statistical methods, both those based upon summary statistics and phylogenetic information, will likely prove adaptable and useful in this regard.

6. SOFTWARE FOR LEAST SQUARES METHODS

Crandall lab: *http://bioag.byu.edu./zoology/crandall_lab/Vasco/eve.htm*
Fu lab: *http://hgc.sph.uth.tmc.edu/fu/*

ACKNOWLEDGEMENTS

This work was supported in part by NIH grants R29 GM50428 (Y-XF) and R0l HG01708 (Y-XF and W.-H. Li), grant DEB-9707567 (Y-XF), R0l HD34350-01A1 and the Alfred P. Sloan Foundation (KAC).

REFERENCES

Avise, J.C., Ball, R.M. and Arnold, J. 1988. Current versus historical population sizes in vertebrate species with high gene flow: a comparison based on mitchondrial DNA lineages and inbreeding theory for neutral mutations. *Mol. Biol. Evol.* **5**: 331-344.

Braverman, J. M., Hudson, R.R., Kaplan, C.H., Langley, N.L. and Stephen, W. 1995. The hitchhiking effect on the site frequency spectrum of DNA polymorphisms. *Genetics* **140**: 783-796.

Crandall, K.A. 1999. *The Evolution of HIV.* Johns Hopkins University Press, Baltimore, MD.

Crandall, K.A., Vasco, D.A., Posada, D. and Imamichi, H. 1999. Advances in understanding the evolution of HIV. *AIDS* **13A**: 39-47.

Crandall, K.A., Posada, D. and Vasco, D.A. 1999. Effective population sizes: missing measures and missing concepts. *Animal Conservation* **2**:317-319.

Crow, J.F. and Kimura, M. 1970. *An Introduction to Population Genetics Theory*. Harper and Row, New York.

Deng, W.-H., and Fu, Y.-X. 1996. The effects of variable mutation rate across sites on the phylogenetic estimation of effective population size. *Genetics* **134**: 783-796.

Donnelly, P. and Tavaré, S. 1995. Coalescents and genealogical structure under neutrality. *Ann. Rev. Genet.* **29**:401-21.

Donnelly, P. and Tavaré, S. 1997. *Progress in Population Genetics and Human Evolution*. Springer. New York.

Ewens, W.J. 1972. The sampling theory of selectively neutral alleles. *Theor. Pop. Biol.* **3**: 87-112.

Ewens, W.J. 1979. *Mathematical Population Genetics*. Springer-Verlag. Berlin

Feller, W. 1968. *An Introduction to Probability: Theory and Applications. Vol. I.* J. Wiley and Sons, New York.

Felsenstein, J. 1992a. Estimating effective population size from samples of sequences: inefficiency of pairwise and segregation sites as compared to phylogenetic estimates. *Genet. Res.* **56**: 139-147.

Felsenstein, J. 1992b. Estimating effective population size from samples of sequences: a bootstrap monte carlo integration method. *Genet. Res.* **59**: 209-220.

Felsenstein, J., Kuhner, M., Yamato, J. and Beerli, P. 1999. Likelihoods on coalescents: A Monte Carlo sampling approach to inferring parameters from population samples of molecular data. In: *Statistics in Molecular Biology and Genetics* (Seiller-Moisewitcsh, F., ed.) American Mathematical Society, Providence, RI.

Fisher, R.A. 1930. *Genetical Theory of Natural Selection.* Clarendon, Oxford.

Fu, Y.-X. 1994a. A phylogenetic estimator of effective population size or mutation rate. *Genetics* **136**: 685-692.

Fu, Y.-X. 1994b. Estimating effective population size or mutation rate using the frequencies of mutations of various classes in a sample of DNA sequences. *Genetics* **138**: 1375-1386.

Fu, Y.-X. 1995. Statistical properties of segregating sites. *Theor. Pop. Biol.* **48**: 172-197.

Fu, Y.-X. 1996. New statistical tests of neutrality for DNA samples from a population. *Genetics* **143**: 557-570.

Fu, Y.-X. 1997. Statistical tests of neutrality of mutations against population growth, hitchhiking and background selection. *Genetics* **146**: 915-925.

Fu, Y. -X., and Li, W.-H. 1993. Maximum likelihood estimation of population parameters. *Genetics* **134**: 1261-1270.

Fu, Y.-X., and Charkraborty, R. 1998. Simultaneous estimation of parameters for a stepwise mutation model. *Genetics* **150**: 487-497.

Fu, Y.-X., and Li, W.-H. 1997. Estimating the age of the common ancestor of a sample of DNA sequences. *Mol. Biol. Evol.* **14**:195-199.

Fu, Y.-X., and Li, W.-H. 1999. Coalescing into the 21st century: An overview of and prospects of coalescent theory. *Theor. Pop. Biol.* **56**:1-10.

Golding, B. 1994. *Non-Neutral Evolution: Theories and Molecular Data.* Chapman and Hall, London.

Grassly, N., Harvey, P.H. and Holmes, E.C. 1999. Population dynamics of HIV-1 inferred from gene sequences. *Genetics* **151**: 427-438.

Griffiths, R. C. 1989. Genealogical tree probabilities in the infinitely-many-sites model. *J. Math. Biology* **27**: 667-680.

Griffiths, R. C. and Tavaré, S. 1994. Sampling theory for neutral alleles in a varying environment. *Phil. Trans. R. Soc. London* **B 344**: 403-410.

Griffiths, R.C. 1999. The time to the ancestor along sequences with recombination. *Theor. Pop. Biol.* **55**: 137-144.

Hey, J. and Wakley, J. 1997. A coalescent estimator of the population recombination rate. Genetics **145**: 833-846.

Holmes, E.C., Zhang, L.Q., Simmonds, P., Ludlam, C.A. and Brown, A.J. 1992. Convergent and divergent sequence evolution in the surface envelope of glycoprotein of human immunodeficiency virus type 1 within a single infected patient. *PNAS* **89**:4835-4839.

Hudson, R. R. 1990. Gene genealogies and the coalescent process, In *Oxford Surveys in Evolutionary Biology*, (Futuyama, D. and Antonovic, J., eds.), Oxford University Press, Oxford.

Hudson, R. R. 1993. The how and why of generating genealogies, In *Mechanisms of Molecular Evolution* (Takahata, N. and Clark, A.G., eds.). Sinauer Associates, Sunderland, MA.

Kaplan, N.L., Hudson, R.R. and Langley C.H. 1989. The "hitchhiking effect" revisited. *Genetics* **123**: 887-899.

Kimura, M. 1983. *The Neutral Theory of Molecular Evolution.* Cambridge University Press, Cambridge.

Kingman, J. 1982. On the genealogy of large populations. *J. Appl. Prob.* **19A**:27-43.

Kuhner, M. K., Yamato, J. and Felsenstein, J. 1995. Estimating effective population size and mutation rate from sequence data using Metropolis-Hastings sampling. *Genetics* **140**: 1421-1430.

Kuhner, M. K., Yamato, J. and Felsenstein, J. 1998. Maximum likelihood of estimation of population growth rates based on the coalescent. *Genetics* **149**: 429-434.

Li, W.-H. 1997. *Molecular Evolution.* Sinauer Associates, Sunderland, MA.

Li, W.-H. and Fu, Y.- X. 1994. Estimation of population parameters and detection of natural selection from DNA sequences, In *Non-Neutral Evolution: Theories and Molecular Data* (Golding, B., ed.) Chapman and Hall, London.

Li, W.-H. and Fu, Y.- X. 1999. Coalescent theory and its applications in population genetics, In *Statistics in Genetics* (Halloran, M. E. and Geisser, S. eds.) Springer. New York.

Majoram, P. and Donnelly, P. 1994. Pairwise comparisons of Mitochondrial DNA sequences in subdivided populations and implications for early human evolution. *Genetics* **136**:673-683.

Malecot, G. 1941. Etude mathematique des populations mendeliennes. *Ann. Univ. Lyon. Sec.* **A 4**:45-60.

Maynard Smith, J. and Haigh, J. 1974. The hitch-hiking effect of a favourable gene. *Genet. Res.* **23**: 23-35.

Neuhauser, C. and Krone, S.M. 1997. The genealogy of samples in models with selection. *Genetics* **145**: 519-534.

Rodrigo, A.G. and Felsenstein, J. 1999. Coalescent approaches to HIV population genetics, In *The Evolution of HIV.* (Crandall, K.A., ed.) Johns Hopkins Univ. Press, Baltimore, MD.

Rodrigo, A.G., Shpaer, E.G., Delwart, E.L, Iverson, A.K.N., Gallo, M.V., Brojatsch, J., Hirsch, M.S., Walker, B.D. and Mullins, J.I. 1999. Coalescent estimates of HIV-1 generation time in vivo. *Proc. Natl. Acad. Sci. USA* **96**: 2187-2191.

Rogers, A. R. and Harpending, H. 1992. Population growth makes waves in the distribution of pairwise genetic differences. *Mol. Biol. Evol.* **9**: 552-569.

Roughgarden, J. 1979. *Theory of Population Genetics and Evolutionary Ecology*. MacMillan, New York.

Searle, S.R. 1971. *Linear Models.* J. Wiley and Sons. New York.

Simonsen, K.I., Churchill, G.A. and Aquadro, C.F. 1995. Properties of statistical tests of neutrality for DNA polymorphism data. *Genetics* **141**: 413-420.

Slatkin, M. and Hudson, R.R. 1991. Pairwise comparisons of mitochondrial DNA sequences in stable and exponentially growing populations. *Genetics* **129**: 555-562.

Slatkin, M. and Maddison, W.P. 1989. A cladistic measure of gene flow inferred from the phylogenies of alleles. *Genetics* **123**: 603-613.

Strobeck, C. 1983. Estimation of the neutral mutation rate in a finite population from DNA sequence data. *Theor. Popul. Biol.* **24**: 160-172.

Strobeck, C. 1987. Average number of nucleotide difference in a sample from a single subpopulation: a test for population subivision. *Genetics* **117**: 149-153.

Tajima, F. 1983. Evolutionary relationship of DNA sequences in finite populations. *Genetics* **105**: 437-460.

Tajima, F. 1989a. The effect of change in population size on DNA polymorphism. *Genetics* **123**: 597-601.

Tajima, F. 1989b. Statistical method for testing the netural mutation hypothesis by DNA polymorphism. *Genetics* **123**:585-595.

Tajima, F. 1989c. DNA polymorphism in a subdivided population: the expected number of segregating sites in the two-subpopulations models. *Genetics* **123**: 229-240.

Tajima, F. 1993. Measurement of DNA polymorphism, In *Mechanisms of Molecular Evolution: Introduction to Molecular Paleopopulation Biology,* (Takahata, N. and Clark, A. G., eds.). Sinauer Associates, Sunderland, MA.

Tajima, F. 1997. Estimation of the amount of DNA polymorphism and statistical tests of the netural mutation hypothesis based on DNA polymorphism, In *Progress in Population Genetics and Human Evolution,* (Donnelly, P. and Tavaré, S., eds.) Springer, New York..

Takahata, N. 1991. A trend in population genetics theory, In *New Aspects of Genetic of Molecular Evolution,* (Kimura, M. and Takahata, N., eds.) Springer, New York..

Tavaré, S. 1984. Lines of descent and genealogical process and their applications in population genetics models. *Theor. Pop. Biol.* **26**: 119-164.

Tavaré, S. Balding, D.J., Griffiths, R.C. and Donnelly, P. 1997. Inferring coalescence times from DNA sequence data. *Genetics* **145**: 505-518.

Vasco, D.A. 1999. *The EVE v1.0 Software Package*. Brigham Young University, Provo, UT, *http://bioag.byu.edu/zoology/crandall_lab/Vasco/eve.htm*.

Wakeley, J. 1999. Segregating sties in Wright's Island Model. *Theor. Pop. Biol.* **53**: 166-174.

Wakeley, J. and Hey, J. 1997. Estimating ancestral population parameters. *Genetics* **145**: 847-855.

Watterson, G.A. 1975. On the number of segregating sites in genetical models without recombination. *Theor. Popul. Biol.* **7**: 256-276.

Weiss, G., Henking, A. and Von Haesler, A. 1997. Distribution of pairwise differences in growing populations, In *Progress in Population Genetics and Human Evolution,* (Donnelly, P. and Tavaré, S. eds.) Springer, New York..

Weiss, G., and Von Haessler, A. 1996. Estimating the age of the common ancestor of men form the ZFY intron. *Science* **272**: 1359-1360.

Wright S. 1931. Evolution in Mendelian populations. *Genetics* **16**: 139-156.

APPENDIX

Sampled sequences of an HIV patient studied by Holmes *et al.* (1992) in Paup Nexus Format:

```
#NEXUS
[!Imported PHYLIP file "holmes.3.DAT" (Friday, February 5, 1999 11:31
AM)]
Begin data;
        Dimensions ntax=12 nchar=234;
        Format datatype=nucleotide gap=- missing=? matchchar=.;
        Matrix
M84250_3.2
TTTCACAGACAATGCTAAAACCATAATAGTACAGCTGAAGGAATCTGTAGAAATTAATTGTACAAGACCC
AACAACAATACAAGAAAAAGTATACATATAGGACCAGGGAGAGCATTTTATACAACAGGAGACATAATAG
GAGATATAAGACAAGCACATTGTAACATTAGTAGAGCAAAATGGAATGACACTTTAGGAAAGATAGTTAT
GAAATTAGGAGAACAATTTGGGAA
M84249_3.3
TTTCACAGACAATGCTAAAACCATAATAGTACAGCTGAAGGAATCTGTAGTAATTAATTGTACAAGACCC
AACAACAATACAAGAAAAAGTATACATATAGGACCAGGGGGAGCATTTTATACAACAGGACAAATAATAG
GAGATATAAGACAAGCACATTGTAACCTTAGTAGAGCAAAATGGAATGAAACTTTAAGACAGATAGTTAT
AAAATTAGGAGAACAATTTGGGAA
M84248_3.4
TTTCACAGACAATGCTAAAACCATAATAGTACAGCTGAAGGAATCTGTAGAAATTAATTGTACAAGACCC
AATAACAATACAAGAAAAAGTATACCTATAGGACCAGGGAGAGCATTTTATACAACAGGACAAATAATAG
GAGATATAAGACAAGCACATTGTAACCTTAGTAGAGCAAAATGGAATGAAACTTTAAGACAGATAGTTAT
AAAATTAGGAGAACAATTTGGGAA
M84241_3.2
TTTCACAAACAATGCTAAAACCATAATAGTACAGCTGAATGAATCTGTAGAAATTAATTGTACAAGACCC
AACAACAATACAAGAAAAAGTATACCTATAGGACCAGGGAGAGCATTTTATACAACAGGACAAATAATAG
GAGATATAAGACAAGCACATTGTAACCTTAGTAGAGCAAAATGGAATGAAACTTTAAGACAGATAGTTAT
AAAATTAGGAGAACAATTTGGGAA
M84255_3.7
TTTCACAGACAATGCTAAAACCATAATAGTACAGCTGAAGGAATCTGTAGTAATTAATTGTACAAGACCC
AACAACAATACAAGAAAAAGTATACCTATAGGACCAGGGAGAGCATTTTATACAACAGGACAAATAATAG
GAGATATAAGACAAGCACATTGTAACCTTAGTAGAGCAAAATGGAATGAAACTTTAAGACAGATAGTTAT
AAAATTAGGAGAACAATTTGGGAA
M84247_3.5
TTTCACAGACAATGCTAAAACCATAATAGTACAGCTGAAGGAATCTGTAGAAATTAATTGTACAAGACCC
AACAACAATACAAGAAAAAGTATACCTATAGGACCAGGGAGAGCATTTTATACAACAGGAGACATAATAG
GAGATATAAGACAAGCACATTGTAACCTTAGTAGAGCAAAATGGAATGAAACTTTAAGACAGATAGTTAT
AAAATTAGGAAAACAATTTGGGAA
M84246_3.6
TTTCACAGACAATGCTAAAACCATAATAGTACAGCTGAAGGAATCTGTAGTAATTAATTGTACAAGACCC
AACAACAATACAAGAAAAAGTATACCTATAGGACCAGGGAGAGCATTTTATACAACAGGAGACATAACAG
GAGATATAAGACAAGCACATTGTAACCTTAGTAGAGCAAAATGGAATGAAACTTTAAAACAGATAGTTAT
AAAATTAGGAGAACAATTTGGGAA
M84245_3.1
TTTCACAGACAATGCTAAAACCATAATAGTACAGCTGAAGGAATCTGTAGAAATTAATTGTACAAGACCC
AACAACAATACAAGAAAAAGTATACATATAGGACCAGGGAGAGCATTTTATACAACAGGAGACATAACAG
GAGATATAAGACAAGCACATTGTAACCTTAGTAGAGCAAAATGGAATGAAACTTTAAAACAGATAGTTAT
AAAATTAGGAGAACAATTTGGGAA
M84243_3.1
TTTCGCAGACAATGCTAAAACCATAATAGTACAGCTGAATGAATCTGTAGAAATTAATTGTACAAGACCC
AACAACAATACAAGAAAAAGTATACATATAGGACCAGGGGGAGCATTTTATACAACAGGACAAATAATAG
GAGATATAAGACAAGCACATTGTAACATTAGTAGAGCAAAATGGAATGAAACTTTAAAACAGATAGTTAT
AAAATTAAGAGAACAATTTGAGAA
M84242_3.1
TTTCACAAACAATGCTAAAACCATAATAGTACAGCTGAATGAATCTGTAGAAATTAATTGTACAAGACCC
```

```
AACAACAATACAAGAAAAAGTATACATATAGGACCAGGGGGAGCATTTTATACAACAGGACAAATAATAG
GAGATATAAGACAAGCACATTGTAACCTTAGTAGAGCAAAATGGAATGAAACTTTAAGACAGATAGTTAT
AAAATTAAGAGAACAATTTGAGAA
M84251_3.a
TTTCACAGACAATGCTAAAACCATAATAGTACAGCTGAATGAATCTGTAGTAATTAATTGTACAAGACCC
AACAACAATACAAGAAAAGGTATACATATAGGACCAGGGAGTGCATTTTATGCAACAGGAGACATAATAG
GAGATATAAGACAAGCACATTGTAACCTTAGTAGAGCAAAATGGAATGAAACTTTAAAACAGATAGTTAT
AAAATTAGGAGAACAATTTAGGAA
M84254_3.8
TTTCACAGACAATGCTAAAACCATAATAGTACAGCTGAATGAATCTGTAGTAATTAATTGTACAAGACCC
AACAACAATACAAGAAAAAGTATACATATAGGACCAGGGAGAGCATTTTATACAACAGGAGGAATAATAG
GAGATATAAGACAAGCACATTGTAACATTAGTAGAGCAAAATGGAATGAAACTTTAAAACAGATAGTTAT
AAAATTAAGAGAACAATTTGAGAA
        ;End;
```

POPULATION GENETICS OF HIV: PARAMETER ESTIMATION USING GENEALOGY-BASED METHODS

Peter Beerli*, Nicholas C. Grassly[§], Mary K. Kuhner*, David Nickle[£], Oliver Pybus[§], Matthew Rain[£], Andrew Rambaut[§], Allen G. Rodrigo[‡], and Yang Wang[£#]

Departments of *Genetics and [£]Microbiology, University of Washington, Seattle, WA 98195 USA

[§]Department of Zoology, University of Oxford, South Parks Rd, Oxford, OX1 3PS, UK

[‡]School of Biological Sciences, University of Auckland. Private Bag 92019, Auckland, New Zealand

[#]*Authors listed in alphabetical order.*

1. INTRODUCTION

Regardless of where viruses sit on the Tree of Life, if indeed they sit there at all, the evolutionary biology of HIV is no different from that of organisms with which we are more familiar. Each virion inherits its genetic material from some parent, and ultimately all virions (and proviruses) within a host are related by virtue of a shared evolutionary history and a common ancestor. Along the way, the HIV population is subject to the forces of evolution operating in its own microcosm, the host environment. The virus population grows in the host, it moves to establish subpopulations in different parts of the host or in different hosts in much the same manner — to use a paradigm of evolutionary biology — as Darwin's finches colonized the Galapagos Islands. The problem facing evolutionary virologists studying HIV populations is easily stated: how can one make inferences about the evolutionary dynamics of a viral population numbering up to 10^{10} virions and 10^{7} infected cells within an infected individual, and hundreds of thousands more globally, from a sample of only

10, 20, even 100 viral sequences? Fortunately, the same tools that evolutionary biologists use to understand how prokaryotes and eukaryotes evolve can also be used to understand how HIV evolves, both within and between hosts. What is required of such methods is that they take account of the stochastic effects that act on small samples of sequences.

In the chapter by Vasco, Crandall and Fu (Chapter 9), readers were introduced to the coalescent. The coalescent is a mathematical construct developed by Kingman (1981a; 1981b) to describe the genealogy of a sample of sequences drawn from a large population. In this chapter, we will assume these to be nucleotide sequences. If we consider a population of constant size, discrete generations and a Poisson birth-death process, it is easy to see that after some length of time all individuals in the population will be the descendants of a single individual (Figure 1). Therefore, if a sample of sequences is obtained and the genealogy of the sample is reconstructed, we will discover, as we move back in time, the common ancestors of pairs or groups of these sequences, culminating in a single common ancestor, the most recent common ancestor (MRCA) of all sequences in our sample. The genealogy of the sequences that is so constructed is best represented as a tree (and indeed, we use phylogenetic methods to reconstruct this genealogy). Lineages on the tree, starting with branches with sequences from our sample at the terminal nodes, are said to coalesce as we move back in time and down the tree towards the root that represents the MRCA. As Vasco and colleagues (Chapter 9) show, and as we revisit

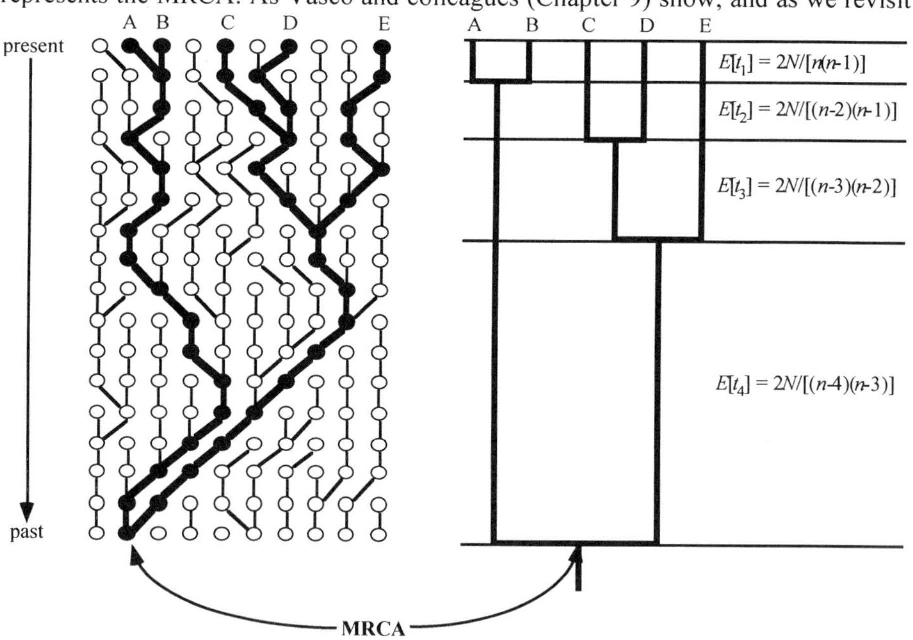

Figure 1. Hypothetical genealogy of a sample of n = 5 sequences. Under a Wright-Fisher model, with a constant-sized population, discrete generations and a Poisson birth-death process, all sequences coalesce ultimately to a most recent common ancestor (MRCA). On the right, the genealogy is depicted as a phylogenetic tree, with time counted in generations from present to past. The expected time of the ith interval, $E[t_i]$, is $2N/k(k-1)$ where k is the number of lineages at the top of the interval i.

in the next section, the times separating coalescent events, the coalescent intervals, are determined by population size, structure and evolutionary processes that operate on the population. Consequently, if we can define the appropriate mathematical relationship between coalescent times and these factors/processes, it becomes possible to characterize the evolutionary dynamics of the population quantitatively. In the following sections, we describe exploratory and statistical methods in which inferences about population growth or decline, population subdivision and migration, and recombination requires the genealogy of a sample of sequences to be explicitly accounted for.

2. THE COALESCENT

Since the methods discussed in this chapter are based on genealogies of sequences, it is appropriate to describe the statistical properties of such genealogies. To do this, we must rely on Kingman's mathematical framework that describes the statistical distribution of times between coalescent events in a sample of sequences. This mathematical framework, the n-coalescent, or coalescent for short, allows one to model the expected time between coalescent events. Consider an ideal Wright-Fisher population of individuals that has remained constant at size N. From n of these individuals (where n is much smaller than N), a sample of sequences is obtained one from each individual. Several other assumptions need to be made. First, the sequences have evolved neutrally, or at least, if they are under selection, the intensity of selection is not strong enough to distort the timing of coalescent events. It is also assumed, for now, that the sequences have not recombined, and that the generations are discrete. If the process of births and deaths in the population is Poisson, i.e., if each birth and death occurs independently and all individuals in the population have equal probabilities of giving birth or dying, the probability that there is a coalescent event in the previous generation is simply the probability that two of n sequences will share a common ancestor in that generation. (Since N is assumed to be much larger than n, the probability that there will be more than one coalescent event in any given generation is assumed to be so small that it is ignored.) For a haploid population, this probability, P, is:

$$P = \frac{\binom{n}{2}}{N} = \frac{n(n-1)}{2N} \tag{1}$$

Conversely, the probability that there is no coalescent event in the preceding generation is (1-P). At this point, we need to orient ourselves with respect to time: with the coalescent, we model genetic change as time moves backward from the present. Hence, in the coalescent framework, t=1 is one generation in the past. So, what is the probability of seeing a coalescent event k generations in the past? We compute this probability simply as the product of two other probabilities, the probability that no coalescent even has occurred from t=0 to t=k-1 and the probability that the first coalescent event occurs at t=k:

$$P(t = k) = P(1 - P)^{k-1} \tag{2}$$

The probability density function for continuous-valued times using the exponential distribution can be approximated by:

$$P(t=k) \approx \int_{k-1}^{k} \frac{n(n-1)}{2N} \exp\left(-\frac{n(n-1)}{2N} t\right) dt = \frac{n(n-1)}{2N} \exp\left(-\frac{n(n-1)}{2N} k\right) \tag{3}$$

From this, the expected time it takes to coalesce from n to n-1 lineage, as well as its variance is (see also Equations 7 and 8 in Vasco *et al.*, Chapter 9 in this book):

$$\mathrm{E}[t_{n \to n-1}] = \frac{2N}{n(n-1)} \tag{4a}$$

$$\mathrm{Var}[t_{n \to n-1}] = \frac{4N^2}{n^2(n-1)^2} \tag{4b}$$

This, of course, only gives us the expected time for the first coalescent event. With a sample of n sequences, and a rooted genealogy, there are n-1 coalescent events. To obtain the expected times for each subsequent coalescent interval, it is important to note that the number of lineages decreases by 1 after each event. Therefore, the second coalescent interval has an expected time of $2N/(n-1)(n-2)$ generations, and so on down to the last coalescent interval, which has an expected time of N generations. The total time it takes a sample of n to coalesce can be obtained by summing all the expected times, and equals $2N(n-1)/n$ or approximately $2N$, for a large enough n.

2.1 Adding Mutations

Up to this point, we have timed the coalescent intervals in generations, but we have no direct way of measuring the number of generations that elapses between coalescent events. Instead, what we do have is a molecular metronome that is based on the clock-like accumulation of mutations along the fragment we have chosen to sequence. Of course, there is no guarantee that our sequences evolve according to a molecular clock, but there are statistical methods one can use to test this (see Posada *et al.*, Chapter 7). With HIV, at least some studies (Leitner and Albert, 1999; Shankarappa *et al.*, 1999) have documented a molecular clock operating for parts of the genome, notably, in *env* and *gag*. If it can be shown that the sequences sampled evolve in a clock-like manner, then time in generations, t, can be re-expressed as a new variable, $t' = \mu t$ where μ is the mutation rate per site per generation. Now the coalescent intervals are measured in numbers of mutations. Using the appropriate mathematical machinery, t' replaces t in Equation 3, and the probability that a coalescent event occurs k' mutations in the past is:

$$P(t' = k') \approx \frac{n(n-1)}{2N\mu} \exp\left(-\frac{n(n-1)}{2N\mu} k'\right) \tag{5}$$

The composite parameter $2N\mu$ is designated Θ, and is a fundamental quantity in population genetics (although Θ is more usually defined by expressing μ as the mutation rate per locus per generation). The mean and variance of the coalescent intervals, in terms of the number of mutations are:

$$\mathrm{E}[t'_{n \to n-1}] = \frac{\Theta}{n(n-1)} \tag{6a}$$

$$\mathrm{Var}[t'_{n \to n-1}] = \frac{\Theta^2}{n^2(n-1)^2} \tag{6b}$$

2.2 Simulating Genealogies

Note that the coalescent says nothing about the topology of the genealogy, only about the distribution of the lengths of the coalescent intervals. However, by once again relating the coalescent intervals to the size of the population, we are in a position to model the genealogy of the population in concert with models of population size change. Consider, for instance, a population that has been growing. At present the size of the population N is large relative to the size of the population sometime in the past. Therefore, the first few coalescent intervals will have long expected times consistent with a large N, but as one moves back in time and N decreases, we expect the intervals to become shorter. The result is a genealogy with a cluster of coalescent nodes at the base of the tree; in fact, with a very rapid rate of growth, the genealogy will look like the familiar star-topology frequently seen with HIV subtype phylogenies.

If the population declines, the converse is true. Since the population size at present is small relative to its size in the past, the more recent coalescent events occur very rapidly, but these get longer towards the base of the tree. The apparent picture is that of a clonal population. However, this picture is deceptive, since any "clonality" is not due to any extrinsic selective mechanism, but is a consequence of population size and sampling.

This is a particularly important point, because the superimposition of coalescent intervals on any random phylogenetic topology can give the appearance that there is some kind of phylogenetic structure to the tree. It is equally tempting to invoke some kind of functional explanation for such structure. However, simulations of genealogies based on the coalescent can often convince a researcher that what the eye detects as pattern or structure is not unusual but the outcome of stochastic processes (Figure 2). In this respect, the coalescent is a very powerful tool for generating null hypotheses using simple population models.

It is indeed easy to simulate genealogies using the coalescent. For example, to simulate a genealogy of samples drawn from a constant sized population,

a) Obtain a random value from an exponential distribution with parameter $n(n-1)/2N$ (where n is the number of remaining lineages) and use this as the number of generations between coalescent events;
b) Select two different sequences/lineages at random and let them coalesce after the interval obtained in (a);
c) Repeat steps (a) and (b) until all lineages have coalesced.

Once the branch-lengths and topology of the genealogy have been obtained, we can now superimpose mutations on the branches. Along each branch, mutations are acquired according to some model of evolution, but with a mean of μb, where μ is the mutation rate per site per generation and b is the branch length in generations. It is easy enough to work from the base of the tree towards the tips adding mutations to a simulated sequence (or simply counting the number of mutations) along the way. Methods for simulating genealogies when the population size varies are also

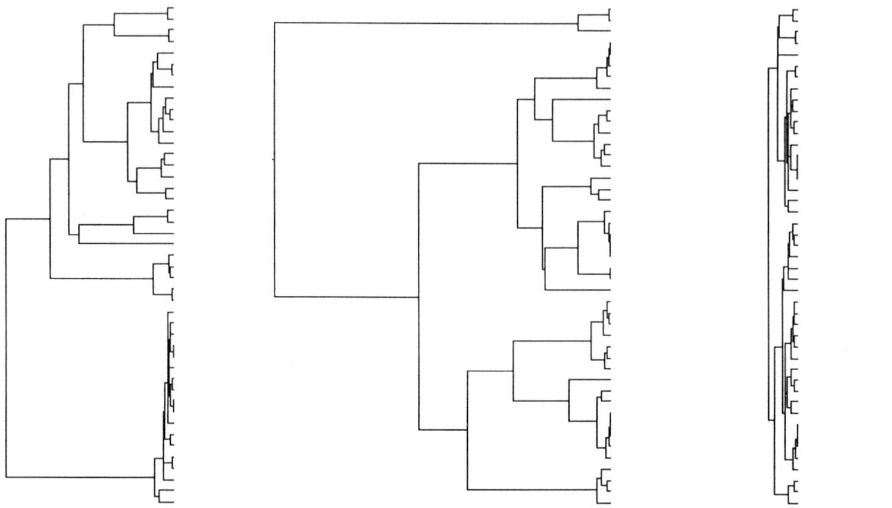

Figure 2. Simulations of samples of 45 sequences from exponentially-declining, constant-sized and exponentially growing populations. Each population has a current effective size of 10^4 individuals. The rate of change in population size was set at –0.00005 and 0.002 for the exponentially declining and growing populations respectively. Note the very striking patterns of phylogenetic "structure". However, such structure is not due to any underlying functional differences between sequences, but is due simply to the effects of sampling sequences from populations with different dynamics.

available (Hudson, 1990; Slatkin and Hudson, 1991). Computer programs are available to perform these simulations assuming constancy of population size (*PEBBLE:* Drummond, Goode and Ewing; all software mentioned in this chapter are described below in Section 7), or under a variety of conditions including fluctuations in population size and population subdivision (*TREEVOLVE:* Grassly and Rambaut).

2.3 A Note on Effective Population Size

Throughout the discussion above we have taken N as the number of individuals in the population, i.e., the census population size. If the population behaves according to our assumptions, i.e., with non-overlapping and discrete generations, births and deaths mediated through a stochastic Poisson process and in the absence of any selective forces acting on the population, using the census population size is appropriate. However, it is pertinent to ask how the use of the model is affected when these assumptions are violated. With HIV, it is perhaps most convenient to think of the infected cells as the individuals in our population, and the cell-free virions as the gametes these individuals produce. (Note that this is not a theoretical requirement, only an intuitively appealing hook on which to hang our concepts. Nonetheless, when thinking about such things, it is useful to remember the witticism that eggs have chickens to produce more eggs). But we know, for instance, that there are several different cell types that are infected, each with its own average life span. Even within a cell type, cells may exist in different states, e.g., $CD4^+$ T-cells may be activated or latent, they may be memory or naive cells. Not only is the assumption of non-overlapping generations violated, each infected cell may have a different generation time. In addition, each cell or variant may produce different numbers of newly infected cells. With a Poisson distribution, the variance in production of newly infected cells per productively infected cell is expected to equal the average number of new infections each infected cell gives rise to. However, this may not be the case, thus violating the assumption that the birth-death process is Poisson. Sewall Wright (Wright, 1938) introduced the notion of the "effective population size" to deal with populations that fail to satisfy the assumptions of the ideal model. Vasco, Crandall and Fu (Chapter 9) offer some mathematical definitions of effective population size, typically denoted N_e in the population genetics literature. Essentially, a population with effective size N_e experiences the same intensity of genetic drift (i.e., stochasticity due to sampling effects) as does an ideal population of the same size, and therefore acquires or loses diversity at the same rate.

What is the effective size of the HIV population in vivo? Leigh Brown (1997) estimated the effective size of the HIV population in vivo, and obtained estimates on the order of 10^3 individuals, certainly far lower than the estimated 10^7 infected cells. However, all this means is that the HIV population has the same rate of change of diversity as would be seen in an ideal population with 10^3 individuals. This can be due to any number of reasons, many of which were noted by Leigh Brown. Episodic crashes of the viral population, mediated perhaps by selective sweeps, lead to very dramatic losses of diversity. Latent infection slows down the loss of diversity, as does random activation and clonal expansion of infected cells. Overlapping generations increase the rate at which diversity is lost, whereas tissue compartmentalization tends to decrease it. For all these reasons, it is not unreasonable to obtain an N_e, which is different from the census population size, although the size of the difference certainly is puzzling. More recently, however, Rouzine and Coffin (1999), using a different method, estimated that the effective size of the HIV population in vivo to be on the order of 10^4 to 10^5 individuals, and argue cogently that that the HIV effective population size is at the very least, close to the deterministic boundary. However, perhaps most importantly, Rouzine and Coffin accepted that there will be instances when stochastic factors may well be the determinants of HIV evolution. Under highly active antiretroviral therapy, for instance, when the numbers of infected cells drop by two or more orders of magnitude, the HIV population may find itself in the stochastic zone. Similarly, the bottleneck at transmission or after the initial phase of acute viremia (which occurs within the few weeks after infection), viral populations may evolve stochastically.

3. INFERRING THE EPIDEMIC HISTORY OF HIV USING COALESCENT BASED GRAPHICAL METHODS

Coalescent methods as applied to HIV molecular sequences are not confined to processes that operate within the infected individual only. At a higher level, we are also interested in samples of sequences obtained from several hosts. In such cases, the inferences made are not about the evolutionary processes that act within the host, but those that act on the population of hosts. Is the number of infected hosts growing and, if so, at what rate? Is there a significant movement of hosts from one geographical region to another? Coalescent theory can, quite naturally, be used to describe the relationship between a genealogy of HIV sequences each obtained from a different individual and the historical dynamics of the epidemic. Here, we discuss methods that graphically display the demographic information contained in genealogies.

The theoretical potential of graphical coalescent methods was first noted by Felsenstein (1992), who suggested that one could "detect whether there is a trend in the [internode intervals] that would indicate that the effective population sizes had changed through time." The first implementation of this idea was the lineages-through-time (LTT) plot (Nee *et al.*, 1995). On its abscissa, a LTT plot displays time in units of substitutions or, if the mutation rate is known, in units of years or generations. Its ordinate shows the log of the number of lineages present in the genealogy at each point in time. A LTT plot therefore describes the rate of coalescence in a genealogy through time. Different demographic histories produce distinctive curves: constant-sized populations tend to generate concave LTT plots, whilst exponentially-increasing populations tend to produce convex plots. Nee and colleagues (1995) also developed two transformations and a statistical test that permit more detailed investigation of population history.

The LTT approach was first applied to a genealogy of 24 HIV-1 group M *env* gene sequences. The resulting plot was convex, suggesting that the number of group M infections had grown exponentially through time (Nee *et al.,* 1995). Holmes and his coworkers (1995) subsequently analyzed a larger phylogeny, containing 72 group M *gag* gene sequences, and obtained the same result. Although reasonable, these analyses were too general to provide detailed information about epidemic history. HIV-1 is not a single population, on the contrary, it is highly subdivided with individual subtypes associated with different transmission routes and geographical regions (McCutchan, 1999).

More recent studies have therefore investigated the demographic history of HIV at the subtype level. In the most extensive LTT plot analysis to date Holmes, Pybus and Harvey (1999) compared two epidemiologically distinct subtypes of HIV-1. Subtype A is mostly found in sub-Saharan Africa where the majority of transmissions occur through heterosexual intercourse. In contrast, subtype B mainly circulates in the developed world and has been predominantly spread via intravenous drug use and homosexual intercourse (UNAIDS, 1998). Genealogies were reconstructed for both subtypes from alignments of *env*, *gag* and *pol* gene sequences. Each alignment was carefully compiled to minimize the effects of recombination and non-random sampling. Despite the epidemiological differences between subtypes A and B, their LTT plots were similar, both indicating a constant-rate exponential increase.

The failure of LTT plots to distinguish between subtypes A and B is puzzling, especially as other coalescent analyses of the same data have consistently suggested different demographic histories for these subtypes. Using a simple test of tree shape, Pybus, Holmes and Harvey (1999) inferred that subtype B's exponential growth rate was significantly greater than A's. Grassly, Harvey and Holmes (1999) investigated the same problem using a pairwise-difference-distribution method and concluded that the current population size of subtype A was larger than that of B.

These inferences are in broad agreement with epidemiological evidence: subtype A predominates in many parts of Sub-Saharan Africa, where ~70% of the world's HIV-1 infected individuals can be found (Rayfield *et al.*, 1998; Neilson *et al.*, 1999; UNAIDS, 1999). Furthermore, the high growth rate inferred for subtype B is probably the result of its rapid transmission in the developed world within closely-connected networks of high-risk individuals (Robertson *et al.*, 1986; Jacquez *et al.*, 1994).

LTT plots are useful in determining the general mode of population change, but can be insensitive to differences in demographic parameters such as exponential growth rate. Also, they must be interpreted subjectively by visual comparison of observed and expected curves. In an attempt to overcome these problems, Pybus, Rambaut and Harvey (2000) developed a new graphical method called the skyline plot. Skyline plots are easy to interpret because they transform observed genealogies into plots of estimated effective population size against time (implemented in *GENIE*: O. Pybus and A. Rambaut).

Skyline plots are constructed as follows. Under the variable-size coalescent process (Griffiths and Tavaré, 1994), the time intervals between successive internal nodes of a genealogy (called internode intervals) are determined by the demographic history of the sampled population. Demographic history is described by $N_e(t)$, a function that represents the effective population size at time t. In many applications, $N_e(\cdot)$ takes the form of constant population size or exponential growth. A genealogy with n tips contains n-1 internode intervals, which have sizes $\gamma_2, \gamma_3, \ldots, \gamma_n$ measured in units of generations. The subscripts refer to the number of lineages present in the genealogy during each interval. Each γ defines an interval in time $[\tau_i, \tau_i + \gamma_i]$, where τ_i is the time at which the ith internode interval begins. It is straightforward to show that the term $M_i = i(i\text{-}1)\gamma_i / 2$ is an estimate of the harmonic mean of $N_e(\cdot)$ during $[\tau_i, \tau_i + \gamma_i]$. Plotting M_i against time therefore defines a non-parametric estimate of demographic history (Figure 3; Pybus *et al.*, 2000). If the reconstructed genealogy has branch lengths in units of substitutions per site then M_i is used to estimate $N_e(\cdot)\mu$, where μ is the substitution rate.

Skyline plots can reconstruct population history under a variety of demographic scenarios (Figure 3). If population size is constant, then the average of the M_i is equal to the maximum likelihood estimate of effective population size (Felsenstein, 1992). However, if $N_e(\cdot)$ is variable then skyline plots can underestimate population size, because a harmonic mean is always smaller than its corresponding arithmetic mean. As illustrated in Figure 3b, the severity of this bias depends on the relative rates of coalescence and population change. The bias is small when the rate of coalescence is large compared to the rate of population change (i.e., when the harmonic and arithmetic means of $N_e(\cdot)$ during an internode interval are similar).

The skyline plots of HIV subtypes A and B are substantially different (Figure 4). Both subtypes initially grew exponentially, with subtype B growing faster than A. However, in the recent past the growth rate of subtype B appears to decline, suggesting a logistic model of population growth. A logistic scenario is partially

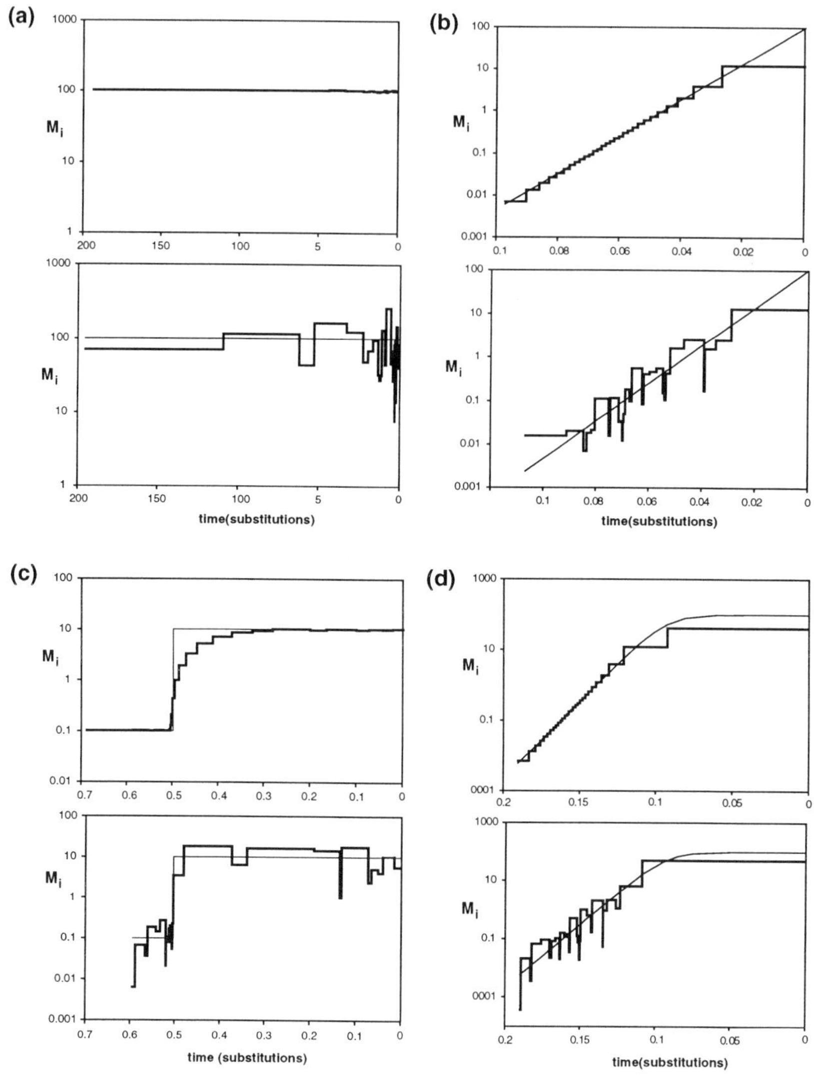

Figure 3. Skyline plots can reconstruct population history under different demographic scenarios: A. Constant population size. B. Exponential growth. C. A 100-fold instantaneous increase in population size at time 0.5. D. Logistic growth. The vertical axis shows estimated $N_e(\cdot)\mu$. The horizontal axis represents time in units of substitutions; time is zero at the present. On each of (A-D), the top graph shows the expected skyline plot, obtained by calculating the mean of 5000 plots. The bottom graph shows the skyline plot of a single genealogy simulated under the same conditions. Both the estimated (thick lines) and true (thin lines) demographic histories are shown. From Pybus *et al.* (2000).

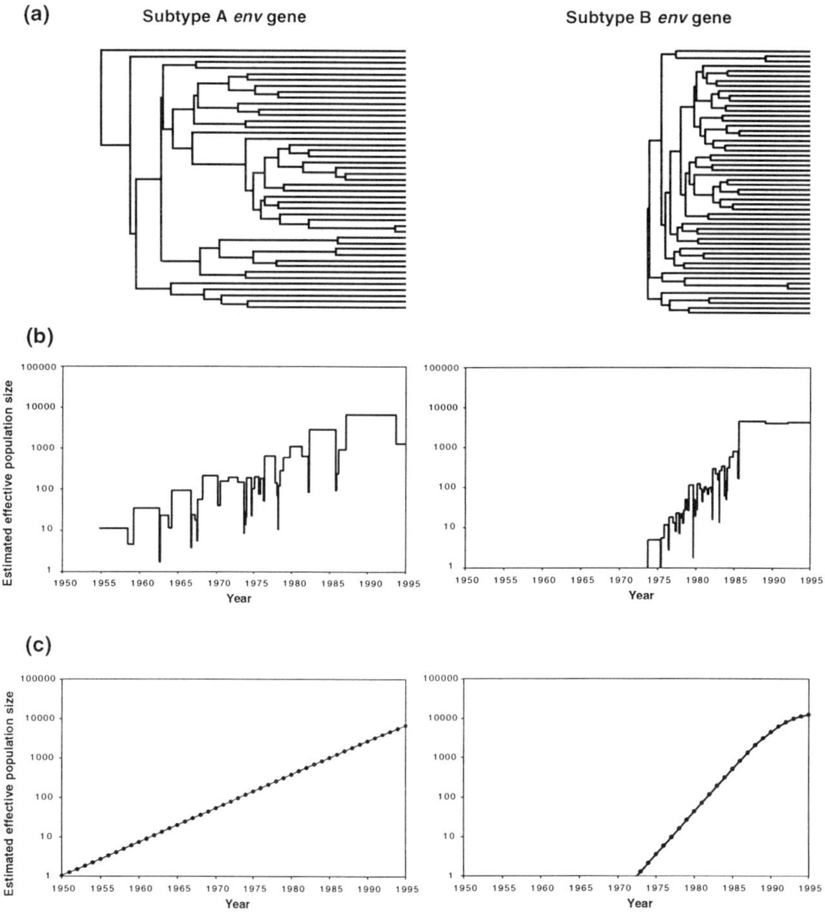

Figure 4. The epidemic history of HIV-1 subtypes A and B. (a) Subtype A and B genealogies reconstructed from env gene sequences. (b) Skyline plots obtained from the env genealogies. Time has been rescaled into years using a substitution rate of 0.0028 substitutions/site/year (Korber *et al.*, 2000). (c) Parametric maximum likelihood estimates of $N_e(\cdot)$ obtained using exponential and logistic demographic models.

consistent with epidemiological evidence, as the introduction of behavioral intervention and antiviral therapies in western Europe has led to a stabilization in the incidence of new infections (UNAIDS, 1998). The skyline plots give estimates of

$N_e(\cdot)$ that are very similar to those obtained using explicitly parametric likelihood-based methods (Figure 4).

Most current methods for the inference of demographic history from genetic data make several strong assumptions, such as the absence of recombination, selection and subdivision. A thorough consideration of these assumptions in the context of HIV is not possible here, but interested readers are directed to the discussions elsewhere (Holmes *et al.*, 1999; Grassly *et al.*, 1999). Suffice to say, the propensity of HIV to recombine will significantly complicate the interpretation of any coalescent-based analysis.

LTT and skyline plots use a single genealogy that is assumed to have been reconstructed without error. Although this assumption would be foolish in some circumstances, the variability and abundance of HIV sequence data permits phylogenies to be estimated quite accurately (Leitner *et al.*, 1996). Moreover, preliminary results suggest that inferences about HIV demographic history based on reconstructed trees are no less accurate than those based on correct simulated trees (Pybus *et al.,* 2000). Despite these assurances, it would be preferable to incorporate phylogenetic error into the skyline plot and we (specifically, OP and AR) are currently working to achieve this goal.

4. INFERRRING HIV COMPARTMENTALIZATION PATTERNS FROM MOLECULAR SEQUENCE DATA

When HIV subpopulations exist in different cellular types, tissues or systemic compartments within a single host, and the movement of infected cells or virions amongst these "locales" is restricted, we say that the population is compartmentalized. Several studies noted intra-host compartmentalization when they observed striking dissimilarities among HIV sequences obtained from different tissue compartments. As examples, such observations have indicated compartmentalization between blood and semen for HIV-1 subtype B (Coombs *et al.*, 1998; Delwart *et al.*, 1998; Kiessling *et al.*, 1998), genital tract and peripheral blood for HIV-1 subtype A (Poss *et al.*, 1998) and subtype B (Shaheen *et al.*, 1999), blood and cerebrospinal fluid (Keys *et al.*, 1993), brain and blood (Korber *et al.*, 1994) brain and lymph node (Haggerty and Stevenson, 1991), brain, spleen, and lymph node (Wong *et al.*, 1997), and within different regions of brain (Shapshak *et al.*, 1999). Distinct variants have also been found in the blood and brain of an HIV-2 infected individual (Sankale *et al.*, 1996). Since the compartmental structure of HIV is likely to affect viral evolution, understanding it could provide useful insights about the pathology of HIV infection, as well as the prospects of evolutionary escape from antiretroviral therapies. In the total absence of compartmentalization, viruses would be expected to flow freely in and out of potential tissue compartments and an HIV sample obtained from any infected tissue should be a random sample of the entire population at that time. In a sense, this assumption underlies every study where HIV subpopulations are sampled from one compartment (e.g., from the peripheral blood compartment) and the results extrapolated to the larger HIV population within a host.

When HIV compartmentalization is extreme, and follows a simple pattern, it may be easy to detect by visual inspection of a sequence alignment or a phylogenetic tree. For example, one might observe that a distinguishable sequence variant occurs exclusively in one tissue compartment, or that phylogenetic clades correspond perfectly with different tissues from which the sequences were sampled (e.g., Figure 5A; Korber *et al.*, 1994). However, it is not hard to imagine more complex scenarios where compartmentalization exists, but is harder to recognize. For example, if a patient has several distinct tissue compartments and multiple distinct HIV variants, even extreme compartmentalization might produce patterns that are indistinguishable to the eye. Also, it is not implausible that some barriers to viral migration act as imperfect screens, allowing some variants to pass unhindered while stopping most, but not all, of another variant. In that case, it can be very difficult to define an objective threshold of significance for discerning a compartmental pattern from random variation.

Slatkin and Maddison (1989; 1990) reported an analytical procedure that addresses that specific problem. Their approach considers biogeographic distribution in a phylogenetics/coalescent framework and, consequently, provides a method for inferring restriction of gene flow from molecular sequence data. This characteristic is useful since nucleotide sequences provide one of the most accessible and information-rich windows into HIV population processes.

The Slatkin-Maddison test employs a parsimony-based approach to questions of biogeographic distribution. To evaluate HIV compartmentalization, one must know the compartment from which each HIV sequence was sampled. For example, in order to consider the hypothesis that HIV in the brain is genetically disjunct from HIV in blood (e.g., Korber *et al.*, 1994), multiple samples from each compartment are required to apply this test. Beyond that, it is convenient to consider the Slatkin-Maddison method as four separate steps:

a) Estimate the phylogenetic relationship of the viral sequences sampled. All subsequent steps are based on this phylogeny, so the accuracy of the results can depend on the accuracy of the phylogeny (Slatkin and Maddison, 1989, 1990).
b) Assign to each taxon in the phylogeny a code that indicates the tissue from which it was sampled (e.g., A for brain and B for blood). As illustrated by the different branch textures in Figures 5A and 6A, this can be used simply to add compartment information to a phylogenetic tree.
c) Estimate the minimum number of migration events that one must postulate to reconcile the observed spatial distribution of the HIV sequences (e.g., the distribution between brain and blood in this example) with the phylogeny that was inferred in step (a).
d) Statistically evaluate whether the number of migration events estimated in step (c) is significantly smaller than one would expect by chance in a population that is not compartmentalized.

Slatkin and Maddison (1989; 1990) provide an accessible explanation of the mathematical/theoretical foundation of their method; we will focus, instead, on the application of the method in HIV studies.

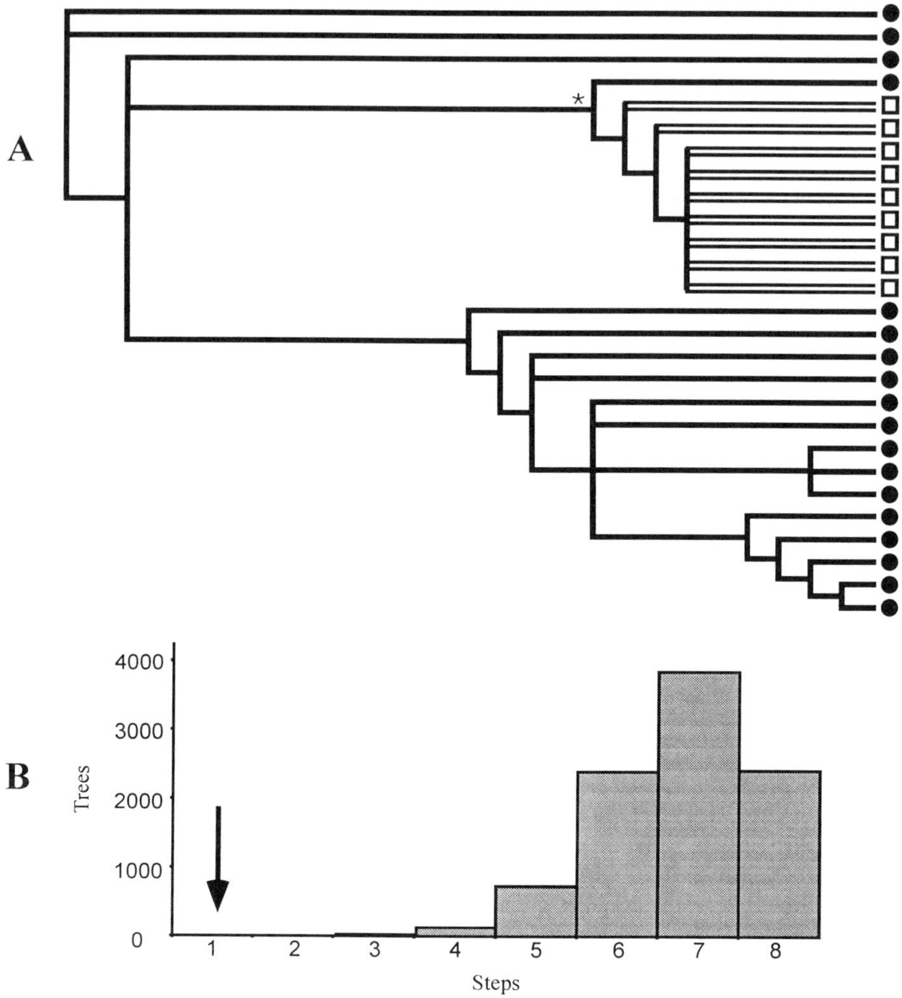

Figure 5. A. Cladogram adapted from Korber *et al.* (1994) showing estimated relationships among HIV sequences, which were obtained from the brain (open squares) and blood (filled circles) of "patient 1." Brain sequences (double lines) form a monophyletic group, and Blood sequences form a separate monophyletic group, suggesting compartmentalized population structure of the virus. MacClade counts one migration event (step) as the most parsimonious reconciliation of evolutionary history and tissue origin. The location of that step is marked with an asterisk. B. Histogram (from MacClade) shows the distribution of migration-event counts (steps) among the 10,000 random trees made by random joining/splitting of the tree shown in A. Most of the random trees had 7 steps, but none had fewer than 3 steps. Since the presumed true tree (A) has 1 step, we say the probability of observing a tree that extreme by chance, in a population without restrictions to gene flow, is $p < 0.0001$.

A

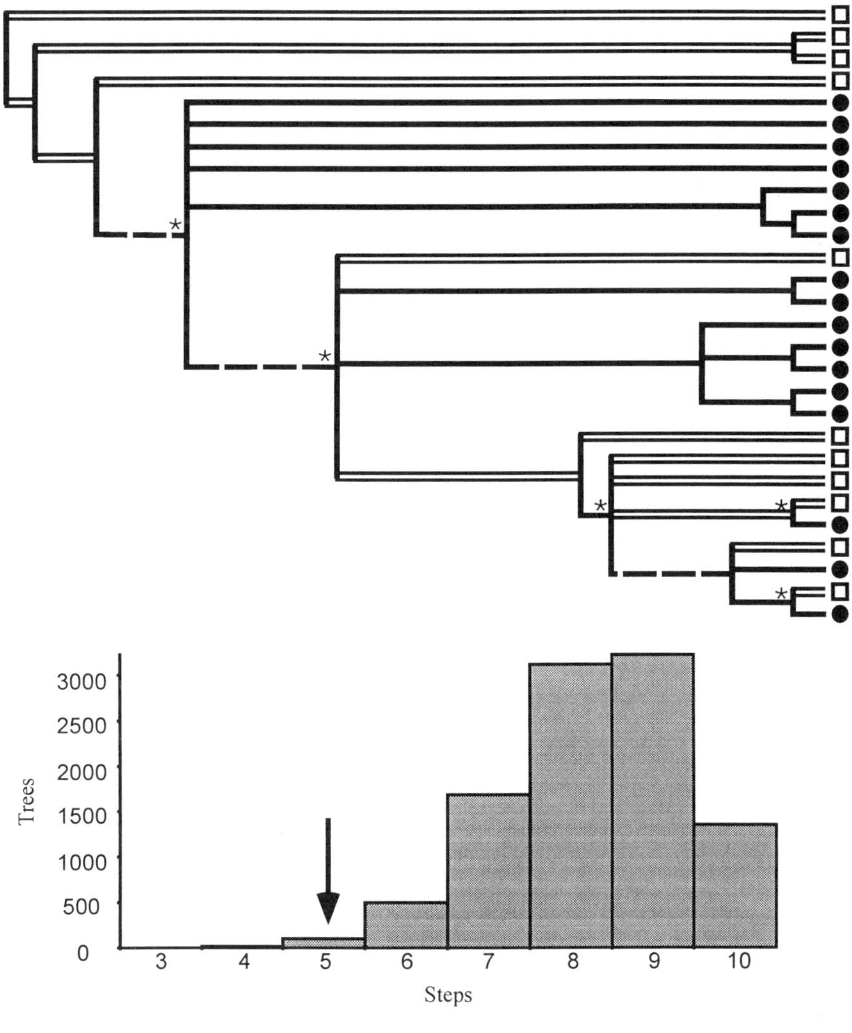

Figure 6. A. Cladogram adapted from Korber *et al.* (1994) showing estimated relationships among HIV sequences, which were obtained from the brain (open squares) and blood (filled circles) of "patient 2." Brain sequences (double lines) form clades that intermingle with the clades of blood-derived sequences. *MacClade* counts 5 migration events (steps) as the most parsimonious reconciliation of evolutionary history and tissue origin. The locations of those steps are marked on the tree with asterisks. Dashed lines represent branches whose ancestral compartment status is equivocal by parsimony state reconstruction. B. Histogram (from *MacClade*) shows the distribution of migration-event counts (steps) among the 10,000 random trees made by random joining/splitting of the tree shown in B. Most of the random trees had 9 steps, but just 172 had 5 steps or fewer. Since the presumed true tree (B) has 5 steps, we say the probability of observing a tree that extreme or more extreme by chance, in a population without restrictions to gene flow, is $p < 0.0172$.

4.1 Estimating the Phylogeny

The Slatkin-Maddison method infers geographic restrictions to gene flow, or compartmentalization, from the topology of a phylogenetic tree that is assumed to be known. To that extent, inferences about viral compartmentalization may only be as reliable as the topology of the phylogenetic tree of the sequences obtained (see Posada *et al.*, Chapter 7, for details on phylogenetic reconstruction). The Slatkin-Maddison method does not incorporate branch-length information in any way, so it is important to be cautious when using trees where unresolved polytomies might be arbitrarily inflated to non-zero lengths. Although some methods for estimating phylogeny will only recover a single tree (e.g., neighbor joining and UPGMA) others will return as many trees as the method finds to be equally optimal for the data (e.g., trees with equal maximum likelihood scores, or trees that are maximally parsimonious). When the phylogeny of sequences can not be narrowed down to just one tree, it is possible to summarize the available trees before applying the Slatkin-Maddison method using a consensus tree. The consensus tree often represents conflicting cladistic relationships amongst equally optimal trees as polytomies (or multifurcating nodes). This approach preserves information about those clades that occur on every tree (for strict-consensus methods) or on a majority of trees (for majority-rule methods) while explicitly indicating the clades that are not present on every tree or a majority of trees, respectively. Thus, the phylogenetic tree that is produced for the subsequent Slatkin-Maddison test is not affected by arbitrary rules for breaking ties in a distance matrix (as sometimes occurs with neighbor-joining, for example), and unequivocal relationships are distinguished from equivocal ones before the Slatkin-Maddison analysis is performed. Although phylogenetic methods that discover more than one tree to explain the phylogeny of HIV sequences have the benefit of preserving more information about ambiguous nodes, for some data sets the extra information may reveal that a tree is simply not suited for a rigorous compartmentalization analysis. Consensus trees frequently have more polytomies than their component trees. The use of consensus trees may, therefore, confound the process of enumerating the minimum number of hypothesized migration events (discussed below). Because the Slatkin-Maddison test assumes that the phylogenetic trees inferred at the start of the process are true, we will refer to the trees obtained in step (a) as "true" for simplicity, and in order to distinguish them from random trees in the subsequent discussion. Once the phylogenetic tree has been estimated, subsequent steps can be performed using the *MacClade* software package (Maddison and Maddison, 1992).

4.2 Encoding the Data with Compartment Information.

It is simple to encode each sequence with compartment information. That information is just a multistate unordered character — where each state represents the tissue from which the sequence was obtained. Each state name can be added to a multiple sequence alignment as a one-letter code after the phylogenetic analysis is completed (but that change to the alignment must not appear in the data during phylogenetic analysis) with an alignment editor such as the one in *MacClade*. After a "compartment" character is added to the sequence alignment *MacClade*'s "trace character" function allows compartmental information to be superimposed onto the tree as in Figures 5A and 6A.

4.3 Estimating the Minimum Number of Migration Events in the True Tree

MacClade's "trace character" function counts the minimum number of migration events, or 'steps' that one would have to postulate in order to account for the distribution of compartment characters across a phylogenetic tree, given the genealogical relationships in that tree. Figures 5 and 6 illustrate this general idea: in Figure 5, the relationship between HIV clades identified in blood and brain samples can be explained by one migration event — either from blood to brain or vice versa (note that this method does not indicate the *direction* of migration). For comparison, it would take at least 5 migration events to reconcile the phylogenetic topology in Figure 6A with the compartment information shown on that tree. In complex trees the number of migration events can be difficult to count at a glance; in fact, the counting rules comprise the most theoretically complex aspect of the procedure. Briefly, the minimum-number-of-migrations approach is based on a maximum-parsimony reconstruction of ancestral states, except that in this case, the ancestral states are the compartments occupied by the hypothetical ancestors of sequences in the sample (Farris, 1970; Fitch, 1971; Slatkin and Maddison, 1989, 1990). First the compartment status is estimated for each internal node in the tree. Then the number of migration events, m^*, are estimated by counting as steps, which are summed for all changes in the phylogenetic tree according to the formula:

$$m^* = \sum_b \begin{cases} 1 & \text{if the nodes at either end of branch } b \\ & \text{have different compartmental states,} \\ 0 & \text{otherwise} \end{cases} \tag{7}$$

where b indexes the branches on the phylogeny, and $\sum_b$ indicates the sum over all branches on the tree. One migration event is counted whenever a node has a different state than its immediate ancestor within the tree. Any node whose compartment status is equivocal, because the ancestral compartment status can not be estimated for that node, gets a value of zero (e.g., dashed branches in Figure 6A). This avoids redundant counting, since the migration events will be counted at the point in the tree where they are not ambiguous under these rules. Slatkin and Maddison (Slatkin and Maddison, 1989) have shown that m^* is a non-linear monotonically increasing function of Nm, the effective population size for each subpopulation (assuming that this is the same for both subpopulations) multiplied by the migration rate, and n, the number of genes sampled from each population. Although Slatkin and Maddison (Slatkin and Maddison, 1989) were unable to provide an analytic formula for this function, they used computer simulations to construct a look-up table for different values of n and Nm that correspond to m^*. To use their table, we need to interpolate between tabulated values of m^* and n, and for unequal sample sizes, randomly delete sequences from the phylogenetic tree; details of how these manipulations can be performed are given in their paper.

Polytomies pose special problems for counting migration events. Under the assumption of soft polytomies, polytomous nodes can not exceed a count of 1 but when polytomies are considered hard, as Maddison and Maddison (1992) write in the manual to *MacClade*, "every branch in a polytomy is assumed to have acquired any character state changes from the ancestor independently from other branches in the polytomy". Unfortunately, hard polytomies may err toward higher counts (i.e., higher estimates of the minimum number of migration events); conversely, soft polytomies can err toward lower counts, and there is usually no objective way to judge where the truth lies between those extremes. Fortunately, polytomies do not affect every tree. In those cases where polytomies are unavoidable, treating these as hard polytomies is the conservative option because, as we shall discuss in the next section, the null hypothesis that compartmentalization is absent is favored when there are more rather than fewer migration events.

Although the examples outlined here discussed counting rules for two-compartments data sets for simplicity, the formula can be expanded to accommodate more than two character states for experiments where viruses are obtained from more than two compartments.

4.4 Statistical Evaluation

The final step of the Slatkin-Maddison method involves a randomization procedure to test whether sequences from the same compartment are more closely related to each other than to sequences from other compartments. An appropriate null model is one in which all HIV move freely amongst compartments. Under such a null model, sequences from one compartment would be as likely to be evolutionarily related to sequences from another compartment as from the same compartment. The frequency distribution of the number of migration events under this null model can be obtained by constructing a large number of random trees, each with the same numbers of sequences from each compartment as in the original sample set. The number of migration events can be estimated for each random tree as it is done for the original genealogy (Figures 5B and 6B). With this null distribution in hand, it is possible to compare the number of migration events on the true tree to the null distribution and estimate the probability that the true tree came from a population lacking compartmentalization. Recall that just one migration event is required to explain the distribution of brain and blood designations in figure 5A, then note Figure 5B, which shows that none of the 10,000 random trees generated by *MacClade* had such an extremely low score (the lowest step count among the random trees was 3 steps). In this case, we can say that the probability of achieving the level of clade-compartment correspondence that was observed in this tree entirely by chance is less than 1 in 10,000 (i.e., $p < 0.0001$).

4.5 Applying the Method: Revisiting an Interesting Case

The topologies in Figures 5 and 6 were reconstructed from trees originally published side by side and discussed in the context of tissue compartmentalization (Korber *et al.*, 1994). The first tree was said to represent a general pattern seen in 4 patients, where blood- and brain-derived HIV populations clustered strongly in their respective trees. The second tree was said to represent a pattern seen in two other patients whose blood- and brain-derived sequences were described as "intermingled." While we also recognize the visual differences between those trees, and while we agree with the characterizations of the original authors, we sought to enhance the resolution for identifying compartmentalization in ambiguous cases by applying the Slatkin-Maddison test. Using the same trees, we set out to determine whether the observed topological difference represented a threshold between an HIV population with compartmental structure from a population without. Out of 10,000 random trees derived from the "true" tree shown in Figure 6A, only 172 had as many (five) or fewer steps as the true tree (see Figure 6B). This suggests that the probability of finding a clade-compartment association like that shown in Figure 6A, in a population that lacks barriers to gene flow, entirely by chance is $p = 0.0172$. Since this tree has polytomies at 2 migration-critical nodes in the tree, we also performed the analysis assuming unresolved nodes to be hard polytomies. In that case there were 6 steps, and 570 out of 10,000 random trees had as many or fewer steps. So another estimate of the same p value, based on an interpretation of the phylogeny that may be excessively conservative, yields $p = 0.0570$. As we mentioned above, without the ability to resolve some polytomies in the tree there is no objective way to determine where the actual probability lies between $0.0172 < p < 0.0570$, but at least we can say that the probability is fairly low. Hence there is reason to believe that both of the trees originally published, and perhaps all of the patients in the study by Korber and her coworkers (1994), had significant compartmental structure, although it seems compartmentalization was more extreme in some patients than in others.

5. LIKELIHOOD-BASED ESTIMATORS OF POPULATION PARAMETERS

In the previous sections, a single reconstructed phylogeny was taken as the best estimate of genealogy, and used in all subsequent analyses. However, the reconstructed phylogeny is never guaranteed to be a true representation of evolutionary relationships. Tree reconstruction is a potentially error-prone affair, contingent on the sequences used and the amount of phylogenetic signal in the data, the particular choice of method and (where appropriate) the model of evolution. Consequently, it is hard to be certain that the reconstructed tree — the estimate of genealogy — is the true tree.

One approach is to acknowledge this uncertainty, and integrate it into our estimation of population parameters. A time-honored statistical approach, advocated appropriately by Fisher, is to associate with each possible genealogy the probability that it will give the observed sequence data (mediated through some model of mutation and evolution). The probability, $P(D \mid G)$, of obtaining the sequences, D, given the genealogy, G, is called the likelihood of G. Note that we are not introducing a new term here — the likelihood of G is precisely the likelihood associated with the phylogenetic tree of the sample of sequences, and is calculated according to the standard method developed by Felsenstein (Felsenstein, 1981) and described by Posada, Crandall and Hillis in Chapter 7 of this book.

Of course, the coalescent dictates that each genealogy, as described by the times of the coalescent intervals, is itself a function of the processes that act on that population from which the sample is obtained. If this process is one of population growth, for instance, then the instantaneous growth rate of the population, g, is a parameter that can take an infinite number of values each more or less likely to account for a particular genealogy. Consequently, for any given genealogy G, we may also obtain the probability $P(G \mid \Phi)$ where Φ is the vector of parameter values associated with one or more processes acting on the population and on the genealogy of the sampled sequences. The form of $P(G \mid \Phi)$ indicates that it is also a likelihood, this time the likelihood of Φ with respect to the genealogy G. Knowing these two likelihoods allows us to compute the likelihood of Φ with respect to the sequence data D as:

$$P(D \mid \Phi) = \sum_G P(D \mid G) P(G \mid \Phi) \tag{8}$$

where $\sum_G$ indicates the sum of the products of the two likelihoods $P(D \mid G)$ and $P(G \mid \Phi)$ over all possible genealogies. Summing over all genealogies allows us to include all genealogies in our calculations weighting each by a measure of uncertainty, the likelihood $P(D \mid G)$. The best estimate of Φ is taken to be the vector of values that maximizes the probability $P(D \mid \Phi)$. Thus we have achieved our stated aim of taking account, in our estimation procedures, of the uncertainty inherent in our estimate of genealogy.

5.1 The *LAMARC* Samplers

Of course, it is one thing to write down how the likelihood of Φ is calculated, quite another actually to devise a method for obtaining the maximum-likelihood estimator of Φ. For a start, there are infinitely many genealogies (remember that a genealogy is both the topology and coalescent times). How do we derive a feasible and sound method for estimating Φ? That is the focus of this section. In particular, the *LAMARC* ("Likelihood Analysis with Metropolis Algorithm using Random Coalescence") samplers (Kuhner *et al.*, 1995; 1998; Beerli and Felsenstein, 1999) that utilize a Metropolis-Hastings algorithm, are described. These samplers attempt to estimate population parameters by making a combined estimate over a large sample of plausible genealogies.

To illustrate the application of the *LAMARC* samplers, we use a simple example, the estimation of Θ. Rewriting equation (8), the likelihood of Θ is given by:

$$L(\Theta) = \sum_G P(D|G)\, P(G|\Theta) \tag{9}$$

Here, $P(G|\Theta)$ is the probability, based on Kingman's coalescent, of the genealogy given the parameter Θ (Equation 5 above). The summation, as before, is over all possible genealogies with all possible branch lengths. For a sample size of n individuals, we would need to investigate $n!(n-1)!/2^{n-1}$ labeled histories (Edwards, 1970) and for each of them we would need to solve an n-1 dimensional integral. For ten individuals we would need to consider 2.571×10^9 labeled histories, each with an associated nine-dimensional integral. With today's and even tomorrow's computers this endeavor is hopeless. We resort to Markov chain Monte Carlo sampling and use a method devised by Metropolis (1955) and refined by Hastings (1972). We can approximate equation (9) by sampling genealogies according to a distribution as close to it as possible, and then making a weighted sum over the sampled genealogies, correcting for the influence of our sampling distribution. Samplers of the *LAMARC* family sample according to the distribution of (9), except that they replace the unknown true parameter Θ with an assumed value Θ_0. The final likelihood then corrects for the sampling bias introduced by Θ_0:

$$\frac{L(\Theta)}{L(\Theta_0)} = \sum_G \frac{P(G \mid \Theta)}{P(G \mid \Theta_0)} \tag{10}$$

Here the summation is over genealogies sampled from the distribution $P(D|G)\ P(G|\Theta_0)$. This is a form of importance sampling: the distribution $L(\Theta_0)$ is used to focus sampling on genealogies that should be informative for the unknown distribution $L(\Theta)$. The value of Θ that maximizes (10) is an approximate maximum likelihood estimate of Θ.

LAMARC works by starting with an initial genealogy, modifying it in a way proportional to $P(G|\Theta_0)$, and accepting or rejecting it based on $P(D|G)$: this process is repeated many times to produce a sample. The sampled genealogies are used to construct a likelihood curve according to (10). The maximum of this curve is a maximum likelihood estimate (MLE) of Θ, and its curvature allows construction of approximate confidence intervals around the estimate. This is an improvement over most non-likelihood estimators, which provide no information about their probable error.

This form of estimation is most effective at finding the maximum likelihood estimate of Θ when the sampling is done at a value of Θ_0 reasonably close to the (initially unknown) actual MLE of Θ. This can be accommodated by repeating the sampler several times, using the final estimate from each run as the starting estimate for the next. We call each run of the sampler with a given Θ_0 a "chain". Our standard approach is to run 5-10 short chains in order to get the starting values into a good range, and then 1-2 long chains to refine the final estimate. The last of these chains is then used to construct final parameter estimates. Currently we are investigating if combining information from multiple chains, using a method of Geyer (1994; Geyer and Moeller, 1996), helps to improve estimates.

One drawback of Metropolis-Hastings samplers is that it is difficult to tell when the sampler has been run long enough. Standard tests for autocorrelation of results can tell whether the sampler has explored its current search area well, but they cannot determine whether there are other important areas in the space of genealogies that it has not searched at all. In the Section 5.4 A Practical Example we suggest some rules-of-thumb for deciding if the sampler has been run long enough.

All of the *LAMARC* samplers estimate Θ along with additional parameters, as appropriate for their model, because Θ determines the fundamental scale of the coalescent genealogy.

5.2 Applications

5.2.1 Data Types

Currently *Fluctuate* can use DNA or RNA sequences; *Recombine* can use DNA, RNA, or SNP data; and *Migrate* can use DNA, RNA, SNPs, microsatellites, or electrophoretic polymorphisms. In the near future all of these samplers should be able to use any of these types of data. Any form of data for which a model of $P(D|G)$ can be developed is theoretically suitable for use in the *LAMARC* samplers (for example, protein data could in principle be used) although the more complex models lead to programs that run slowly.

5.2.2 Population size and growth rate

LAMARC samplers have been developed for several sets of population parameters. The simplest model involves a single, constant-size, non-recombining population and estimates Θ. One extension adds estimation of an exponential rate of growth g. This parameter is defined by the equation $\Theta_t = \Theta_n e^{-gt}$, which gives the value of Θ at time t given its value now (Θ_n). Time is measured in units of μ generations. These two models are available in the program *Fluctuate* (Kuhner *et al.*, 1995, 1998).

Simulation results show that the constant-size case (originally distributed as program *Coalesce*, now part of *Fluctuate*) produces nearly unbiased estimates of Θ and is somewhat more powerful than pairwise estimators of Θ such as Watterson's estimator (Watterson, 1975). Efficiency is greatly improved if multiple unlinked loci are analyzed at the same time. In the growing-population case, however, a pronounced upwards bias in the estimate is observed when only one locus is analyzed, even if sequence data are abundant. This bias is inherent to the method, rather than being an artifact of the *LAMARC* sampler. It has two causes:

a) The relationship between branch lengths estimated from the genealogy and the final estimate of the growth parameter g is non-linear; thus, error in estimating branch lengths becomes bias in g.
b) There is a severe lack of information about the most ancient parts of the genealogy, since only a few lineages from that period are still present in the population: this allows for pathological hypotheses that explain the observations of the distant past very well, but are highly unlikely *a priori* — in particular, the hypothesis that the population was of size 1 at the common ancestor of the sampled sequences.

The first problem can be somewhat ameliorated by using longer and more numerous sequences; the second responds only to use of multiple unlinked loci. This is a difficulty for HIV researchers, given the small size and linkage of the entire viral genome. One potentially fruitful direction for future research would be putting a prior distribution on the growth rate, or in some other way attempting to restrain the influence of spurious hypotheses about the base of the genealogy. In the meantime, it should be noted that in simulation results (Kuhner *et al.*, 1998) the 95% confidence intervals were fairly reliable (they included the truth an appropriate fraction of the time) even when the maximum likelihood estimator was badly biased.

Fluctuate is normally used to co-estimate modern-day Θ and historical *g*. These parameters are strongly correlated, with the likelihood surface often taking the form of a long, narrow ridge. If information about one parameter is available from another source, however, a much more powerful estimate can be made of the other. This is particularly important in HIV data. HIV samples from within a patient often approximate a star phylogeny — every sequence is about equally diverged from every other sequence. Co-estimation of Θ and *g* on such data does not succeed, because the likelihood surface in Θ is flat for high values of Θ — it does not come to a maximum. In practice the estimator will fail computationally since it is not capable of returning an infinite estimate. If an external estimate of Θ is available, however, a good estimate of *g* can be made. *Fluctuate* is a reasonably fast program and large runs are feasible.

5.2.3 Recombination Rate.

The *LAMARC* approach can be extended to the estimation of recombination rate. Current models require recombination to be uniformly distributed across the sequence, and estimate two parameters: Θ and the per-site recombination rate *r*. This model is implemented in the program *Recombine* (Kuhner, Yamato and Felsenstein, submitted). The rearrangement model used in the case of recombinant genealogies is necessarily both more complex and less efficient than that used in *Fluctuate*, and so *Recombine* runs more slowly and must produce a larger sample in order to obtain equal efficiency. It is also much more memory-intensive, because very large recombinant genealogies are sometimes produced. However, uniquely among the *LAMARC* programs it does not require multiple unlinked loci for the most efficient estimate, but can produce good estimates using a single long locus with some internal recombinations.

Preliminary simulation results suggest that Θ can be accurately estimated from recombinant data, but that relatively little power for estimating *r* is available. Accurate estimation of *r* requires a long sequence: with short sequences the estimate varies wildly and tends to show an upward bias. The ability to estimate *r* depends strongly on the value of Θ, since Θ determines the degree of polymorphism in the data. Nearly monomorphic data are consistent with a very wide range of recombination rates, since most recombinations in such data will be invisible: thus, the confidence intervals for *r* will be very wide. In contrast, highly polymorphic data will tend to indicate a fairly narrow range of possible recombination rates.

5.2.4 Migration Rates and Subpopulation Sizes

Takahata and Slatkin (1990) found that Kingman's results could not be extended to populations with subdivision. They were unable to obtain the distribution of coalescence times in the presence of migration. Beerli and Felsenstein (1999) have avoided this difficulty by having the genealogy *G* specify not only the coalescences, but also the times and places of migration events. With this information in *G*, its probability density for 2 populations becomes a product of terms for the intervals between events in the genealogy:

$$P(G \mid \Theta_1, \Theta_2, M_1, M_2) = \prod_{i=1}^{2} \left(\left[\frac{2}{\Theta_i} \right]^{v_i} M_i^{w.i} \right) \prod_{j=1}^{T} p_j \tag{11}$$

where

$$p_j = \exp\left(-u_j \sum_{i=1}^{2} \left[k_{ji} M_i + \frac{k_{ji}(k_{ji} - 1)}{\Theta_i} \right] \right) \tag{12}$$

M_i is the number of migrants moving into the ith population per unit time, where time is measured in $1/\Theta$ units. The variable v_i is the total number of coalescences in subpopulation i and $w_{.i}$ is the sum of the numbers of migration events to subpopulation i over all time intervals T. p_j is the probability that no event occurs during time interval j, u_j is the length of time interval j, and k_{ji} is the number of lineages in subpopulation i during time interval j. Using this approach, it now becomes possible to estimate a different Θ and migration rate for each subpopulation. This approach for migration rate estimation has a drawback: the number of parameters increases with the square of the number of populations. This makes it rather difficult to achieve reliable results from real data from systems with many populations. The search space in the migration directions becomes very large, and good results can only be achieved by summarizing over many independent loci (Beerli and Felsenstein, 1999, Felsenstein *et al.*, 1999, Kuhner *et al.*, 1998). Table 1 compares a 2-parameter and a 16-parameter model for 1 and 10 loci, and shows the substantial advantage given by additional loci.

Table 1: Reduction of variance by increasing number of loci and reducing number of estimated parameters. Data set had 40 individuals from 4 populations. Coefficient of variation is the standard deviation divided by the mean. The standard deviations were approximated through profile likelihoods.

Loci	Coefficient of Variation			
	2-parameter model		16-parameter model	
	Θ	$2Nm$	Θ^{a}	$2Nm^{b}$
1	0.20	0.25	0.26	0.22
10	0.06	0.04	0.11	0.15

[a] Averages over all individual Θ
[b] Averages over all individual $2Nm$

If a researcher has external evidence about migration scenarios s/he can reduce the number of parameters and try to estimate fewer than the n^2 possible interactions (Figure 7). For example, in understanding insect migration among the Hawaiian Islands it may be sufficient to consider migrations between adjacent islands only and set all other rates to zero. This will greatly increase the speed and accuracy of the algorithm, though of course the results will be only as good as the initial simplifying assumption.

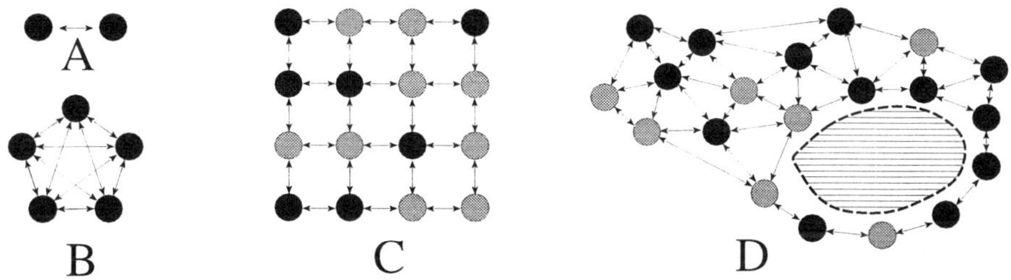

Figure 7 Population structures that can be analyzed using *Migrate*. (A, B) n-island models, (C) Stepping stone model, (D) Arbitrary migration models. Black disks mark sampled populations, Arrows mark migration considered in the program. The program is able to estimate parameters from a minimal n-island model with two parameters (the scaled population size Θ and the scaled migration parameter *M*) to a *n* population structure with maximally n^2 parameters (n Θ_i and $n(n-1)\ M_{ji}$.)

5.3 Comparison with Other Methods

The preference of the *LAMARC* approach for highly polymorphic data can be contrasted with the other major coalescent-based likelihood estimator, developed by Griffiths, Tavaré and colleagues (for population growth: Griffiths and Tavaré, 1994; for recombination: Griffiths and Marjoram, 1996; and for migration and growth: Bahlo and Griffiths, 2000). The Griffiths/Tavaré (GT) estimator performs well on relatively monomorphic data, which can be approximated by an infinite-sites mutation model, but tends to bog down on more polymorphic data. The two methods may thus complement one another, with GT efficiently extracting information from small, relatively monomorphic data sets and *LAMARC* from larger, more polymorphic ones.

The coalescent likelihood methods contrast sharply with methods using an intermediate statistic that is based on the variance of gene frequencies in and among populations, the F_{ST}-statistic (Wright, 1951) and its relatives. The translation of F_{ST} into a migration rate often assumes very restrictive population structures, such as all migration rates the same and all population sizes equal. Real data often heavily violate these assumptions and therefore estimates using F_{ST} based methods are often wrong (Beerli, 1998; Whitlock and McCauley, 1999).

Comparisons with other coalescence based methods, such as those of Wakeley (2000) who bases the estimation of gene flow on the analysis of variances on segregating sites, or Nielsen and Slatkin (2000), who devise a Bayesian method without mutation for the estimation of divergence time and migration rates between two populations, have not yet been made. But given the specificity of the model of the population structure used, the current implementations of *Migrate* or *Genetree* (Bahlo and Griffiths, 2000) are more versatile.

5.4 A Practical Example

To illustrate the method we selected 46 HIV sequences from the Los Alamos HIV Sequence database. These sequences were obtained from individuals in the US and Botswana. We chose the *gag* locus (1564 bp) for this example. A maximum likelihood clock-like tree of the sequences is shown in Figure 8. We do not present this analysis as conclusive — much more remains to be done with these data — but it illustrates use of the *LAMARC* samplers.

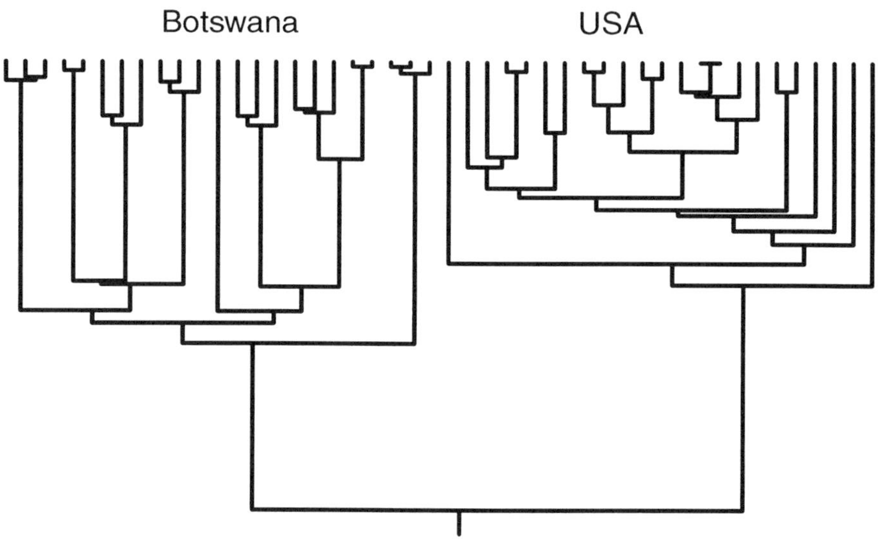

Figure 8. 46 partial *gag* sequences, obtained from individuals from Botswana and USA.

We obtained aligned sequences and did a visual check for adequacy of alignment, since the final results will depend on the quality of the starting alignment. While quite diverse, the sequences appeared well aligned with no dubious regions. It is best to delete doubtful alignment columns. If *Recombine* is being used, any deleted columns should be indicated in the spacefile so that the map of the sequence will not be distorted.

5.4.1 Estimating Recombination.

We were interested in the question of whether recombination can be detected in *gag*. The literature asserts that there is nearly free recombination between *gag* and *env*, but does this also hold within a single locus? This question can be addressed with *Recombine*, as long as we are willing to wave aside the issue of population subdivision (ideally we should use a sampler that combines *Recombine* and *Migrate*, but as of the printing of this book such a sampler was still in development).

For an initial investigation we used nucleotide frequencies calculated from the data, a transition/transversion rate of 2.0, and assumed that all sites evolve at the same rate. It would be appropriate to use a more complex model in subsequent runs, such as assuming two rates, a higher one for third positions and introns and a lower one for first and second positions. We used Watterson's estimate of Θ as a starting point, and an arbitrary value of r=0.01.

This is quite a large data set, but it is worthwhile to do a relatively short run initially to make sure that everything is working. We chose 10 short chains of 2000 steps each and 2 long chains of 50,000 steps each; this should be enough to give us some feedback. Part of the resulting outfile is shown in Figure 9.

```
There are 690 variable sites in the data
Watterson estimate of theta is 0.10038263


Intermediate chains (only)
chain# theta rec-rate
-----------------------------
 1 0.126114 0.047275 accepted 1372/3000 trees, dropped 161
 2 0.179966 0.021247 accepted 692/3000 trees, dropped 9
 3 0.212383 0.017482 accepted 399/3000 trees
 4 0.227373 0.012415 accepted 306/3000 trees
 5 0.246564 0.010072 accepted 101/3000 trees
 6 0.271848 0.009434 accepted 150/3000 trees
 7 0.294451 0.007965 accepted 56/3000 trees
 8 0.309872 0.006051 accepted 36/3000 trees
 9 0.313204 0.006693 accepted 55/3000 trees
 10 0.338996 0.006726 accepted 70/3000 trees
 11 0.369408 0.004474 accepted 942/51000 trees
Point estimate of parameters for locus 1
Theta = 0.409663 Recombination = 0.003588
Log(Likelihood) at maximum 0.382937
```

Figure 9. Partial output of *Recombine* run

There are several points to consider here. After the first few short chains (during which the sampler was finding its way to a reasonable genealogy) only about 2% of proposed trees were accepted. These data are highly polymorphic and will actually fit only a limited number of trees, so this is not surprising, but it does suggest that we will want fairly long runs. A stronger indication that the program has not run long enough is the visible trend in both Θ and r. The estimates of Θ are climbing steadily and the estimates of r are dropping; the program has not reached equilibrium. We could add more short chains, or lengthen each chain. This is true for all the *LAMARC* samplers: if the estimates have not settled down during the short chains, the amount of short-chain effort should be increased.

Some trees were "dropped" initially because they tried to exceed the program's upper limit on recombinations (imposed to keep the computer from running out of memory). This is not a problem as long as it does not continue throughout the run: if it had, we would rerun using the "final coalescence" strategy option. This strategy reduces the number of recombinations in the trees without affecting the final estimate. It is normally not used because it runs slowly, but for very high values of r it may be the only way to keep the program within memory limits.

Since the estimate of *r* appeared to be dropping steadily, we did a second run (results not shown) with a starting value of *r*=0.00001. The estimate of *r* never became visibly higher than 0, and we received a warning that going into the last long chain it was necessary to artificially force a recombination. (The program must not be allowed to settle on an estimate of 0, because it can never leave 0 once it arrives there.) It is likely that these results reflect acceptance of few or no recombinations throughout the run. This might be because the data strongly reject recombination, but it's likely that our starting value of *r* was too low and not enough recombinations were tried: this is especially worrisome because of the low overall acceptance rate. It is best to start with parameter values that are too high, rather than too low, as the sampler moves more freely with higher values.

The output in Figure 10 was produced using the "interactive design of output" option to allow us to focus on the most interesting area of the likelihood surface. The asterisks indicate the MLEs of Θ and *r*.

```
=====================================================================
 ln(Likelihood) for various values of Theta and rec

=====================================================================
                                         rec

Theta 0.0000000 0.0005000 0.0035881* 0.0036667 0.0068333 0.0100000
=====================================================================
0.2800    ----     -8.4981   -3.4634    -3.4640   -4.6804   -7.0579
0.3013    ----     -7.0913   -2.0622    -2.0630   -3.2832   -5.6640
0.3227    ----     -6.0894   -1.0619    -1.0627   -2.2868   -4.6712
0.3440    ----     -5.4040   -0.3752    -0.3761   -1.6040   -3.9923
0.3653    ----     -4.9690    0.0632     0.0622   -1.1693   -3.5617
0.3867    ----     -4.7340    0.3030     0.3019   -0.9332   -3.3296
0.4080    ----     -4.6601    0.3825     0.3814   -0.8572   -3.2577
0.4097*   ----     -4.6602    0.3829*    0.3818   -0.8570   -3.2579
0.4293    ----     -4.7169    0.3319     0.3307   -0.9111   -3.3156
0.4507    ----     -4.8804    0.1749     0.1737   -1.0714   -3.4797
0.4720    ----     -5.1312   -0.0695    -0.0708   -1.3188   -3.7308
0.4933    ----     -5.4539   -0.3859    -0.3872   -1.6380   -4.0536
0.5147    ----     -5.8358   -0.7619    -0.7632   -2.0166   -4.4356
0.5360    ----     -6.2668   -1.1871    -1.1884   -2.4443   -4.8665
0.5573    ----     -6.7382   -1.6530    -1.6545   -2.9127   -5.3378
0.5787    ----     -7.2430   -2.1528    -2.1543   -3.4147   -5.8426
0.6000    ----     -7.7754   -2.6805    -2.6820   -3.9445   -6.3751

Dashed lines indicate lnL values less than -1000;
Asterisks indicate the MLEs of Theta and rec
```

Figure 10. Log-likelihood results from *Recombine* run

A few points are worth noting about tables such as this. The values given are relative log-likelihoods, $\ln(L(\Theta)/L(\Theta_0))$. As such, their values have no fixed meaning; they can be greater than 1, which simply indicates that the genealogies fit the MLE much better than they fit the parameter values under which they were generated. However, relative likelihood values much greater than the number of parameters being estimated (here, 2) are a warning sign that the sampler had not yet settled on good working values of the parameters when entering the final chain, and that more or longer short chains would therefore be desirable. We do not see this problem in the table shown in Figure 10 above: the highest relative likelihood is much less than 2. For two parameters, a ln$L$ difference of about 3 units indicates an approximate 95% confidence limit. From the table we can interpolate that this limit runs, in the Θ direction, from approximately 0.29 to 0.63, and in the r direction from approximately 0.001 to 0.01. The same can be seen, in less detail, from the diagram of the likelihood surface (not shown).

What can we conclude from these results? There is clear evidence (the continuing upward trend in Θ and downward trend in r) that the program had not reached its final state. As such, both the maxima and the confidence intervals are questionable. Indeed, when we reran the program for the same amount of time but with starting r=0.00001 the estimate of r was essentially zero, whereas in this run zero is clearly excluded. For this large data set a much more intensive search is required.

5.4.2 Estimating Migration

When we look at Figure 7 we can immediately see that a most parsimonious interpretation would require only one migration between Botswana and USA. Analyzing this data set with *Migrate* reveals a similar pattern (Table 2), but rather than making an arbitrary assumption about which branch had the migration, the outcome of *Migrate* reflects our uncertainty and has a rather large confidence interval. The analysis does confirm our prejudice that not much migration has happened between these two populations.

Table 2. Analyses of population parameters based on 23 sequences of the *gag* locus from Botswana and 23 sequences from the USA, using *Migrate*. Values in parenthesis are the approximate 95% confidence intervals.

Population	Θ	2*Nm*	
		From Botswana	From USA
Botswana	0.140 (0.099 – 0.206)	—	0.003 (0 – 0.566)
USA	0.211 (0.121 – 0.404)	0.009 (0 – 1.705)	—

If we ignore the confidence intervals for the migration rates then we could even suggest that the gene flow was more likely from Botswana to the USA than from the USA to Botswana.

6. DISCUSSION: LOOSE ENDS AND FUTURE APPLICATIONS

In this chapter, we describe analyses that rely on genealogical information to make inferences about population processes. These analyses include a graphical method for identifying changes in population size, the Slatkin-Maddison randomization test for compartmentalization, and likelihood-based estimators of population growth, recombination and migration rates. For each of these analyses, several assumptions apply (Table 3).

One assumption that is common to all methods, with the possible exception of the Slatkin-Maddison test (see Slatkin and Maddison, 1989), is that of neutrality, i.e., that no selection acts on the sequences sampled. It is likely that the HIV population within each host is subject to a range of selective pressures imposed by the immune system, different therapeutic strategies, the availability of new cellular, tissue and systemic niches, and the need to preserve functionality in the face of an unrelenting mutational drive. Between hosts, there are the added pressures of overcoming a variety of barriers to transmission, and adapting to new host environments. How, then, can one justify the assumption of neutrality?

There are several lines of evidence suggesting that even if selection is operating on the HIV genome, neutrality is not a bad working assumption. First, as noted earlier, there is evidence that at least some regions of the HIV genome evolve in a clock-like fashion, so that even in different individuals, the rate of substitutions in parts of *env* (Shankarappa *et al.*, 1999) and *gag* (Leitner and Albert, 1999) are remarkably consistent. As Kimura (1968) pointed out in his seminal paper on molecular evolution, a widespread molecular clock in which substitutions accumulate at the same rate for different protein molecules is most easily explained by assuming that most of the observed variation is selectively neutral. The alternative explanation — that most of the variation is due to selection — must account for the fact that presumably different selective pressures acting on very different molecules can still lead to consistent rates of evolution. With HIV, the same argument can be brought to bear: that the same rates of substitutions occur in HIV populations across hosts makes Kimura's argument more appealing, since it is difficult to imagine how different hosts can exert the same selective pressures on their resident HIV populations.

Of course, a molecular clock will be obtained, and neutrality will be a good approximation, if selection is weak relative to genetic drift. Nearly neutral evolution (Ohta and Gillespie, 1996) operates when the product of effective size and the selection coefficient of a certain allele, $N_e s < 1$. $N_e s$ is essentially a measure of the differential reproductive success of that allele, and when this quantity is less than 1, the presence or absence of an allele in successive generations is determined more by sampling effects than its fitness. This brings us back to our discussion of HIV effective population size. If N_e is small, as Leigh Brown suggests, then any new substitution must have a considerable selective advantage to be fixed in the population, whereas a larger effective size will allow a variant with a smaller selective advantage to persist. This may account for the frequent inability to statistically detect selection operating on HIV sequences (Leigh Brown, 1997; Rodrigo and Mullins, 1996; Rodrigo *et al.*, 1999; Yamaguchi and Gojobori, 1997). In addition, as Nielsen (Chapter 11) points out, only a few sites may be under strong selection, so that when all sites of the sampled sequences are considered collectively, the "average" effect is one of neutral or nearly neutral evolution.

Finally, recent results by Golding (1997) and Neuhauser and Krone (Krone and Neuhauser, 1997; Neuhauser and Krone, 1997) indicate that the distributions of coalescent times appear to remain close to that expected under neutral models of evolution even when moderate levels of purifying selection operate on the population. At this point, we do not know how positive selection influences coalescent times. Perhaps the safest approach is to test for selection (Nielsen, Chapter 11 in this book) before applying any of the genealogy-based methods discussed in this chapter.

Table 3 also highlights the modular nature of existing methods. It is not possible, at present, to jointly estimate the rates of migration, growth and recombination, for instance. There is no theoretical impediment to doing this; the computational complexity of the task, however, is daunting. This is likely to be a research imperative in the near future. In fact, with *LAMARC*, work is ongoing to combine all of the forces discussed so far into a single sampler. In addition, Krone and Neuhauser (Krone and Neuhauser, 1997; Neuhauser and Krone, 1997) provide one possible approach to incorporating natural selection into the coalescent, by modeling a selective advantage of one allele over another as a pseudo-branch that can only be traversed if the allele is in the favored state. A possible alternative is to model the favored and disfavored alleles as defining pseudo-populations: information about the selection coefficient is then found in the relative growth rate of the favored class.

There is a second imperative that we believe will play a key role in focusing research and development of population genetics methods in the next few years. HIV, unlike eukaryotes and most prokaryotes, evolves so rapidly that over the entire course of infection in an individual (lasting possibly 10 years or more), the HIV *env* gene will diverge as much from that of the founder population as small-subunit ribosomal RNA has over the entire history of metazoan evolution (500 – 1000 million years). This means that it is possible to obtain samples of sequences from the HIV population within infected individuals, as well as from the population of hosts as a whole, at several times and statistically detect a measurable accumulation of substitutions. But how should one analyze these serial samples? Analyzing each sample separately ignores the fact that there is considerable shared historical information amongst samples. Consider, for instance, two samples each of 10 sequences, obtained a year apart from an HIV population with a constant effective size of 1000. Assume that there are 300 generations a year (Rodrigo *et al.*, 1999). The expected time to the MRCA of both samples is $2N_e = 2000$ generations, which is approximately 6.7 years from the time of sampling, counting backwards in time. If the samples are only a year apart, that still leaves over 5 years of shared history. Most of the historical information in the genealogies of these two samples of sequences is not independent and neither are any estimates derived separately for each sample. For this reason, methods must be developed that use a single genealogy of all serial samples obtained from the population. These methods are only now appearing: a method to reconstruct serial genealogies under the assumption of a molecular clock is available (*sUPGMA*: A. Drummond and A. Rodrigo), as is a method that returns the maximum likelihood estimate of mutation rate when serial samples have been obtained (*TipDate*, Rambaut, 2000). Rodrigo and Felsenstein (1999) also derived the probability $P(G|\Theta)$ for samples obtained serially from a constant-sized population; this probability, readers will recall, is used to obtain the likelihood estimate of Θ (Equation 9). Likelihood estimators of population parameters based on serial-sample genealogies are not far off from being developed.

Table 3. Methods discussed in this chapter, and their assumptions.

Method	Assumptions
LTT/Skyline plots	Genealogy known Neutral evolution No recombination Molecular clock No subdivision
Slatkin-Maddison test	Genealogy known Neutral evolution (?) No recombination
Fluctuate (likelihood estimator of growth rate)	Neutral evolution No recombination Molecular clock No subdivision
Recombine (likelihood estimator of recombination rate)	Neutral evolution Constant population size Molecular clock No subdivision
Migrate (likelihood estimator of subpopulation sizes and migration rates)	Neutral evolution No recombination Constant population size Molecular clock

7. SOFTWARE

Genetree (Bahlo and Griffiths, 2000) implements an alternative Markov Chain Monte Carlo method to that described for *LAMARC*. The coalescent-derived estimates of mutation and time to MRCA are likelihood-based. *Genetree* does however allow for models that incorporate population subdivision and size change, but excludes recombination and selection and needs an infinite sites model. The program is available at *http://www.maths.monash.edu.au/~mbahlo/mpg/gtree.html.*

GENIE (Genealogy Interval Explorer; O. Pybus and A. Rambaut) is a program for the inference of demographic history from reconstructed phylogenies. It implements the methods by Pybus, Rambaut and Harvey (2000), including the Skyline plots described above. The program is available at *http://evolve.zoo.ox.ac.uk.*

LAMARC is a suite of programs including *Migrate* (Beerli and Felsenstein, 1999), *Fluctuate* and *Recombine* (M. Kuhner, J. Yamato, and J. Felsenstein). The programs implement a Markov Chain Monte Carlo algorithm to derive maximum-likelihood estimators of population parameters. The programs in the *LAMARC* package are available as C-source code or executable programs for several platforms, including Windows and Macintosh from *http://evolution.genetics.washington.edu/lamarc.html.*

MacClade (Maddison and Maddison, 1992) is a MacOS-based program that allows users to manipulate phylogenetic trees and explore the properties of character-state distributions, including the number of state changes on one or more topologies. Random trees can be constructed, and character states of hypothetical ancestors can reconstructed. The program and a comprehensive manual are available from Sinauer Associates, Inc.

PEBBLE (A. Drummond, M. Goode and G. Ewing) builds a graphical user interface around a functional language specifically designed for evolutionary analyses. Methods presently implemented in *PEBBLE* include the construction of serial sample phylogenies, estimation of population parameters including substitution/mutation rates using pairwise distances, ML estimation of divergence between serial samples assuming constant or varying mutation rates, and the simulation of genealogies and sequences under a constant-sized population model with or without serial sampling. The program is written in JAVA and can run on any platform with JAVA Runtime Environment 1.1 or later installed. The program can be obtained from *http://www.cebl.auckland.ac.nz.*

sUPGMA (A. Drummond and A. Rodrigo) takes a distance matrix of serially sampled sequences and reconstructs a phylogenetic tree under the assumption of a molecular clock. The program can be run either as a JAVA applet or application, and is available at *http://www.cebl.auckland.ac.nz.*

TipDate (Rambaut, 2000) is an application for estimating the rate of molecular evolution (and hence a time-scale) for a phylogeny consisting of dated tips. These will most frequently be from viruses or other fast-evolving pathogens that have been isolated over a range of dates. The program can also return the likelihood for the simple molecular clock model (i.e., assuming that all sequences are contemporary) or the non-clock model. These are useful for likelihood ratio tests of the fit of the model to the data. The program is available at *http://evolve.zoo.ox.ac.uk.*

Treevolve simulates the evolution of DNA sequences under a coalescent model, which allows exponential population growth, population subdivision according to an island model, migration and recombination. In addition different periods of population dynamics can be enforced at different times. For example, a period of exponential growth can be followed by a period of stasis where the population is subdivided into demes. Multiple sets of such simulated sequence data can then be compared to sequence data sampled from a population of interest using suitable statistics, and various evolutionary hypotheses concerning the evolution of this population tested. The program is available at *http://evolve.zoo.ox.ac.uk.*

ACKNOWLEDGMENTS

PB and MKK were supported by National Health Institutes USA grants GM51929 and GM01989, both to Joseph Felsenstein. AR and OGP are funded by a grant (number 050275) from the Wellcome Trust. Research by AGR was funded by a NIH grant GM59174. MWR, DN, and YW were supported by grants from the US Public Health Service to James I. Mullins.

REFERENCES

Bahlo, M. and Griffiths, R. C. 2000. Inference from gene trees in a subdivided population. *Theor. Popul. Biol.* **57**: 79-95

Beerli, P. 1998. Estimation of migration rates and population sizes in geographically structured populations, In *Advances in Molecular Ecology* (Carvalho, G., ed.). NATO Science Series A: Life Sciences, IOS Press, Amsterdam.

Beerli, P., and Felsenstein, J. 1999. Maximum likelihood estimation of migration rates and effective population numbers in two populations using a coalescent approach. *Genetics* **152**: 763-773.

Coombs, R. W., Speck, C. E., Hughes, J. P., Lee, W., Sampoleo, R., Ross, S. O., Dragavon, J., Peterson, G., Hooton, T. M., Collier, A. C., Corey, L., Koutsky, L. and Krieger, J. N. 1998. Association between culturable human immunodeficiency virus type 1 (HIV- 1) in semen and HIV-1 RNA levels in semen and blood: evidence for compartmentalization of HIV-1 between semen and blood. *J. Infect. Dis.* **177:** 320-30.

Delwart, E. L., Mullins, J. I., Gupta, P., Learn, G. H., Jr., Holodniy, M., Katzenstein, D., Walker, B. D. and Singh, M. K.. 1998. Human immunodeficiency virus type 1 populations in blood and semen. *J. Virol.* **72:** 617-623.

Farris, J. S. 1970. Methods for computing Wagner trees. *Syst. Zool.* **18:** 83-92.

Felsenstein, J. 1981. Evolutionary trees from DNA sequences: a maximum likelihood approach. *J. Mol. Evol.* **17**: 368-376

Felsenstein, J. 1992. Estimating effective population size from samples of sequences: inefficiency of pairwise and segregating sites as compared to phylogenetic estimates. *Genet. Res.* **59**: 139-147.

Fitch, W. M. 1971. Toward defining the course of evolution: minimum change for a specific tree topology. *Syst. Zool.* **20:** 406-416.

Fu, Y-X. 1994. A phylogenetic estimator of effective population size or mutation rate. *Genetics* **136**: 685-692.

Geyer, C. J. 1994. Estimating normalizing constants and reweighting mixtures in Markov chain Monte Carlo. *Technical Report No. 568r4*, School of Statistics, University of Minnesota, St. Paul, MN.

Grassly, N. C., Harvey, P. H.and Holmes, E. C. 1999. Population dynamics of HIV-1 inferred from gene sequences. *Genetics* **151**: 427-438.

Griffiths, R. C. and Marjoram, P. 1996. Ancestral inference from samples of DNA sequences with recombination. *J. Computational Biol.* **3**: 479-502.

Griffiths, R. C. and Tavaré, S. 1994. Sampling theory for neutral alleles in a varying environment. *Phil. Trans. R. Soc. Lond. B* **344**: 403-410.

Golding, B. 1997. The effect of purifying selection on genealogies, In *Progress in Population Genetics and Human Evolution, Vol. 87.* (Donnelly, P. and Tavaré, S., eds.) Springer-Verlag, New York.

Haggerty, S. and Stevenson, M. 1991. Predominance of distinct viral genotypes in brain and lymph node compartments of HIV-1-infected individuals. *Viral Immunol.* **4:** 123-131.

Hastings, W. 1970. Monte Carlo sampling methods using Markov chains and their applications. *Biometrika* **57**: 97-109.

Holmes, E. C., Nee, S., Rambaut, A., Garnett, G. P. and Harvey, P. H. 1995. Revealing the history of infectious disease epidemics through phylogenetic trees. *Phil. Trans. R. Soc. Lond. B.* **349**: 33-40.

Holmes, E. C., Pybus, O. G. and Harvey, P. H. 1999. The molecular population genetics of HIV-1, In *The Evolution of HIV* (Crandall, K. A., ed.). John Hopkins University Press, Baltimore, MD.

Hudson, R. R. 1990. Gene genealogies and the coalescent process, In *Oxford Surveys in Evolutionary Biology, Vol. 7* (Dawkins, R. and Ridley, M.) Oxford University Press, Oxford, UK.

Jacquez, J. A., Koopman, J. S., Simon, C. P. and Longini, I. M., Jr. 1994. Role of primary infection in epidemics of HIV infection in gay cohorts. *J. Acquir. Immune Defic. Syndr.* **7**: 1169-1184.

Keys, B., Karis, J., Fadeel, B., Valentin, A., Norkrans, G., Hagberg, L. and Chiodi, F. 1993. V3 sequences of paired HIV-1 isolates from blood and cerebrospinal fluid cluster according to host and show variation related to the clinical stage of disease. *Virology*. **196:** 475-483.

Kiessling, A. A., Fitzgerald, L. M., Zhang, D., Chhay, H., Brettler, D., Eyre, R. C., Steinberg, J., McGowan, K. and Byrn, R. A. 1998. Human immunodeficiency virus in semen arises from a genetically distinct virus reservoir. *AIDS Res. Hum. Retroviruses*. **14 Suppl. 1:** S33-41.

Kingman, J. F. C. 1982a. The coalescent. *Stochastic Processes and Their Applications* **13**: 235-248.

Kingman, J. F. C. 1982b. On the genealogy of large populations, In *Essays in Statistical Science* (Gani, J. and Hannan, E. eds.) Applied Probability Trust, London.

Korber, B., M., Muldoon, J., Theiler, J., Gao, F., Gupta, R., Lapedes, A., Hahn, B. H., Wolinsky, S. and Bhattacharya, T. 2000. Timing the ancestor of the HIV-1 pandemic strains. *Science*. **288**: 1789-1796.

Korber, B. T. M., Kunstman, K. J., Patterson, B. K., Furtado, M., McEvilly, M. M., Levy, R. and Wolinsky, S. M. 1994. Genetic differences between blood- and brain-derived viral sequences from HIV-1-infected patients: Evidence of conserved elements in the V3 region of the envelope protein of brain-derived sequences. *J. Virol.* **68:** 7467-7481.

Krone, S. M. and Neuhauser, C. 1997. Ancestral processes with selection. *Theor. Pop. Biol.* **51**: 210-237.

Kuhner, M., Yamato, J. and Felsenstein, J. 1995. Estimating effective population size and mutation rate from sequence data using Metropolis-Hastings sampling. *Genetics* **140**: 1421-1430.

Kuhner, M., Yamato, J. and Felsenstein, J. 1998. Maximum likelihood estimation of population growth rates based on the coalescent. *Genetics* **149**: 429-434.

Leigh Brown, A. J. 1997. Analysis of HIV-1 env gene sequences reveals evidence for a low effective number in the viral population. *Proc. Natl. Acad. Sci. USA* **94**: 1862-1865.

Leitner, T. and Albert, J. 1999. The molecular clock of HIV-1 unveiled through analysis of a known transmission history. *Proc. Natl. Acad. Sci. USA* **96**: 10752-7.

Leitner, T., Escanilla, D., Franzen , C., Uhlen, M. and Albert, J. 1996. Accurate reconstruction of a known HIV-1 transmission history by phylogenetic tree analysis. *Proc. Natl. Acad. Sci. USA* **93**: 10864-10869.

Maddison, W. P. and Maddison, D. R.. 1992. *MacClade - Analysis of Phylogeny and Character Evolution - Version 3.* Sinauer Associates, Sunderland, MA.

McCutchan, F. E. 1999. Global diversity in HIV, In *The Evolution of HIV* (Crandall, K. A., ed.) John Hopkins University Press, Baltimore, MD.

Metropolis, N., Rosenbluth, A., Rosenbluth, M., Teller, A. and Teller, E. 1953. Equation of state calculations by fast computing machines. *J. Chem. Phys.* **21**: 1087-1092.

Nee, S., Holmes, E. C., Rambaut, A. and Harvey, P. H. 1995. Inferring population history from molecular phylogenies. *Phil. Trans. R. Soc. Lond. B* **349**: 25-31.

Nielsen, R. and Slatkin, M. 2000. Likelihood analysis of ongoing gene flow and historical association. *Evolution* **54**: 44-50.

Neilson, J. R., John, G. C., Carr, J. K., Lewis, P., Kreiss, J. K., Jackson, S., Nduati, R. W., Mbori-Ngacha, D., Panteleeff, D. D., Bodrug, S., Giachetti, C., Bott, M. A., Richardson, B. A., Bwayo, J., Ndinya-Achola, J. and Overbaugh, J. 1999. Subtypes of human immunodeficiency virus type 1 and disease stage among women in Nairobi, Kenya. *J. Virol.* **73**: 4393-4403.

Neuhauser, C. and Krone, S. M. 1997. The genealogy of samples in models with selection. *Genetics* **145**: 519-534.

Ohta, T. and Gillespie, J. H. 1996. Development of neutral and nearly neutral theories. *Theor. Popul. Biol.* **49**: 128-142.

Poss, M., Rodrigo, A. G., Gosink, J. J., Learn, G. H., de Vange Panteleeff, D., Martin, H. L., Jr., Bwayo, J., Kreiss, J. K. and Overbaugh, J. 1998. Evolution of envelope sequences from the genital tract and peripheral blood of women infected with clade A human immunodeficiency virus type 1. *J. Virol.* **72:** 8240-8251.

Pybus, O. G., Holmes, E. C. and Harvey, P. H. 1999. The mid-depth method and HIV-1: a practical approach for testing hypotheses of viral epidemic history. *Mol. Biol. Evol.* **16**: 953-959.

Pybus, O. G., Rambaut, A. and Harvey, P. H. 2000. An integrated framework for the inference of viral population history from reconstructed genealogies. *Genetics* **155**: in press.

Rambaut, A. 2000. Estimating the rate of molecular evolution: Incorporating non-contemporaneous sequences into maximum likelihood phylogenies. Bioinformatics **16**: 395-399.

Rayfield, M. A., Downing, R. G., Baggs J., Hu, D. J., Pieniazek, D., Luo, C. C., Biryahwaho, B., Otten, R. A., Sempala, S. D. and Dondero, T. J. 1998. A molecular epidemiologic survey of HIV in Uganda. *AIDS* **12**: 521-527.

Robertson, J. R., Bucknall, A. B., Welsby, P. D., Roberts, J. J., Inglis, J. M., Peutherer, J. F. and Brettle, R. P.1986. Epidemic of an AIDS related virus (HTLV-III/LAV) infection among intravenous drug abusers. *Br. Med. J.* **292**: 527-529.

Rodrigo, A. G. and Felsenstein, J. 1999. Coalescent approaches to HIV population genetics. In *The Evolution of HIV* (Crandall, K. A., ed.). John Hopkins University Press, Baltimore, MD.

Rodrigo, A. G., and Mullins, J. I. 1996. HIV-1 molecular evolution and the measure of selection. *AIDS Res. and Hum. Retrovir.* **12**: 1681-1685.

Rodrigo, A. G., Shpaer, E. G., Delwart, E. L., Iversen, A. K. N., Gallo, M. V., Brojatsch, J., Hirsch, M. S., Walker, B. D. and Mullins, J. I.. 1999. Coalescent estimates of HIV-1 generation time in vivo. *Proc. Natl. Acad. Sci. USA* **96**: 2187-2191.

Rouzine, I. M., and Coffin, J. M. 1999. Linkage disequilibrium test implies a large effective population number for HIV in vivo. *Proc. Natl. Acad. Sci. USA* **96**: 10758-10763.

Sankale, J. L., De La Tour, R. S., Marlink, R. G., Scheib, R., Mboup, S., Essex, M. E. and Kanki, P. J. 1996. Distinct quasi-species in the blood and the brain of an HIV-2-infected individual. *Virology* **226:** 418-423.

Shaheen, F., Sison, A. V., McIntosh, L., Mukhtar, M. and Pomerantz, R. J. 1999. Analysis of HIV-1 in the cervicovaginal secretions and blood of pregnant and nonpregnant women. *J. Hum. Virol.* **2:** 154-166.

Shankarappa, R., Margolick, J. B., Gange, S. J., Rodrigo, A. G., Upchurch, D., Farzadegan, H., Gupta, P., Rinaldo, C. R., Learn, G. H., He, X., Huang, X.-L. and Mullins, J. I.. 1999. Consistent viral evolutionary dynamics associated with the progression of HIV-1 infection. *J. Virol.* **73**: 10489-10502.

Shapshak, P., Segal, D. M., Crandall, K. A., Fujimura, R. K., Zhang, B. T., Xin, K. Q., Okuda, K., Petito, C. K., Eisdorfer, C. and Goodkin, K. 1999. Independent evolution of HIV type 1 in different brain regions. *AIDS Res. Hum. Retroviruses* **15:** 811-820.

Slatkin, M. and Hudson, R. R. 1991. Pairwise comparisons of mitochondrial DNA sequences in stable and exponentially growing populations. *Genetics* **129**: 555-562.

Slatkin, M. and Maddison, W. P. 1989. A cladistic measure of gene flow inferred from the phylogenies of alleles. *Genetics* **123:** 603-613.

Slatkin, M., and Maddison, W. P. 1990. Detecting isolation by distance using phylogenies of genes. *Genetics* **126:** 249-260.

Takahata, N. and Slatkin, M.. 1990. Genealogy of neutral genes in two partially isolated populations. *Theor. Popul. Biol.* **38**: 331-50.

UNAIDS. 1998. Report on the global HIV/AIDS epidemic. *http://www.unaids.org*.

UNAIDS. 1999. AIDS epidemic update: December 1999. *http://www.unaids.org*.

Wakeley, J. 1998. Segregating sites in Wright's island model. *Theor. Popul. Biol.* **53**: 166-174.

Watterson, G. A., 1975. On the number of segregating sites in genetical models without recombination. *Theor. Popul. Biol.* **7**: 256-276.

Whitlock, M. C. and McCauley, D. E. 1999. Indirect measures of gene flow and migration: FST $\neq$ 1/(4Nm + 1). *Heredity* **82**: 117-125.

Wong, J. K., Ignacio, C. C., Torriani, F., Havler, D., Fitch, N. J. S. and Richman, D. D. 1997. In vivo compartmentalization of human immunodeficiency virus: evidence from the examination of *pol* sequences from autopsy tissues. *J. Virol.* **71:** 2059-2071.

Wright, S. 1938. Size of population and breeding structure in relation to evolution. *Science* **87**: 430-431.

Wright, S. 1952 The theoretical variance within and among subdivisions of a population that is in a steady state. *Genetics* **37**: 312-321.

Yamaguchi, Y. and Gojobori, T. 1997. Evolutionary mechanisms and population dynamics of the third variable envelope region of HIV within single hosts. *Proc. Natl. Acad. Sci. USA* **94**: 1264-1269.

DETECTING SELECTION IN PROTEIN CODING GENES USING THE RATE OF NONSYNONYMOUS AND SYNONYMOUS DIVERGENCE

Rasmus Nielsen

Department of Organismic and Evolutionary Biology
Harvard University, Cambridge, MA 02138 USA

1. INTRODUCTION

One of the major projects of molecular population genetics and molecular evolutions has been to quantify the role of selection in molecular evolution. Studies by Lewontin and Hubby (1966) and others in the late sixties demonstrated an abundance of variation at the molecular level. This high level of diversity was consistent with models of positive selection acting to maintain polymorphisms within species and causing substitutions between species. However, as argued by Kimura (1968) and King and Jukes (1969), models that include only selectively neutral mutations could also explain the high levels of standing genetic variation and divergence between species. The basic idea was that neutral mutations, having no effect on the fitness of the organism, might be much more common than positively selected mutations. Such mutations may increase in frequency in a population by random genetic drift. Molecular divergence between species and molecular variation within species can therefore be explained solely in terms of mutation and genetic drift, and selection need not be invoked. This theory of molecular evolution was named "the neutral theory" and has since then served as a paradigm of molecular evolution. One important point not to forget is that the neutral theory allows for arbitrary high levels of negative selection. The existence of functional constraints and conserved amino acid sites are in fact quite consistent with the neutral theory. However, large proportions of segregating slightly deleterious mutations or mutations with positive selection coefficients are not accommodated by the neutral theory in its most strict form.

The importance of the neutral theory has as much been to provide a null model used when testing for the presence of selection as a universal theory for the causes of molecular evolution. Statistical tests of the neutral null model are referred to as "neutrality tests" and have played a prominent role in detecting cases of adaptation at the molecular level. Familiar examples include the Major Histocompatibility Complex (MHC) in humans (Hughes and Nei, 1988), the alcohol dehydrogenase (*adh*) locus in *Drosophila* (McDonald and Kreitman, 1991), the lysozyme locus in mammalian foregut fermenters (Stewart *et al.*, 1987; Messier and Stewart, 1997) and the HIV-1 envelope (*env*) gene (Bonhoeffer *et al.*, 1995; Seibert *et al.*, 1995; Yamaguchi and Gojobori, 1997). In all these cases, the neutral null model has been rejected and an adaptive explanation has been provided. For example, in the MHC, the most common adaptive explanation is overdominant selection (heterozygote advantage) to increase the range of possible immune responses, although other types of selection, typically in form of frequency dependent selection, have been invoked. The evidence for positive selection in this system is an excess of amino acid variation, beyond what can be explained by neutrality, in the antigen binding cleft.

The importance in testing neutrality in a particular genetic system lies less in the possibility of providing sweeping statements about the nature of molecular variation as such, than pointing out candidate systems for molecular adaptation. In the HIV-1 *env* gene, the presence of excess amino acid variation has been used to demonstrate the existing of diversifying selection. The inferences regarding the causes of molecular evolution have in this case provided important information regarding the interaction between immune system and virus. In the following, we will briefly provide a review of some of the most common neutrality tests. We will argue that tests based on inferences regarding the ratio of nonsynonymous to synonymous divergence have been more useful and provided the most accurate information regarding selection and adaptation. (Nonsynonymous substitutions are substitutions that alter the amino acid in the protein sequence and synonymous substitutions are substitutions between two codons that code for the same amino acid.) We will take this as a motivation for describing some of the methods used to estimate synonymous and nonsynonymous divergence and some of the statistical methods used to detect selection based on nonsynonymous and synonymous variation. Finally, we will very briefly discuss how these methods have been used to analyze HIV-1 *env* gene evolution.

2. TESTS OF NEUTRALITY

Considering the importance of the neutrality question, it is not surprising that there exists a wealth of different statistical tests of selective neutrality. Many of these tests are based on comparing the observed allelic distribution to the distribution expected under a neutral equilibrium model. One of the first such test, the Ewens-Watterson test (Ewens, 1972; Watterson, 1977; 1978), compares the observed homozygosity in the sample to the expected homozygosity given the number of observed alleles. This test assumes that the data conforms to an infinitely-many alleles model which may be appropriate for some types of allozyme or restriction fragment length polymorphism (RFLP) data. For DNA sequence data, most tests are based on comparing various summary statistics of the variability in a sample. Probably the most famous test of neutrality based on DNA sequence data, Tajima's D test, (1989) compares the number of segregating sites to the average number of pairwise differences in the sample. If the difference in estimates of $\theta = 4N_e\mu$ (N_e = effective population size, μ = mutation rate) based on these two statistics is too large or small, the neutral equilibrium model can be rejected. Several other tests based on similar test statistics have been suggested (e.g., Fu and Li, 1993; Simonsen *et al.*, 1995).

The problem with these tests lies in the interpretation. The null hypothesis being tested is a composite hypothesis involving demographic assumptions such as no population subdivision and no population growth, in addition to assumptions regarding neutrality. Often the tests may have similar power to detect deviations from the demographic model and to detect deviations from selective neutrality (Simonsen *et al.*, 1995). The availability of these tests has therefore done very little to demonstrate the action of selection in real data. The problem is that these tests are based on statistics that summarize information relating to the underlying genealogy of the examined genes. However, the distribution of genealogies is highly dependent on the demographic model. For example it is well known that the shape of the genealogy is strongly dependent on the growth rate of the population (e.g. Slatkin and Hudson, 1991). In contrast, it has been argued on theoretical grounds that some types of selection do not have a very strong effect of the shape of the genealogies (e.g. Neuhauser and Krone, 1997). Significant results of many neutrality tests may just as well be caused by population growth or population subdivision as selection. The use of the term "neutrality" test is therefore a bit misleading.

Somewhat more robust to assumptions regarding demographics is the HKA test (Hudson *et al.*, 1987). In this test, the level of polymorphism within species is compared to the number of fixed differences between species. The test requires DNA sequence data from more than one gene and it tests if the ratio of the number of fixed differences to the number of polymorphic sites is consistent among genes. However, even this test cannot be claimed to be robust to the effects of the demographic model, because it relies on an estimate of the variance in the number of polymorphic sites based on the neutral equilibrium model. Demographic factors such as population growth or population subdivision will change the variance in the number of polymorphic site in the sample relative to its mean.

The fundamental problem in (the interpretation of) these tests can be circumvented by examining different types of variation in the same regions. The test by McDonald and Kreitman (1991) compares the number of nonsynonymous and synonymous substitutions between species to the number of synonymous and nonsynonymous polymorphic sites within species using a test of homogeneity. Since both nonsynonymous and synonymous sites in the same region share the same evolutionary history, this type of test does not rely on strong assumptions regarding the demographics of the examined populations. Under neutrality, the ratio of the expected number of nonsynonymous polymorphisms to the expected number of nonsynonymous substitutions is identical to the ratio of the expected number of synonymous polymorphisms to the expected number of synonymous substitutions. If this relationship between synonymous and nonsynonymous divergence is rejected by the test, the interpretation is quite unambiguously the presence of natural selection.

This McDonald-Kreitman test illustrates quite well the power of using comparisons of nonsynonymous to synonymous mutations. Under neutrality, both synonymous and nonsynonymous mutations will occur approximately as a Poisson process on the underlying genealogy. The number of nonsynonymous mutations given the total number of mutations in the genealogy is therefore binomially distributed and independent of the shape of the genealogy. By comparing nonsynonymous and synonymous mutations, the McDonald-Kreitman test does not rely on assumptions regarding the structure of the genealogy, and is therefore making no assumptions regarding the demographics of the populations.

Selection may not have as strong an influence on the shape of genealogies as previously thought (Neuhauser and Krone, 1997). In contrast, the distribution of nonsynonymous mutations on a genealogy is strongly influenced by selection (e.g. Nielsen and Weinreich, 1999). Therefore, it seems that tests using information in the data regarding the distribution of nonsynonymous and synonymous mutations on the genealogy provide the most direct method for examining hypotheses regarding natural selection. In contrast, test based on genealogical information or allelic distributions alone may be difficult to interpret and are very sensitive to assumptions regarding demographics and population structure.

The McDonald-Kreitman test (1991) belongs to a larger class of tests in which constancy of the ratio of nonsynonymous to synonymous divergence, in different parts of a gene genealogy, is tested. We will return to this point when describing statistical methods for analyzing the ratio of nonsynonymous to synonymous divergence. For now it should be noted that while a significant result in this type of test does indicate the presence of selection, it does not necessarily demonstrate the action of positive selection. Changes in the ratio of nonsynonymous to synonymous divergence do not alone demonstrate which type of selection has been involved. For example, in the McDonald-Kreitman test, an increased level of nonsynonymous variability within species over nonsynonymous divergence between species may be caused both by balancing selection and by selection against weakly deleterious mutations.

To clearly demonstrate the action of positive selection, the footprints of molecular adaptation, it is necessary to identify values of $\omega > 1$. ω can loosely be defined as the ratio of the rate of nonsynonymous substitutions per nonsynonymous site (d_N) to the rate of synonymous substitution per synonymous site (d_S) (a more precise mathematical definition is provided later). The assumption is that the evolution of synonymous mutations is approximately neutral and ω therefore provides a measure of the strength of selection for or against new nonsynonymous mutations (but see Akashi, 1993; 1995). Most of the cases where the action of positive selection has been unambiguously demonstrated, involve showing that $\omega > 1$ in some part of a molecule or along a lineage in a genealogy. Some examples were mentioned in the introduction, including the MHC and the HIV-1 *env* gene.

In most cases where positive selection has been demonstrated, there has been a prior expectation regarding which part of the molecule should be examined or which part of the phylogeny may be relevant. For example, in the MHC the antigen binding cleft had been specified a priori as a candidate for positive selection. When researchers investigated this particular region, they found in fact that $\omega > 1$ in this area of the molecule. In the case of the lysozyme gene in foregut fermenters, researchers knew the lineages in the genealogy in which foregut fermentation was thought to have evolved. When they investigated the lysozyme locus in these particular lineages of the mammalian phylogeny, the found values of $\omega > 1$, demonstrating positive selection and adaptation at the molecular level (Stewart *et al.*, 1987; Messier and Stewart, 1997).

In many systems we do not know a priori which part of a molecule or in which part of the phylogeny to look for positive selection. In fact, in many molecules there may be a reasonable proportion of the sites in which $\omega > 1$, but this goes undetected because these sites are dispersed among sites in the molecule in which $\omega << 1$. It has therefore often been claimed that using $\omega > 1$ is a very conservative criterion for detecting positive selection.

Another problem is that results based on ω may often be questioned on statistical grounds. The major problem is that the number of synonymous and nonsynonymous substitutions or polymorphisms cannot be directly observed but must be estimated. Many results have been questioned because of problems relating to the statistical methodologies involved (e.g. Maynard Smith, 1994; Rodrigo and Mullins, 1996; Nielsen, 1997; Zhang *et al.*, 1997). It is therefore worthwhile to carefully consider the statistical methodology used when estimating ω and testing neutrality based on inferences regarding ω.

In the following, we will discuss some of the statistical issues related to the estimation of the level of nonsynonymous and synonymous divergence. In particular we will be concerned with the estimation of ω, which is the parameter of interest in many studies because it is a measure of the relative strength of selection on nonsynonymous mutations versus synonymous mutations. As mentioned, values of ω significantly larger than one is usually interpreted as evidence of the action of positive selection.

3. ESTIMATION OF ω: HEURISTIC METHODS

The early methods for estimating ω were based on estimating the average number of nonsynonymous and synonymous substitutions per site (d_N respectively d_S) separately. The parameter of interest, ω, is then estimated as $\omega = d_N/d_S$. Notice that this definition assumes the existence of nonsynonymous and synonymous sites. In real data, some codons contain sites in which a mutation may sometimes be synonymous and sometimes nonsynonymous. For example, in the third codon position of the codon CAA (which codes for the amino acid Glutamine), a transition to CAG is synonymous, but a transversion to CAC is nonsynonymous since CAC codes for the amino acid Aspargine. A synonymous site represents therefore at best a possibility of a nonsynonymous change and not a real physical unit. Some skepticism has therefore been expressed regarding the utility of the concept of nonsynonymous and synonymous sites (e.g. Muse, 1996). However, the use of these terms is pervasive in the literature, maybe because most of the early methods for estimating ω operated with these terms. In these methods, the numbers of nonsynonymous and synonymous sites in the sequences are first estimated. Thereafter, the numbers of synonymous (S_d) and nonsynonymous (N_d) differences between a pair of sequences are estimated. In this part of the inference procedure there may be some ambiguity regarding the identity of a nucleotide difference. For example, if the two codons AAG and ACC are compared, the difference between the codons could have been caused by two nonsynonymous mutations, a nonsynonymous mutation and a synonymous mutation or by more than two mutations (Fig. 1).

All the heuristic methods only consider the two "parsimonious" pathways that include only two mutations and each of these two parsimonious pathways are given a weight. For example, if the two parsimonious pathways are weighted equally, the two nucleotide differences in this codon will be scored as 1.5 nonsynonymous nucleotide difference and 0.5 synonymous nucleotide difference. However, notice that in most real cases, the two parsimonious pathways will in fact not be equally likely because of $\omega \neq 1$, unequal base frequencies, etc. Also, it is usually quite difficult to correctly incorporate well-known features of molecular evolution, such as unequal nucleotide frequencies or transition/transversion biases in these approaches for estimating nonsynonymous nucleotide sites and nucleotide differences.

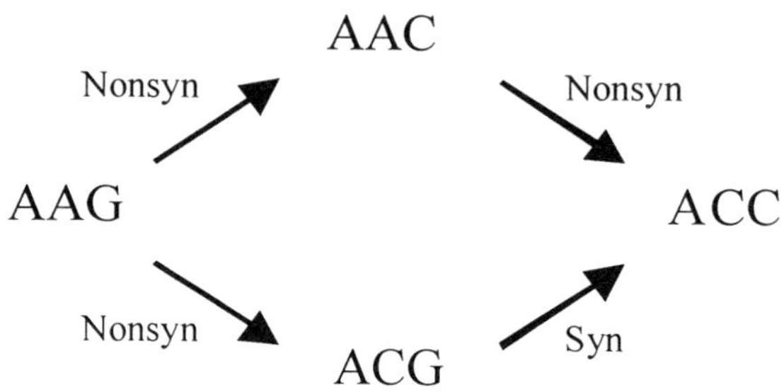

Figure 1. The two parsimonious pathways along which the codon AAG can change to ACC.

After the number of sites and the number of nucleotide differences have been estimated a correction for multiple hits in the same site is performed. The need for such a correction arises because only the most "parsimonious" pathways have been considered when estimating the number of nucleotide differences. The correction is done separately for nonsynonymous and synonymous substitutions and provides the final estimate of d_S and d_N between the two sequences. Usually the correction will be based on a continuos time Markov chain model that does not take into account the complexity of the genetic code; for example, it will ignore the fact that many synonymous sites can change into only one other synonymous state and not three other synonymous states.

The first of these methods are those of Miyata and Yasunaga (1980) and Perler and colleagues (1980), and a simplified version was subsequently provided by Nei and Gojobori (1986). This latter method is today the most commonly applied method for estimating d_N and d_S. The model used to correct for multiple hits in this method is the Jukes and Cantor (1969) model. This model assumes that all four nucleotides are of equal frequency and that all potential mutations are equally frequent. Furthermore, when estimating the number of sites and the number of differences, it is assumed that all mutations are equally likely and all parsimonious pathways are weighted equally. It is implicitly assumed that nonsynonymous and synonymous mutations are equally likely, transitions and transversions are equally likely, etc. In real data, the results of these assumptions may be a quite substantial bias in the estimation of $\omega = d_N / d_S$ (e.g. Yang and Nielsen, 2000).

Other common methods for estimating d_N and d_S based on separate estimation of the number of sites and the number of nucleotide differences include the methods by Li, Wu and Luo (1985), Li (1993), Pamilo and Bianchi (1993), Comron (1995) and Ina (1995; 1996). These methods attempt to various degrees to correct for the biases in the Nei and Gojobori (1986) method. However, it should be obvious from the above discussion that it is not trivial to establish a reliable method for estimating ω based on an estimation of nucleotide sites and nucleotide differences separately. Part of the problem is that synonymous and nonsynonymous sites do not exist as real physical units. Fortunately, it is now computationally possible to use maximum likelihood methods that can incorporate the full complexity of the genetic code when estimating d_N and d_S.

4. ESTIMATION OF ω: LIKELIHOOD METHODS

The first real statistical approaches for estimating ω were developed independently by Muse and Gaut (1994) and Goldman and Yang (1994). In these methods, it is assumed that the evolution of a DNA sequence in a coding region can be described as a continuos time Markov chain with state space on the possible codons (excluding stop codons), i.e. the size of the state space is 61 assuming the universal genetic code. Using such an approach, it is possible to obtain maximum likelihood (ML) estimates of ω and of other parameters of the model.

A popular version of the Goldman and Yang (1994) model used in Nielsen and Yang (1998), Yang and Nielsen (1998) and Yang and cowrkers (2000) assumes the following instantaneous rates of substitution from codon i to j

$$Q_{ij} = \begin{cases} 0, & \text{if the two codons differ at more than one position,} \\ \mu\pi_j, & \text{for synonymous transversion,} \\ \mu\kappa\pi_j, & \text{for synonymous transition,} \\ \mu\omega\pi_j, & \text{for nonsynonymous transversion,} \\ \mu\omega\kappa\pi_j, & \text{for nonsynonymous transition,} \end{cases} \quad , \ i \neq j \qquad (1)$$

$Q = \{Q_{ij}\}$ is the infinitesimal generator of the process with diagonal elements given by the mathematical requirement $\sum_j q_{ij} = 0$ for all i. Here κ is the transition/transversion bias and π_j is the stationary frequency of codon j. The reason for disallowing substitutions that involve more than one position in the codon is that the mutational process is supposed to occur primarily as point mutations in individual nucleotide sites.

The definition of the assumed substitution process given in Equation 1 provides a more precise definition of ω. We see from Equation 1 that ω is the relative increase or decrease in the rate on nonsynonymous substitution in comparison to the rate of synonymous substitution. A treatment of the problem in this statistical framework does not rely on the assumption of the existence of nonsynonymous and synonymous sites as physical entities. However, re-interpreting the proportion of nonsynonymous and synonymous sites as the relative proportion of changes that would be nonsynonymous or synonymous without selection, it is possible to get estimates of d_S and d_N using the ML method (Goldman and Yang, 1994; Yang and Nielsen, 1998).

Usually, ω and κ are estimated jointly from the data using maximum likelihood. The stationary base frequencies are typically estimated separately using a method of moments estimator to reduce the computational complexity of the problem. For example, the observed codon frequencies may be used as estimates of the stationary codon frequencies. For very long sequences, where there is a reasonably large count of each type of codon, this method will work well. However, for short sequences various methods for estimating the codon frequencies based on the observed nucleotide frequencies may work better.

Given a particular rate matrix (Q), there exist standard numerical methods for calculating the corresponding transition probabilities of the substitution process. The transition probabilities, $P_{ij}(t)$, are the probabilities of observing a change from codon i to codon j in time t, and $P(t) = \{P_{ij}(t)\}$ is the matrix of transition probabilities of all possible substitutions. They are calculated as $P(t) = e^{Qt}$, which usually is obtained either by diagonalization of Q or by expanding e^{Qt} in a Taylor's series (e.g. Karlin and Taylor, 1975, pp. 150-152).

For two sequences, with observed codons c_1 and c_2 at site k, the likelihood function at site k is given by

$$\ell_k(\Theta) = \ln\left(\pi_{c_1}\right) + \ln\left(P_{c_1c_2}(t)\right) \quad (2)$$

and the log likelihood function for the entire sequences is given by $\ell(\Theta) = \sum_{k=1}^{s} \ell_k(\Theta)$. Θ is the set of parameters, which in the model described by Equation 1 is given by $\Theta = \{\kappa, \omega, t\}$ and s is the number of codon sites in the sequence (equal to the number of nucleotide sites divided by 3). Joint ML estimates of these parameters can then be obtained by optimizing this likelihood function. Notice that it does not matter if the likelihood is calculated as $\ln\left(\pi_{c_1}\right) + \ln\left(P_{c_1c_2}(t)\right)$ or as $\ln\left(\pi_{c_2}\right) + \ln\left(P_{c_2c_1}(t)\right)$ because it is ensured by the definition of the Markov chain (Equation 1) that the process is time reversible. The computational time required to optimize the likelihood function for a pair of sequences is typically less than 30 seconds on an average workstation. In cases where the number of sequences is so large (>> 100 sequences) that it may not be feasible to obtain all pairwise comparison using ML, an efficient approximation to the ML estimator can be used (Yang and Nielsen, 2000). This approximation is based on weighting pathways according to their probability given by a rate matrix such as Equation 1.

Some of the statistical properties of the ML method are examined in Goldman and Yang (1994), Muse and Gaut (1994), Muse (1996) and Yang and Nielsen (2000). Although the ML estimate often has a lower variance than the heuristic methods (Yang and Nielsen, 2000), a more important advantage of the ML method is that it correctly incorporates important features of molecular evolution such as the transition/transversion bias and unequal codon frequencies. It thereby avoids some of the strong biases in the parameter estimates encountered in other methods.

Another advantage of the likelihood method is that it can be applied to multiple sequences simultaneously. Equation 2 can be generalized to multiple sequences, assuming an underlying phylogeny. The set of parameters Θ will then be larger and contain parameters relating to the topology of the underlying phylogeny and the branch lengths of the phylogeny. In such a model ω can be estimated for all sequences jointly, for each lineage of the phylogeny separately or for specific sets of lineages. Often, it is of interest to estimate ω along a particular lineage (e.g. Gillespie, 1989; Messier and Stewart, 1997; Yang, 1998). The heuristic methods will proceed either by attempting to estimate ancestral sequences at particular nodes in the phylogeny or using methods based on pairwise comparisons. Using ML such heuristics is not necessary. However, there are some computational limitations. While 20 sequences can be analyzed quite fast using ML it may not currently be computationally feasible to analyze 200 sequences jointly using these methods.

Probably the most important advantage of the ML methods is that this framework easily can be used for hypothesis testing. In fact, the emergence of likelihood methods applicable to codon based models are in many ways revolutionizing the way we analyze comparative DNA sequence data. Huelsenbeck and Rannala (1997) provided a general review of the use of likelihood methods for testing evolutionary hypotheses using DNA data sequences. In the following we will discuss some of the specifics relating to codon based models.

5. TESTING HYPOTHESES REGARDING ω

As discussed in the **Tests of Neutrality** section, the McDonald-Kreitman (1991) test belongs to a larger class of tests in which constancy of ω is tested. Variance in ω among lineages is usually interpreted as evidence against strict neutrality (e.g. Gillespie, 1989). The only strictly neutral explanation is variance in the amount of functional constraints among lineages, an explanation that very few have been willing to entertain as a common feature of molecular evolution of closely related species (but see Takahata, 1987). Tests of constant ω can easily be performed in the likelihood framework. In Yang and Nielsen (1998), the case of three species was considered. The null hypothesis was constancy of ω among lineages, i.e. H_0: $\omega_1 = \omega_2 = \omega_3$, where ω_1, ω_2 and ω_3 is the nonsynonymous/synonymous rate ratio on the first second and third lineage of the genealogy. Under the alternative hypothesis (H_1) ω_1, ω_2 and ω_3 are free parameters. A classical likelihood ratio test is performed by first maximizing the likelihood under H_0 and H_1. Then assuming the usual asymptotic properties hold, minus two times the log likelihood ratio is approximately distributed as a χ^2-distribution with 2 degrees of freedom (*df*) for the three taxon case. For example, if the ratio of the maximized log likelihoods is more than approximately 3 log likelihood units larger under H_1 than under H_0, constancy in ω among lineages can be rejected at the 5% level. In the general case, with n taxa and a fixed topology of the phylogeny, there are $2n$-3 *df* for the χ^2 approximation.

Yang and Nielsen (1998) applied this test to 48 loci from humans, artiodactyls and rodents. In 22 of the 48 loci the nonsynonymous/synonymous rate ratio was found to vary significantly. One likely explanation is the fixation of slightly deleterious mutations in combination with differences in the effective population size among species (e.g. Ohta, 1993; 1995). Variation in ω among lineages is indicative of the action of selection, but not necessarily of positive selection.

Another situation where tests of varying ω is of importance is to demonstrate selection in particular lineages of a phylogeny. Previous work by Stewart, Schilling and Wilson (1987) and Messier and Stewart (1997) had demonstrated that episodes of positive selection might have occurred in the lysozyme gene on the lineages of the mammalian phylogeny in which foregut fermentation is thought to have evolved. Stomach lysozyme is important in digestion and some molecular adaptations may therefore be expected to have occurred in this gene during the evolution of foregut fermentation. Yang (1998) addressed this problem in a likelihood framework by using a likelihood ratio test to show that ω was significantly higher in the lineages of the phylogeny in which foregut fermentation had involved than in the other lineages of the phylogeny. Furthermore, he showed, again using a likelihood ratio test, that ω in these lineages was significantly larger than 1. The action of positive selection in these lineages can therefore be firmly established. The likelihood framework proved useful in this case for establishing unambiguous tests of important evolutionary hypotheses.

Hoffmann and colleagues (2000) performed a similar analysis of the *Cytochrome b* gene. Using a likelihood ratio test, they showed that a significant acceleration of the rate of nonsynonymous substitution had occurred in some subterranean rodent species compared to other closely related species. The subterranean atmosphere is very hypoxic and some molecular adaptations may therefore be expected during the invasion of the subterranean niche in genes involved in respiration such as *Cytochrome b*.

The McDonald-Kreitman (1991) test itself can also be performed as a likelihood ratio test. The null hypothesis for such a test is that the lineages in the shared genealogy within and between species all have identical nonsynonymous/synonymous rate ratios. The alternative hypothesis is that ω may differ between lineages in the interspecific and the intraspecific part of the genealogy (Hasegawa *et al.*, 1999).

Another way of testing the neutral hypothesis is to examine if there are regions of a gene in which $\omega > 1$. A well known example is the work on the MHC region by Hughes and Nei (1988), which demonstrated that positive selection is acting on the MHC. This type of test has not been very popular, because it will in most cases only have very little power to detect positive selection. The reason is that ω is calculated as an average over multiple sites, many of which may be strongly functionally constrained and not undergoing any positive selection. However, even in such cases, it is possible to detect positive selection using appropriate statistical techniques. Nielsen and Yang (1998) developed a method in which ω is allowed to vary among sites. In such a model the log likelihood function in a particular site can be written as

$$\ell_k(\Theta) = \ln\left(\int \Pr(x_k \mid \omega, \Theta) f(\omega \mid \Theta) d\omega\right) \tag{3}$$

Where $\Pr(x_k \mid \omega, \Theta)$ is the probability of the observed data in (codon) site k and $f(\omega \mid \Theta)$ is some prior distribution of ω among sites. For example, a representation of the strictly neutral model of molecular evolution might state that there are two categories of codon sites, sites in which mutations are selectively neutral ($\omega_1 = 1$) and invariable sites which are completely functionally constrained ($\omega_2 = 0$). If the proportion of neutral sites is denoted by p, the likelihood function is then given by

$$\ell_k(p) = \ln\left(\Pr(x_k \mid \omega = \omega_1)p + \Pr(x_k \mid \omega = \omega_2)(1-p)\right) \tag{4}$$

Here and in the following other parameters such as κ and parameters relating to the phylogeny have been suppressed in the notation. The parameter p can be estimated using Equation 4 by maximum likelihood jointly with any other parameters of the model.

An alternative model which includes positively selected sites may have an additional category of positively selected sites in which the nonsynonymous/synonymous rate ratio is ω_3, $\omega_3 > 1$. There are then three parameters of the model: p_1, the proportion of neutral sites, p_3, the proportion of positively selected sites, and ω_3. The proportion of invariable sites is then $p_2 = 1 - p_1 - p_3$. The log likelihood function is given by

$$\ell_k(p) = \ln\left(\Pr(x_k \mid \omega = \omega_1)p_1 + \Pr(x_k \mid \omega = \omega_3)p_3 + \Pr(x_k \mid \omega = \omega_2)(1 - p_1 - p_3)\right) \quad (5)$$

By optimizing this likelihood function, the three parameters can be estimated. To test if there are any positively selected sites in the model, the maximized likelihood under the two models is compared. Using the usual asymptotic approximations, minus two times the log likelihood ratio is distributed as a χ^2-distribution with 2 d.f. We see that it is quite easy in this framework to test the null hypothesis that there are no sites in which $\omega > 1$. Such tests may have considerable power to detect positive selection even when the average ω is much less than 1.

Yang and his colleagues (2000) developed a series of similar models, many of them assuming continuous distributions for ω, instead of the very simple discrete distributions discussed above. They also applied tests of the hypothesis $\omega \leq 1$ for all sites, to several genes in which positive selection had not previously been documented. They found significant evidence of positive selection in several genes in which the average ω was much less than one, including β-globin genes, mitochondrial DNA from hominids and several viral genes.

These methods can also be used to obtain an empirical Bayes estimate of ω in each site. Such estimates are useful for identifying the sites undergoing positive selection and may be related directly to the function of the gene. For the model described by Equation (5), the probability that site k belongs to category i, $i = 1, 2, 3$, is

$$\Pr(\omega = \omega_i \mid x_k) = \frac{\hat{p}_i \Pr(x_k \mid \omega = \omega_i)}{\Pr(x_k)} = \frac{\hat{p}_i \Pr(x_k \mid \omega = \omega_i)}{\sum_{j=1}^{3} \hat{p}_j \Pr(x_k \mid \omega = \omega_j)} \quad (6)$$

where $\hat{p}_j$ is the ML estimate of p_j, $j = 1, 2, 3$. An empirical Bayes estimate of the distribution of ω among sites can similarly be obtained for any other prior distribution of ω. It is therefore possible to make probabilistic statements regarding which sites in the sequence are undergoing positive selection. Given sufficient data, it is possible to predict which regions of a protein are being targeted by positive selection.

6. THE HIV-1 GENE: A CASE STUDY

Nielsen and Yang (1998) applied the likelihood method discussed above to the HIV-1 *env* gene. The data used had previously been published by Holmes *et al.* (1992) and contained sequences from the HIV-1 *env* gene from a single infected individual at different points in time after the initial infection. Bonhoeffer, Holmes and Nowak (1995) had also analyzed this data and were able to show that $\omega > 1$ in the hypervariable region of this gene. The type of selection acting on the *env* gene is thought to be immune mediated selection for sequence diversity. Mutants with a new genotype may be able to avoid immune recognition and will therefore have a selective advantage.

Bonhoeffer and his coworkers (1995) further suggested that there was a decrease in ω during the course of infection. They hypothesized that this decrease was due to a weakening of the immune system and was correlated with the $CD4^+$ cell count. However, Rodrigo and Mullins (1996) questioned, on statistical grounds, if there in fact was a significant decrease in ω at different points in time after the infection.

Nielsen and Yang (1998) reanalyzed the data using the likelihood functions in Equation 4 and 5. Some of the estimated parameter values obtained in the study are in Table 1.

Table 1. Estimates of p_3, ω_3 and the log likelihood ratio of the hypothesis of no positively selected sites in the data (Nielsen and Yang, 1998)*.

	Year 3	Year 4	Year 5	Year 6	Year 7
p_3	0.216	0.046	0.157	0.084	0.043
ω_3	29.54	14.89	6.92	6.85	7.71
Log likelihood ratio	-12.88	-5.98	-4.61	-6.08	-2.39

* The data are from the HIV-1 *env* gene from an infected human individual at different stages of the infection (Holmes *et al.*, 1992).

We see that the hypothesis of no positively selected sites can be rejected for all years except year 7 after infection.

Similar strong evidence for positive selection in the *env* gene can be identified in chimpanzees infected with HIV. The occurrence of AIDS in chimpanzees infected with HIV-1 virus has recently been reported (Novembre *et al.*, 1997). The two available sequences of the *env* gene from a chimpanzee with AIDS were obtained from GenBank (accession number: AF049494 and AF049495). These two sequences were from the same infected chimpanzee at the time of AIDS disease and the time of acute infection (Mwaengo and Novembre, 1998). The sequences were aligned with the three inoculating viruses SF2, LAV and NDK. Assuming no positive selection, estimates of the parameters were $p_1 = 0.549$, $p_2 = 0.451$, and the log likelihood was -6334.02. Allowing the presence of positively selected sites, estimates of $p_1 = 0.545$, $p_2 = 0.396$, $p_3 = 0.058$ and $\omega_3 = 8.73$ were found, and the log likelihood was -6289.52. The log likelihood ratio of the hypothesis of no positively selected sites was -44.5, demonstrating that these sequences have evolved under positive selection. Positive selection in the *env* gene seems to be a general property of this system, whether the infection is in humans of chimpanzees.

We also see from Table 1 that there appears to be a change in ω_3 and p_3 between year 3 and the subsequent years, but only little change in the subsequent years. One explanation for the change in ω and p_3, that does not involve a change in the intensity of selection, is that nonsynonymous and synonymous sites accumulate variability at different rates. Shortly after the infection there will be very little sequence variability in the DNA sequences of the infecting HIV strain. Thereafter, selection and mutation will work to build up variability in the sequences. Because of the action of positive selection, nonsynonymous variability will build up faster than synonymous variability. The change in ω is therefore expected during the transient phase in which variability is being built up in the viral population. This effect can be demonstrated easily in simulation studies (Nielsen, 1999).

There may be some cause for warning when interpreting observed changes in ω between different samples. The expected value of ω is a complex function of the demographics of a population, effective population sizes and the distribution of selection coefficients. In addition, a given genetic system may not be in equilibrium. While values of $\omega > 1$ can be interpreted as evidence for positive selection, it may be more difficult to provide unambiguous interpretations of observed changes or difference in ω.

7. CONCLUSION

While many proposed tests of neutrality are highly sensitive to demographic factors, it is possible to establish tests of neutrality based on the nonsynonymous/synonymous rate ratio (ω) that do not rely on assumptions regarding the demographics of the examined populations. Furthermore, the only unambiguous method for identifying positive selection and adaptation at the molecular level is to demonstrate that $\omega > 1$ in parts of a DNA sequence or in certain lineages of a genealogy. Recently developed maximum likelihood methods based on models of codon evolution provide a natural framework for estimating ω and for testing hypothesis regarding ω. Using these methods it is possible to test hypothesis regarding the nonsynonymous/synonymous rate ratio on specific lineages of a phylogeny or to test the hypothesis of no variation in ω along the lineages of a phylogeny. Such methods are useful for identifying occurrences of adaptive molecular evolution. It is also possible to test for the presence of positively selected sites in a DNA sequence without having prior knowledge regarding which part of the molecule is being targeted by selection. In addition, the sites undergoing positive selection may be identified using an empirical Bayes approach. The codon based statistical methods provide a natural framework for investigating hypotheses regarding selection and adaptation at the molecular level.

REFERENCES

Akashi, H. 1993. Synonymous codon usage in *Drosophila melanogaster*: Natural selection and translational accuracy. *Genetics* **136**: 927-935.

Akashi, H. 1995. Inferring weak selection from patterns of polymorphism and divergence at "silent" sites in *Drosophila* DNA. *Genetics* **139**: 1067-1076.

Bonhoeffer. S, Holmes, E. C. and Nowak, M. A. 1995. Causes of HIV diversity. *Nature* **376**: 125.

Comeron, J. M. 1995. A method for estimating the numbers of synonymous and nonsynonymous substitutions per site. *J. Mol. Evol.* **41**: 1152-1159.

Ewens, W. 1972. The sampling theory of selectively neutral alleles. *Theoret. Pop. Biol.* **3**: 78-112.

Fu, Y.-X. and Li., W.-H. 1993. Statistical tests of neutrality of mutations. *Genetics* **133**: 693-709.

Gillespie, J. H. 1989. Lineage effects and the index of dispersion of molecular evolution. *Mol. Biol. Evol.* **6**: 636-647.

Goldman, N., and Yang, Z. 1994. A codon-based model of nucleotide substitution for protein-coding DNA sequences. *Mol. Biol. Evol.* **11**: 725-736.

Hasegawa, M., Cao, Y. and Yang, Z. 1998. Preponderance of slightly deleterious polymorphism in mitochondrial DNA: Nonsynonymous/synonymous rate ratio is much higher within species than between species. *Mol. Biol. Evol.* **15**: 1499-1505.

Hoffmann, F., Tomasco, I., Wlasiuk, G., Lessa, E. P. and Cook, J. A. 2000. Accelerated rates of replacement substitutions in the *Cytochrome b* of subterranean South American Octodontid rodents. *In review*.

Holmes, E. C., Zhang, L. Q., Simmonds, P., Ludlam, C. A., and Leigh Brown, A. J. 1992. Convergent and divergent sequence evolution in the surface envelope glycoprotein of human immunodeficiency virus type 1 within a single infected patient. *Proc. Natl. Acad. Sci. USA* **89**: 4835-4839.

Hudson, R. R., Kreitman, M. and Aguade, M. 1987. A test for neutral molecular evolution based on nucleotide data. *Genetics* **116**: 153-159.

Huelsenbeck, J. P., and Rannala, B. 1997. Phylogenetic methods come of age: Testing hypotheses in a phylogenetic context. *Science* **276**:227-232.

Hughes, A. L. and Nei, M. 1988. Pattern of nucleotide substitution at major histocompatibility complex class I loci reveals overdominant selection. *Nature* **335**: 167-170.

Ina, Y. 1995. New methods for estimating the numbers of synonymous and nonsynonymous substitutions. *J. Mol. Evol.* **40**: 190-226.

Ina, Y. 1996. Patterns of synonymous and nonsynonymous substitutions: an indicator of mechanisms of molecular evolution. *J. Genet.* **75**: 91-115.

Jukes, T. H., and Cantor, C. R.. 1969. Evolution of protein molecules. Pp. 21-123 in *Mammalian Protein Metabolism*, (H. N. Munro, ed.) Academic Press, New York.

Karlin, S. and Taylor, H. M. 1975. *A First Course in Stochastic Processes*. Academic Press, New York.

Kimura, M. 1968. Evolutionary rate at the molecular level. *Nature* **217**: 624-626.

King, J. L. and Jukes, T.H. 1969. Non-Darwinian evolution. *Science* **164**: 788-798.

Lewontin. R. C. and Hubby, J. L. 1966. A molecular approach to the study of genic heterozygosity in natural populations of *Drosophila pseudoobscura*. *Genetics* **54**: 595-609.

Li, W.-H. 1993. Unbiased estimation of the rates of synonymous and nonsynonymous substitution. *J. Mol. Evol.* **36**: 96-99.

Li, W.-H., Wu, C.-I., Luo, C.-C. 1985. A new method for estimating synonymous and non-synonymous rates of nucleotide substitutions considering the relative likelihood of nucleotide and codon changes. *Mol. Biol. Evol.* **2**: 150-174.

McDonald J. H. and Kreitman, M. 1991. Adaptive protein evolution at the *Adh* locus in *Drosophila*. *Nature* **351**: 652-654.

Maynard Smith., J. 1994. Estimating selection by comparing synonymous and substitutional changes. *J. Mol. Evol.* **39**: 123-128.

Messier, W., and Stewart, C.-B. 1997. Episodic adaptive evolution of primate lysozymes. *Nature* **385:** 151-154.

Miyata T., and Yasunaga, T. 1980. Molecular evolution of mRNA: a method for estimating evolutionary rates of synonymous and amino acid substitutions from homologous nucleotide sequences and its applications. *J. Mol. Evol.* **16**: 23-36.

Muse, S. V. 1996. Estimating synonymous and nonsynonymous substitution rates. *Mol. Biol. Evol.* **13**: 105-114.

Muse, S. V., and Gaut, B. S. 1994. A likelihood approach for comparing synonymous and nonsynonymous nucleotide substitution rates, with application to chloroplast genome. *Mol. Biol. Evol.* **11**: 715-724.

Mwaengo, D. and Novembre, F. J. 1998. Molecular cloning and characterization of viruses isolated from chimpanzees with pathogenic human immunodeficiency virus type 1 infection. *J. Virol.* **72**: 8976-8987.

Nei, M., and Gojobori, T. 1986. Simple methods for estimating the number of synonymous and nonsynonymous nucleotide substitutions. *Mol. Biol. Evol.* **3**: 418-426.

Neuhauser, C. and Krone, S. M. 1997. The genealogy of samples in models with selection. *Genetics* **145**: 519–534.

Nielsen, R., 1997 The ratio of replacement to silent divergence and tests of neutrality. *J. Evol. Biol.* **10**: 217-231.

Nielsen, R. 1999. Changes in d_s/d_n in the HIV-1 *env* gene. *Mol. Biol. Evol.* **16:** 711-714.

Nielsen, R. and Yang, Z. 1998. Likelihood models for detecting positively selected amino acid sites and applications to the HIV-1 envelope gene. *Genetics* **148**: 929-936

Nielsen, R. and Weinreich, D. M. 1999. The age of nonsynonymous and synonymous mutations in mtDNA and implications for the mildly deleterious theory. *Genetics* **153**: 497-506.

Novembre, F. J., Saucier, M., Anderson, D. C., Klumpp, S. A., O'Neil, S. P., Brown II, C. R., Hart, C. E., Guenthner, P. C., Swebson, R. B. and McClure, H. M. 1997. Development of AIDS in chimpanzee infected with human immunodeficiency virus type 1. *J. Virol.* **71**: 4086-4091.

Ohta, T. 1993. The nearly neutral theory of molecular evolution. *Ann. Rev. Ecol. Syst.* **23**:263-286.

Pamilo, P., and Bianchi, N. O. 1993. Evolution of the *Zfx* and *Zfy* genes — rates and interdependence between the genes. *Mol. Biol. Evol.* **10**: 271-281.

Perler, F., Efstratiadis, A., Lomedica, P., Gilbert, W., Kolodner, R., and Dodgson, J. 1980. The evolution of genes: the chicken preproinsulin gene. *Cell* **20**: 555-566.

Rodrigo, A. G. and Mullins, J. I. 1996. Human immunodeficiency virus type 1 molecular evolution and the measure of selection. *AIDS Res. Hum. Ret.* **12**:1681-1685.

Seibert S. A., Howell, C. Y., Hughes, M. K. and Hughes, A. L. 1995. Natural selection on *gag*, *pol* and *env* genes of human immunodeficiency virus 1 (HIV-1). *Mol. Biol. Evol.* **12**: 803-813.

Simonsen, K. L. Churchill, G. A. and Aquadro, C. F. 1995. Properties of statistical tests of neutrality for DNA polymorphism data. *Genetics.* **141**: 413-429.

Slatkin, M. and Hudson, R. R. 1991. Pairwise comparisons of mitochondrial DNA sequences in stable and exponentially growing populations. *Genetics* **129**: 555-562.

Stewart, C. B., Schilling, J. W. and Wilson, A. C. 1987. Adaptaive evolution in the stomach lyzosymes of foregut fermenters. *Nature* **330**: 401-404.

Tajima, F. 1989. Statistical methods for testing the neutral mutation hypothesis by DNA polymorphism. *Genetics* **123**: 585-595.

Takahata, N. 1987. On the overdispersed molecular clock. *Genetics* **116**: 169-179.

Watterson, G. A. 1977. Heterosis or neutrality? *Genetics* **85**: 789-814.

Watterson, G. A. 1978. The homozygosity test of neutrality. *Genetics* **88**: 405-417.

Yamaguchi, Y., and Gojobori, T. 1997. Evolutionary mechanisms and population dynamics of the third variable envelope region of HIV within single hosts. *Proc. Natl. Acad Sci. USA* **94**: 1264-1269.

Yang, Z. 1998. Likelihood ratio tests for detecting positive selection and application to primate lysozyme evolution. *Mol. Biol. Evol.* **15**: 568-573.

Yang, Z., and Nielsen, R. 1998. Synonymous and nonsynonymous rate variation in nuclear genes of mammals. *J. Mol. Evol.* **46**: 409-418.

Yang, Z., and Nielsen, R. 2000. Estimating synonymous and nonsynonymous substitution rates under realistic evolutionary models. *Mol. Biol. Evol.* **17**: 32-43.

Yang, Z., Nielsen, R., and Hasegawa, M. 1998. Models of amino acid substitution and applications to mitochondrial protein evolution. *Mol. Biol. Evol.* **15**: 1600-1611.

Yang, Z., Nielsen, R., Goldman, N., and Pedersen, A.-M. K. 2000. Codon-substitution models for variable selection pressure at amino acid sites. *Genetics***155**: 431-449.

Zhang, J., Kumar, S. and Nei, M. 1997. Small-sample tests of episodic adaptive evolution: a case study of primate lysozymes. *Mol. Biol. Evol.* **14**: 1335-1338.

DRUGS TARGETED AT HIV – SUCCESSES AND RESISTANCE

Clare Sansom* and Alexander Wlodawer§

*Department of Crystallography, Birkbeck College, London WC1E 7HX, UK
§Macromolecular Crystallography Laboratory, Program in Structural Biology, NCI-Frederick Cancer Research and Development Center
Frederick, MD 21702 USA.

1. INTRODUCTION

Acquired immunodeficiency syndrome (AIDS) was first identified as a previously unknown disease about twenty years ago, and since that time has become one of the most thoroughly studied human diseases. In the Western countries, it is most often caused by the human immunodeficiency virus type 1 (HIV-1), although in Africa it is commonly associated with another viral variant, HIV-2. The complete nucleotide sequence of HIV-1 was published in 1985 (Ratner *et al.*, 1985), and shows a relatively simple retrovirus, with the genome consisting of three open reading frames, *gag*, *pol*, and *env*. The *gag* open reading frame contains structural proteins such as capsid, nucleocapsid, and matrix, while regulatory proteins are encoded in the multiply-spliced *env* ORF. HIV-1 genome encodes only three unique enzymes, all of them within the *pol* open reading frame. They include reverse transcriptase (RT), a multi-domain enzyme involved in copying the RNA retroviral genome into DNA and in destroying the original RNA strand with its RNase H domain. Another enzyme is protease (PR), a symmetric homodimer related to cellular aspartic proteases such as pepsin, which is necessary for the cleavage of the viral polyproteins into individual mature proteins. Finally, HIV encodes viral integrase (IN), a multifunction processive enzyme which makes staggered cuts of the ends of viral DNA, inserting it in a non-specific manner into the host DNA.

The identification of the pathogen that causes AIDS required several years, and initially no drugs were available to fight the disease. However, since 1987, no fewer than 11 different anti-AIDS drugs have been approved for use in the United States (and in most other countries). Not surprisingly, all of these are inhibitors of the retroviral enzymes, RT and PR, although other drug design targets are still under active investigation and some novel potential drugs are now entering clinical trials. In this chapter, the design, properties, and the development of resistance against these drugs will be discussed.

The past 20 years have seen significant changes to the methodology of pharmaceutical development (Hubbard, 1997). The complementary methods of computer-aided molecular design (Leach, 1996) and combinatorial chemistry (Balkenkohl *et al.,* 1996) are now routinely employed in both the lead identification and the compound development phases of drug design. Since these 20 years have coincided with the emergence of the AIDS pandemic, it is hardly surprising that the complete battery of techniques available to the pharmaceutical scientist have been unleashed on the HIV virus. The development of the protease inhibitors, in particular, owes a lot to the use of rational drug design techniques (Sansom, 1998; Vacca and Condra, 1997; Wlodawer and Vondrasek, 1998).

2. DRUGS TARGETING REVERSE TRANSCRIPTASE

The reverse transcriptase is a heterodimer consisting of two chains, p66 and p51, encoded by the same gene belonging to the *pol* ORF. The only difference in primary structure between these chains is that p51 does not contain the RNase H domain, which is found in p66 only. Although the amino acid sequences of the common parts of the two domains are identical, only p66 contains the active sites for both primary activities of the enzyme (RNA- or DNA-dependent DNA polymerase, and RNase H), while p51 is enzymatically inactive and is presumed to play a structural role only. The structures of RT have been solved in the presence of non-nucleoside inhibitors (see below) (Kohlstaedt *et al.*, 1992); for the apoenzyme (Jager *et al.*, 1994); in a ternary complex with a bound template:primer and Fab (Jacobo-Molina *et al.*, 1993); and as a covalently trapped catalytic complex with a DNA template:primer and a deoxynucleoside triphosphate (Huang *et al.*, 1998), among others. The structure of each domain (Figure 1) can be described as a right hand, with four domains denoted as fingers, palm, thumb, and connection (Kohlstaedt *et al.*, 1992). The residues responsible for the polymerase activity include three aspartates in the palm domain (D110, D185, and D186), which is a counterpart of similar domains in other DNA and RNA polymerases. These aspartate residues act through bound divalent cations, which under physiological conditions consist of a pair of Mg^{2+} ions.

Reverse transcriptase inhibitors can be divided into two general classes. The first class to be discovered are compounds that act as terminators of chain elongation, that is, that can be incorporated by the enzyme into the newly synthesized strand, but which lack groups necessary for further extension of the chain. Chain terminators, which are analogs of the nucleoside substrates and bind in the substrate binding site, can inhibit both HIV-1 and HIV-2 RT, or any other polymerase that uses similar substrates. For these reasons, many were originally developed as cancer drugs, even though none of them found practical use for that purpose. Another class of RT inhibitors are compounds called non-nucleoside inhibitors (NNI's) that are specific to a pocket which is found in the vicinity of the active site in HIV-1 RT, but which does not exist in HIV-2 RT. These are noncompetitive inhibitors that were found largely by serendipity, and their practical application as AIDS drugs was originally subject to considerable controversy. Eventually some NNI's have been found to be excellent drugs if used in combination therapy, even though they are largely useless in monotherapy.

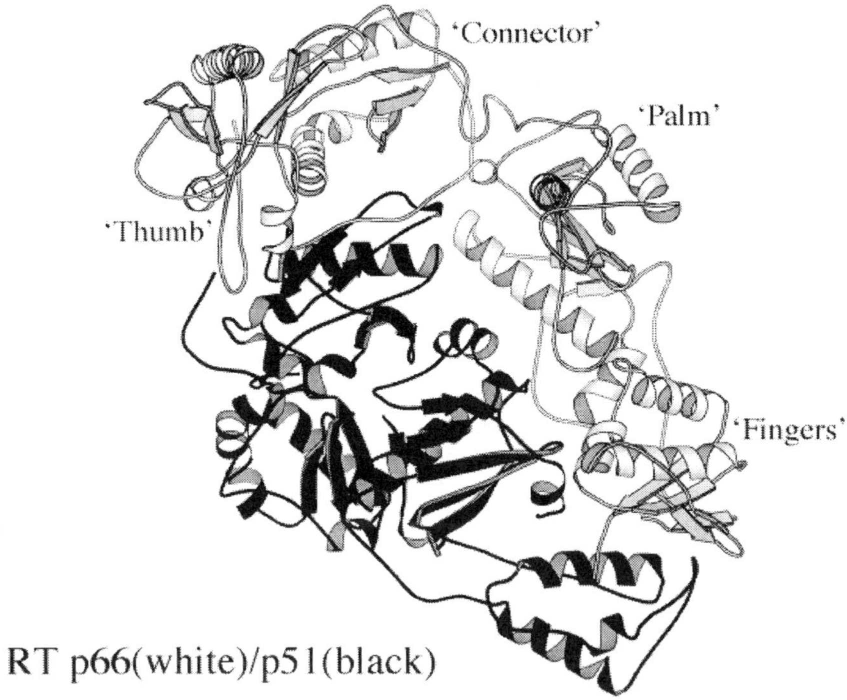

Figure 1. Ribbon structure of RT. The coordinates used to generate this figure are from Ren *et al.* (1995); PDB code 1REV. The active p66 domain is shown in gray, while the p51 domain is black.

2.1 Nucleoside Analogs

The nucleoside analogs are competitive inhibitors of HIV reverse transcriptase activity, even though they are not traditional mechanism-based inhibitors which bind to the enzyme tighter than the substrate (as are protease inhibitors, discussed below), but rather become incorporated into the growing product chains and prevent their further extension. Their inhibitory properties are due either to the lack of 2' or 3' hydroxyl groups, or to their replacement by other functional groups. In the case of AZT (azidothymidine, zidovudine), for example, the presence of the 3'-azido group prevents subsequent creation of a 3'-5' phosphodiester bond and thus terminates the chain (Figure 2).

The introduction of nucleoside analogs as potential AIDS drugs was based on the quickly growing understanding of the mechanism of action of retroviral enzymes such as RT, rather than on any detailed knowledge of their structure, which was still unknown at the time when these compounds were introduced into medical practice. Nucleoside analogs are prodrugs in the sense that they need to be phosphorylated in order to gain inhibitory properties against RT. Such phosphorylation depends on the presence of specific enzymes in the host organism.

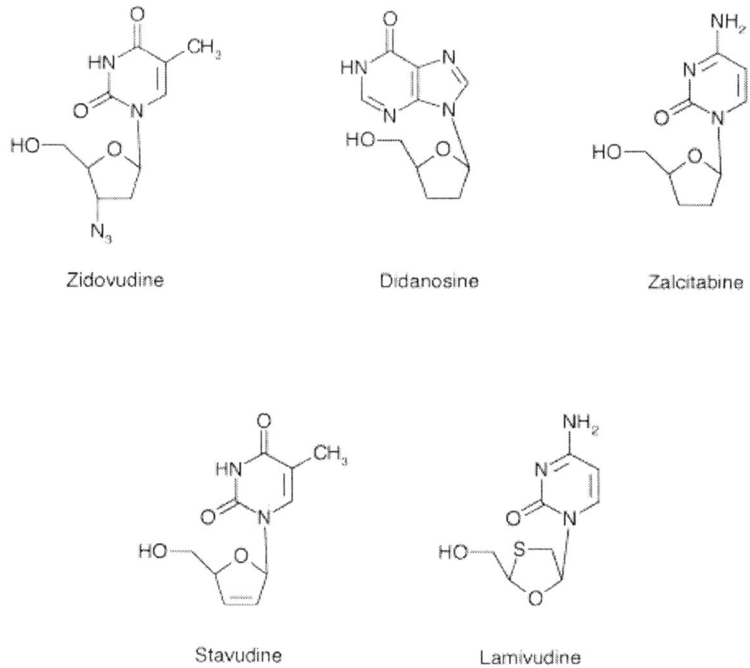

Figure 2. Chemical structures of clinically important nucleoside analog inhibitors of RT.

Five nucleoside analogs (NRTI's) have so far been approved by the U.S. Food and Drug Administration (FDA). The first of these was zidovudine (Retrovir, Glaxo-Wellcome). Its use was initially approved as a monotherapy in 1987, although its efficacy in that mode was shown to be only transitory (Volberding *et al.*, 1990). However, the importance of zidovudine as a therapeutic agent against AIDS cannot be overemphasized, since for several years it was the only generally available approved drug. The activities of all other anti-HIV compounds have been routinely compared to it ever since. In particular, usefulness in preventing transmission of HIV from mother to child during pregnancy was of special note. The drug is delivered orally and has very high bioavailability in that mode. Its affinity for HIV-1 RT is two orders of magnitude higher than for human DNA polymerase, thus minimizing potential side effects, which are, however, still considerable. The most common side effect of the continuous use of zidovudine is anemia.

Another nucleoside analog AIDS drug is didanosine (dideoxyinosine: ddI, Videx), a product of Bristol-Myers Squibb. It is an analog of inosine, lacking both the 2'- and 3'-hydroxyl groups on its ribose moiety. In common with zidovudine, its active form is also a triphosphate, produced by a cellular enzyme. Its intracellular half-life is 8-24 hours, much longer than that of zidovudine, thus allowing once-a-day dosing. Since the compound is inactivated by an acidic environment, it is usually buffered in order to increase gastric pH.

Another drug related to didanosine is zalcitabine (dideoxycytidine: ddC, Hivid), a product of Hoffmann-La Roche. This compound has been FDA-approved since 1992 and its mode of action is the same as that of other nucleoside analogs. This pyrimidine analog is active against HIV in vitro at very low concentrations, although its plasma half-life is rather short, requiring several daily doses of the drug. The principal side effects are pancreatitis and peripheral neuropathy.

Stavudine (d4T: 3'-deoxy-2'-thymidinene, Zerit), manufactured by Bristol-Myers Squibb, is a modification of thymidine. It has been used in clinical practice since 1993, and was approved for the treatment of HIV-infected adults who have received prolonged zidovudine therapy (Spruance *et al.*, 1997). The drug has been found to be generally well tolerated and has minimal side effects.

The latest of the currently approved nucleoside inhibitors of RT is lamivudine (3TC: 3'-thio-2',3'-dideoxcytidine, Epivir), which was discovered by BioChem Pharma and developed by Glaxo-Wellcome. It gained FDA approval in 1995. In addition to being a potent AIDS drug, lamivudine has also been shown to be potent against another viral disease, chronic hepatitis B. Since lamivudine has been shown to elicit very rapid resistance (see below), its initial development was questionable. However, after it was found that the principal mutation due to the exposure to lamivudine, M184V, prevents resistance to zidovudine, a combination of both drugs (Combivir, which consists of 150 mg of lamivudine combined with 300 mg of zidovudine) was developed.

2.2 Non-nucleoside analogs (NNI's)

All members of this heterogeneous class of compounds (Figure 3) are potent inhibitors of HIV-1 RT (but not of HIV-2 RT, or indeed of reverse transcriptase from any other retroviruses). The first few of these compounds, such as HEPT (Baba *et al.*, 1989) or TIBO (Pauwels *et al.*, 1990) were discovered to be active in cell culture before their target was identified. A number of other members of this class, including nevirapine, were identified in screening programs specifically targeting HIV-1 RT (Merluzzi *et al.*, 1990). The NNI's bind in a pocket located about 10 Å from the substrate-binding site, which includes residues such as Val179, Tyr181, Tyr188, and Trp229. When bound into that pocket in HIV-1 RT, most NNI's maintain a similar, butterfly-like shape. Most of them superimpose structurally quite well and appear to act as π-electron donors to the aromatic side chains surrounding the pocket (Kroeger *et al.*, 1997). The mode of action of NNI's is not completely clear, although it has been suggested that they might alter the conformation of the active site due to their proximity, or else restrict the motions of the p66 thumb domain (Kohlstaedt *et al.*, 1992).

So far, two NNI's have gained FDA approval for use against HIV (De Clercq, 1998). Nevirapine (Viramune, manufactured by Boehringer-Ingelheim) was the first of these to gain approval in 1996. This compound is highly bioavailable in oral form. It induces its own metabolism by activating the hepatic cytochrome P450 pathways. The main side effect of its utilization is a rash. The structure of a complex of nevirapine with RT was the first reverse transcriptase structure to be determined by x-ray crystallography (Kohlstaedt *et al.*, 1992).

Another FDA-approved NNI is delaviridine (Rescriptor), manufactured by Pharmacia. This compound is usually prescribed in combination with other anti-HIV drugs, either nucleoside RT inhibitors or PR inhibitors. It has relatively short plasma half-life, requiring several daily doses to maintain its concentration.

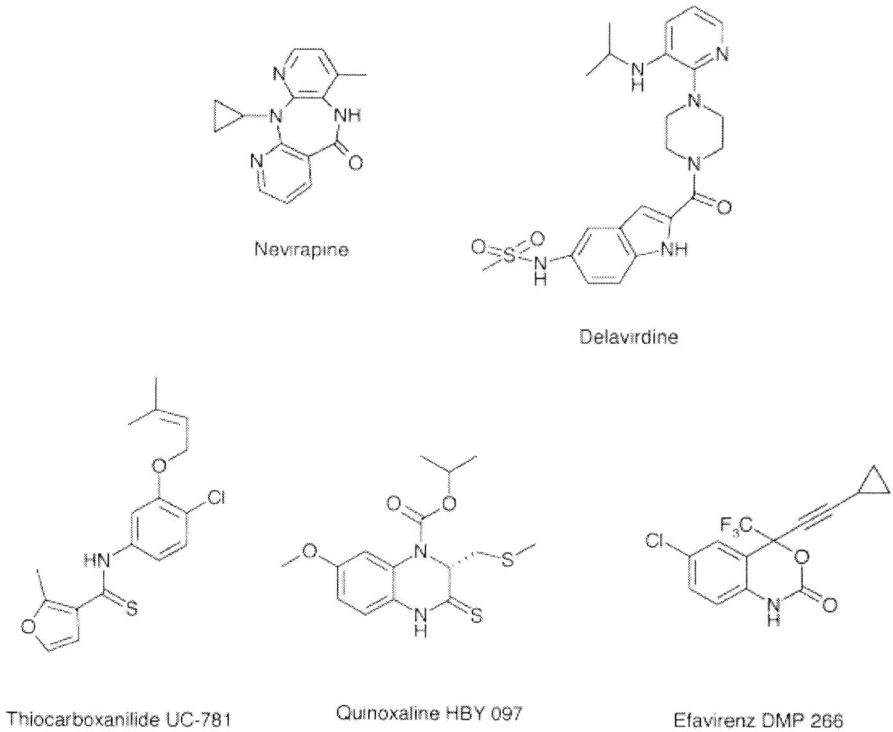

Figure 3. Chemical structures of clinically important non-nucleoside RT inhibitors.

2.3 Development of resistance to drugs targeting RT

Due to the lack of an editing function in retroviral RT, transcription errors during nucleic acid replication are very common, and the viral pool contains species with all conceivable mutations. The presence of drugs provides a powerful selection pressure for virus modifications that produce lower susceptibility to such compounds. This is especially true for monotherapy, in which a single drug is expected to provide suppression of viral replication. Development of resistance is often observed very soon after initiation of therapy, often after only a week or two. Such phenomena were observed early on, when zidovudine was the only approved AIDS drug, and rapid appearance of drug-resistant HIV species was considered a major block in the development of newer therapies, such as PR inhibitors or NNI's.

While nucleoside inhibitors of RT have been in clinical use for almost 15 years, the mechanism of resistance to them has been elucidated only recently in the studies by Huang and colleagues (1998). The structure of a trapped catalytic complex of RT provided data on the exact location of the incoming deoxynucleoside triphosphate and, by extension, of the nucleoside drugs. Not surprisingly, the point mutations sufficient for resistance (K65R, K70R, L74V, Q151M, M184I/V, and T215Y) are all located in the vicinity of the incoming nucleotide and may affect directly the position, stability, or reactivity of the bound analog (Figures 4a, 5a). There appears to be direct correlation between the location of mutation sites and the chemical nature of the analogs, explaining why mutations such as K70R are particularly responsible for the development of resistance to zidovudine with its extra azido group. This mutation is often followed by further ones such as T215Y/F, K210W, which may stabilize the mutation of residue 215, M41L, D67N, and K219Q. Conversely, mutations responsible for the resistance to dideoxynucleotides are located on the opposite surface of the enzyme and include L74V, M174V, K65R, and T69D. The mutation Q151M arises primarily in the patients on dual therapy (zidovudine and didanosine or zalcitabine), affecting both contact with the deoxynucleoside triphosphate and the character of the 3' pocket. An analysis of the steric nature of these mutations can also explain why resistance to one class of inhibitors may sensitize the enzyme to another class, forming the basis of sequential therapy.

The emergence of resistance to NNI's is particularly rapid, since the binding site for these compounds in not a direct part of the active site of RT (De Clerq, 1998; Schinazi *et al.*, 1997). As a rule, mutations leading to resistance to NNRTI's involve residues lining the binding site of these inhibitors (Figures 4b, 5b). Two mutations, K103N and Y181C, are induced by almost all known NNRTI's except the quinoxalines (De Clercq, 1998). Other common mutations which induce resistance to one or more NNRTI's include L100I, V106A/I/L and G190A/T/V. It was originally thought that the rapid emergence of resistant variance would compromise the utility of these drugs in the clinic. However, it is now clear that resistance can be minimized, both by using NNRTI's in combination with other inhibitors and by starting therapy with high concentrations of the drugs. NNRTI's are generally prescribed in combination with drugs from other classes (Coleman and Holtzer, 1998).

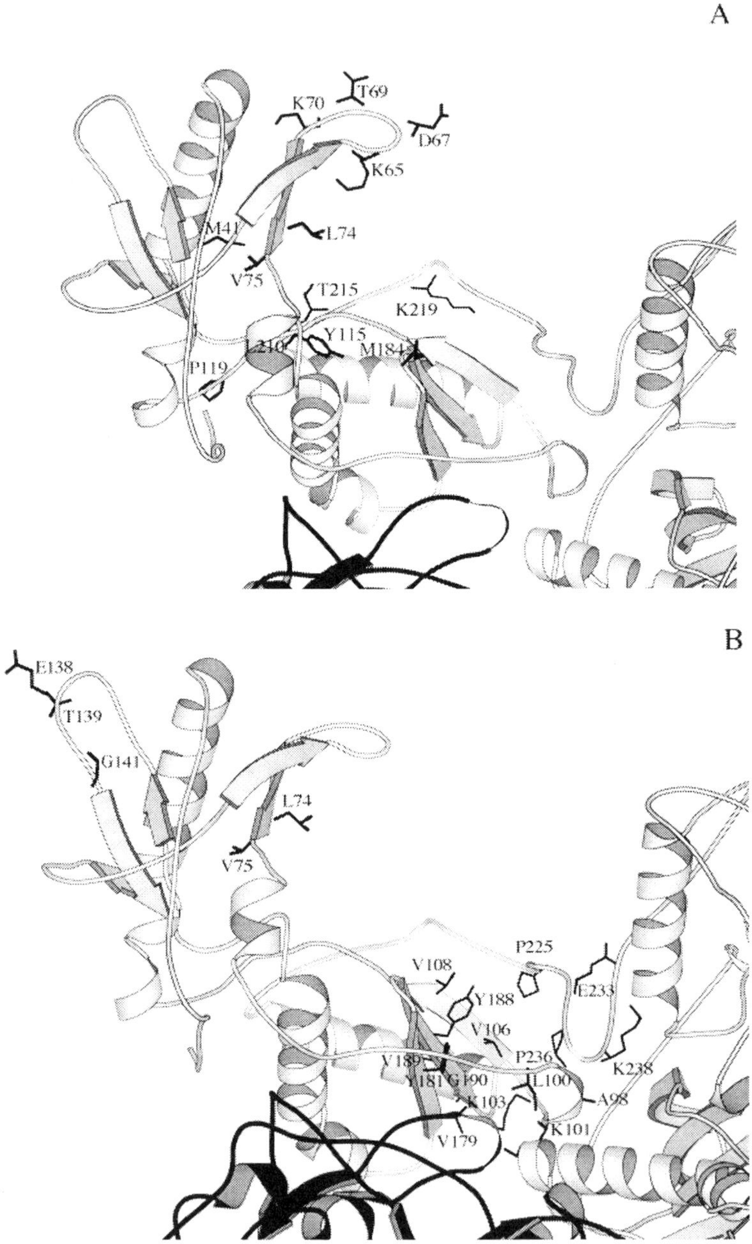

Figure 4. Ribbon structure of a fragment of RT showing amino acids with mutations arising from the use of either **A:** NRTI's or **B:** NNRTI's.

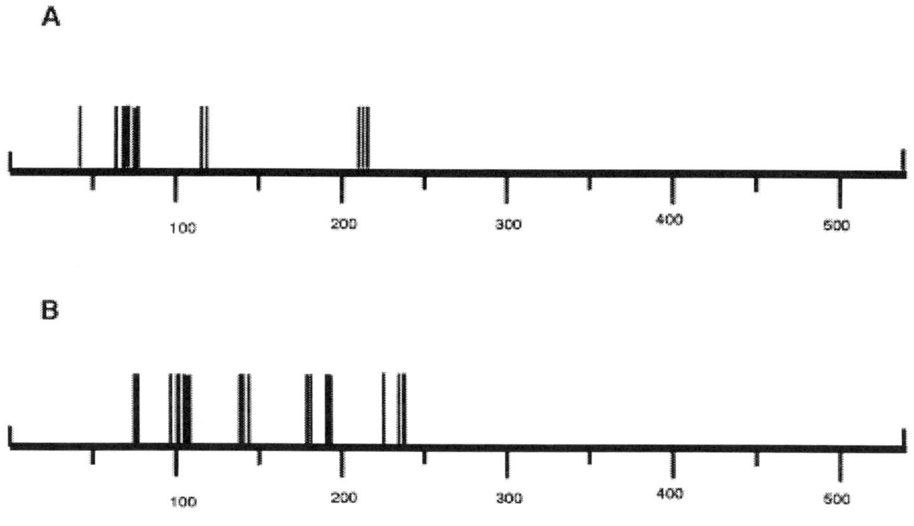

Figure 5. Schematic diagram comparing positions of RT mutations arising from the use of NRTI's and NNRTI's. Data taken from Schinazi *et al.* (1997). **A**: mutations arising from the use of NRTI's; **B**: mutations arising from the use of NNRTI's.

3. DRUGS TARGETING PROTEASE

Retroviral proteases are unique proteins in the sense that they are the only known homodimeric enzymes containing a single, symmetric active site, which includes two adjacent aspartic acid side chains (Figure 6). One aspartate is protonated and the complex is stabilized by a hydrogen bond between them (Pearl, 1987). While each of the two molecules forming a retroviral protease resembles a domain of a single-chain aspartic protease such as pepsin or renin, the latter enzymes retain only approximate twofold symmetry. However, the similarity of the retroviral protease to cellular aspartic proteases was invoked early on as the reason to concentrate on designing inhibitors of that enzyme as potential AIDS drugs. There was a long history of attempts to design renin inhibitors as antihypertensive drugs (Greenlee, 1990). While no such drugs have ever been successfully introduced, the immense experience gained in that work turned out to be invaluable for designing pharmacologically successful inhibitors of HIV PR.

In both HIV-1 and HIV-2 PR, each chain consists of 99 amino acids, with about 50% identity between the enzymes from these two variants of HIV. The principal active site residue is Asp25. Mutation of this residue to any other, including Asn, completely inactivates the enzyme (Kohl *et al.*, 1988). The determination of the crystal structure of the apoenzyme (Navia *et al.*, 1989; Wlodawer *et al.*, 1989) was rapidly followed by the structure of a complex with a substrate-based inhibitor (Miller *et al.*, 1989), with many other similar structures to follow (Wlodawer and Erickson, 1993; Wlodawer and Vondrasek, 1998). These structures, as well as a

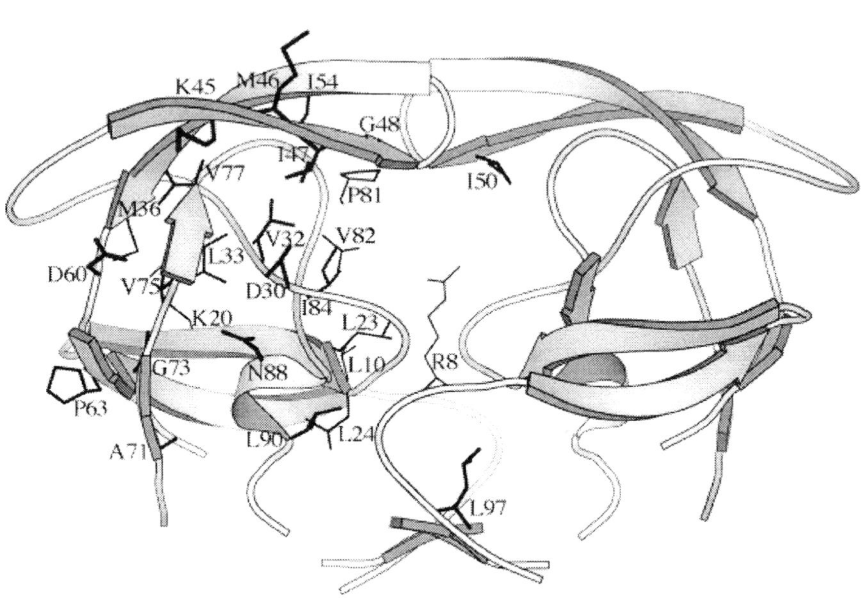

Figure 6. Ribbon structure of protease showing amino acids with drug-resistant mutations found in one of the two identical subunits. Structure based on Swain *et al.* (Swain *et al.*, 1990); PDB code 7HVP.

number of structures solved using NMR (Yamazaki *et al.*, 1996) provided crucial information for the design of new classes of inhibitors, which, ultimately, became an important new category of AIDS drugs (Vacca and Condra, 1997; Wlodawer and Vondrasek, 1998). All publicly available structures of HIV PR and the related simian retrovirus SIV have been made available on the Web in the HIVdb database (Vondrasek and Wlodawer, 1997; Wlodawer and Vondrasek, 1999). The current version of HIVdb, released in early 2000, contains 142 different structures of HIV-1 PR, of which 35 (25%) are mutants, and 24 structures of HIV-2 PR of which 14 (58%) are mutants. The database contains structures of HIV-1 PR in complex with all approved protease drugs, and many other potential drugs and drugs in development; about 35% are "unofficial releases" of structures not available in the Protein Data Bank. This is a useful resource for studying the structural effects of mutations and their influence on drug interactions.

3.1 Drugs interacting with the active site

A very large number of compounds aimed at inhibiting HIV PR by binding to the active site of the enzyme have been created (Figure 7). The initial approach was based on the observation that the smallest oligopeptide substrates that can be efficiently processed by the enzyme consist of about six amino acids, three on the N-terminal side and three on the C-terminal side of the bond to be cleaved. Each amino acid of the substrate occupies a matching subsite in the enzymes, with the subsites on both sides of the scissile bond being symmetric. The early inhibitors were peptide-mimetics: peptides mimicking the specific protease cleavage sites, with a nonscissile insert substituting for the peptide bond which would be normally cleaved. A particular sequence, cleaved effectively by retroviral proteases only, but not by the related cellular enzymes, contained a central sequence Tyr-Pro or its modifications. This dipeptide was used in the design of many early inhibitors.

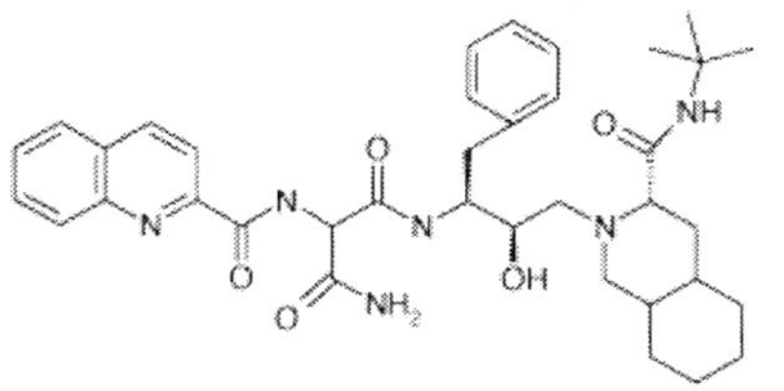

Saquinavir

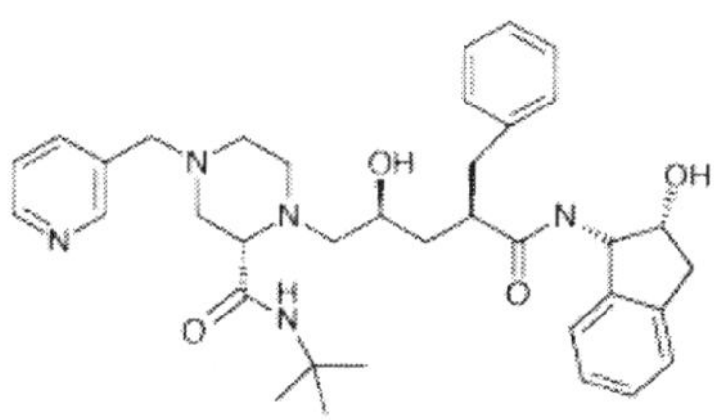

Indinavir

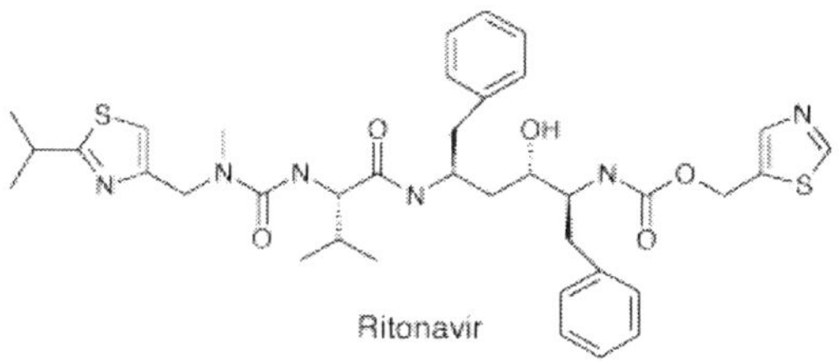

Ritonavir

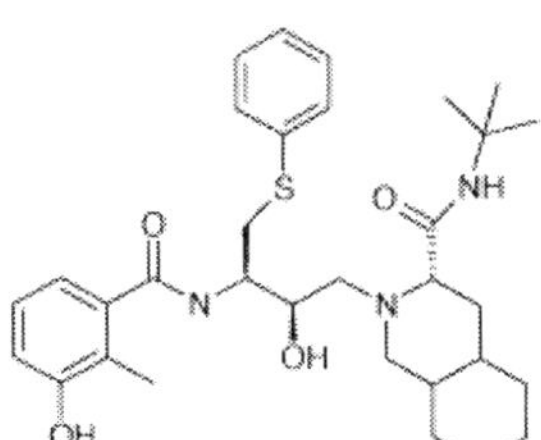

Nelfinavir

Figure 7. Chemical structures of some clinically important protease inhibitors.

Five PR inhibitors have gained FDA approval as AIDS drugs. The first of these, saquinavir (Ro 31-8959; initially formulated by Hoffmann-La Roche as hard gel capsules under the name of invirase, and later as soft gel capsules as fortovase) was approved in 1995. This compound was created from a short peptide by replacing the scissile peptide bond by a hydroxyethylamine moiety, unexpectedly with *R* stereochemistry. The proline residue that followed the nonscissile insert was changed to (S, S, S)-decahydroisoquinoline-3-carbonyl (DIQ), resulting in a considerable increase in potency. While the presence of several chiral centers made the synthesis difficult, and its bioavailability was rather poor, saquinavir turned out to be quite useful, especially due to its almost negligible side effects.

Ritonavir (Norvir) was developed at Abbott Laboratories in a program that initially made significant use of the symmetric nature of HIV PR, as compared with the only quasi-symmetric nature of cellular proteases. This compound also utilizes a hydroxyethylene replacement for the scissile bond, located between two phenylalanines, but in *S* stereochemistry. Unexpectedly, ritonavir was found to be a powerful inhibitor and inducer of the cytochrome P450 metabolic pathway system, and thus the initial dose of the drug needs to be subsequently raised to maintain its proper level. This drug is the least well-tolerated of all the approved protease inhibitors, but its side effects (related to its P450 activity) can be turned into its actual strength, by utilizing it together with new-generation inhibitors such as ABT-378, which is now undergoing Phase II clinical trials (Carillo *et al.*, 1998). This unexpected property of ritonavir may be the main reason for its use in the future.

Indinavir (Crixivan), developed by Merck Pharmaceuticals, was licensed by FDA in 1996. Like saquinavir, it was designed on the peptide-mimetic principle. The drug should be given only under fasting conditions, making its delivery comparatively more difficult. It is, however, relatively safe and well tolerated. In combination with other AIDS drugs such as zidovudine and lamivudine, indinavir was shown to be capable of durable suppression of the virus for periods as long as over 2 years.

Nelfinavir (Viracept) is an Agouron Pharmaceuticals drug, which was approved for adult use in 1997. Subsequently it became the first protease inhibitor to be approved for pediatric use. Chemically it shares parts of its structure (DIQ group) with saquinavir, although it is much smaller, and has much higher oral bioavailability. Since its introduction three years ago it has become the most highly prescribed PR inhibitor, even though it exhibits some side effects, particularly diarrhea.

The Glaxo-Wellcome drug amprenavir (also known as 141W94, or under the designation VX-478 by Vertex, the company where it was first synthesized) is the latest to be approved. Because of its long half-life, it can be administered only twice a day. Several protease inhibitors are in advanced clinical studies and thus widely available; the following list is almost certainly only partial. Pharmacia is conducting an early clinical trial of PNU-140690, a unique PR inhibitor with completely non-peptidic character. Triangle Pharmaceuticals is developing DMP-450, a symmetric cyclic urea derivative with unique properties, originally designed at DuPont Merck.

A number of other inhibitors are in earlier stages of develpoment.

3.2 Resistance to PR inhibitors

The appearance of HIV strains with reduced susceptibility to PR inhibitors has been monitored both in vivo and in vitro, using clinically approved inhibitors, as well as a variety of other ones. A recent study of such mutations has shown that about one third of the residues in HIV-1 PR have been found to be mutated in samples obtained with the help of 21 drugs (Schinazi *et al.*, 1997) (Figures 6, 8). While some of these mutations are in the pocket directly adjacent to the inhibitors, other mutations are observed throughout the protein. The appearance of the mutations is usually sequential, and remote mutations usually develop subsequently to the primary ones.

The most common mutations elicited by saquinavir are G48V and L90M, both in the drug-binding pocket. These mutations decrease the potency of the inhibitor several-fold. The nature of the formulation of the drug (soft or hard capsules) does not modify the pattern of resistance. The discovery of a pattern of multiple resistance mutations in patients subjected to indinavir monotherapy (Condra *et al.*, 1995), as well as cross-resistance with six other PR inhibitors, has raised serious questions about the possible efficacy of the drugs belonging to that category. This initial pessimism, however, has turned out to be unwarranted, since the use of sufficiently high doses of the drugs, and combination therapies, have been shown to be quite successful in delaying or overcoming the appearance of resistance.

The importance of the appearance of drug-resistant mutants of HIV PR is considered to be so high that resistance studies now precede any attempts of introducing such compounds into clinical practice. A good example is provided by ABT-378, a new PR inhibitor from Abbott Laboratories, which is currently in phase II clinical trials. Serial passages of the virus grown in the presence of the inhibitor established a sequential pattern of resistance development The mutations appear in the order I84V, L10F, M46I, T91S, V32I, and I47V. Further selection led to a mutation V47A, followed by reversion of I32 back to V (Carillo *et al.*, 1998). It is likely that similar mutations will also be observed in vivo. Clearly, the development of drug-resistant strains of HIV is quite complicated.

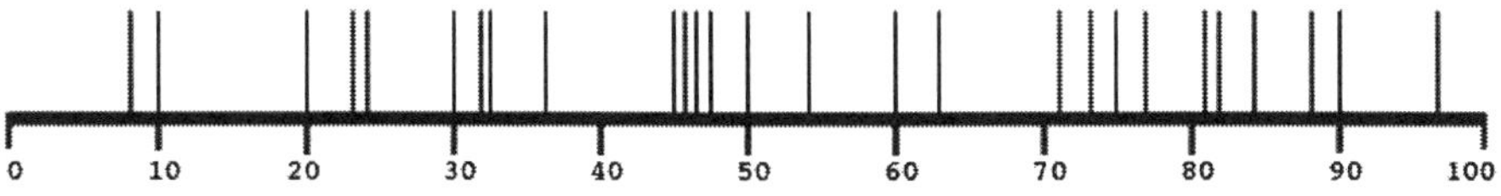

Figure 8. Schematic diagram showing mutations in PR arising from the use of clinically important PR inhibitors.

4. DRUGS TARGETING INTEGRASE

HIV integrase (IN) is the third enzyme found in the viral genome, and thus is a natural target for drug design. This three-domain protein is absolutely required for the support of viral life cycle, with its role being the processing of the DNA copy of the viral genome and its integration into the genome of the host. While the structure of intact integrase is not yet available, crystal and NMR structures of individual domains have been published in the last few years (Cai *et al.*, 1997; Dyda *et al.*, 1994; Eijkelenboom *et al.*, 1995; Eijkelenboom *et al.*, 1997; Goldgur *et al.*, 1998; Lodi *et al.*, 1995; Maignan *et al.*, 1998). These structures provided detailed data which is potentially very useful for drug design, although at this time no drugs have been approved, and only one clinical trial is under way (see below).

The amino- and carboxy-terminal domains of integrase are directly involved in the binding of the DNA substrates, but not directly in the catalytic activity of the enzyme. The central domain (residues ~50-210) (Figure 9) is directly responsible for the catalysis by integrase, and is capable of at least limited catalytic activity in the absence of the two other domains. The structure of that domain (Dyda *et al.*, 1994) elucidated a close relationship between integrase and other nucleotidyl transferases such as RNase H, phage μ resolvase, and RuvC. The enzyme also bears distant relationship to a number of DNA and RNA polymerases. It was puzzling, however, that the active site as seen in the structure did not show a typical divalent cation binding motif, especially since the crystal structure of a related integrase from avian sarcoma virus showed that it could bind divalent cations such as magnesium or manganese (Bujacz *et al.*, 1996). This discrepancy was recently resolved with new structures of the catalytic domain of HIV IN, which showed bound cations and proved that the original structure corresponded to an inactive enzyme (Goldgur *et al.*, 1998; Maignan *et al.*, 1998).

Although extensive efforts to discover inhibitors of integrase have been going on for a number of years (Pommier *et al.*, 1997), the only putative integrase inhibitor currently under clinical development is AR-177 (Zintevir) from Aronex. That compound belongs to the family of guanosine-quartets but while it can interact with integrase (Cherepanov *et al.*, 1997), it might inhibit HIV primarily by a completely different mechanism, involving interactions with the envelope glycoprotein gp120 (Este *et al.*, 1998). Neither the future therapeutic potential of Zintevir nor its mode of action are certain at this time. Interestingly, one of the indications that integrase might not be its primary target comes from an in vitro study of the resistant viral strains, which showed mutations in the gp120 gene, but no mutations in the part of the genome encoding integrase.

5. STRATEGIES TO OVERCOME DRUG RESISTANCE

In the decade and a half since the first anti-HIV drug entered clinical trials, it became absolutely clear that any monotherapy utilizing enzyme inhibitors must lead to the emergence of resistant strains of the virus. While the time until such resistance is apparent varies among different classes of drugs and the individual compounds, it may be extremely rapid. It can be as short as about a week for a non-nucleoside analog nevirapine (Coleman and Holtzer, 1998). Combination therapies promise a

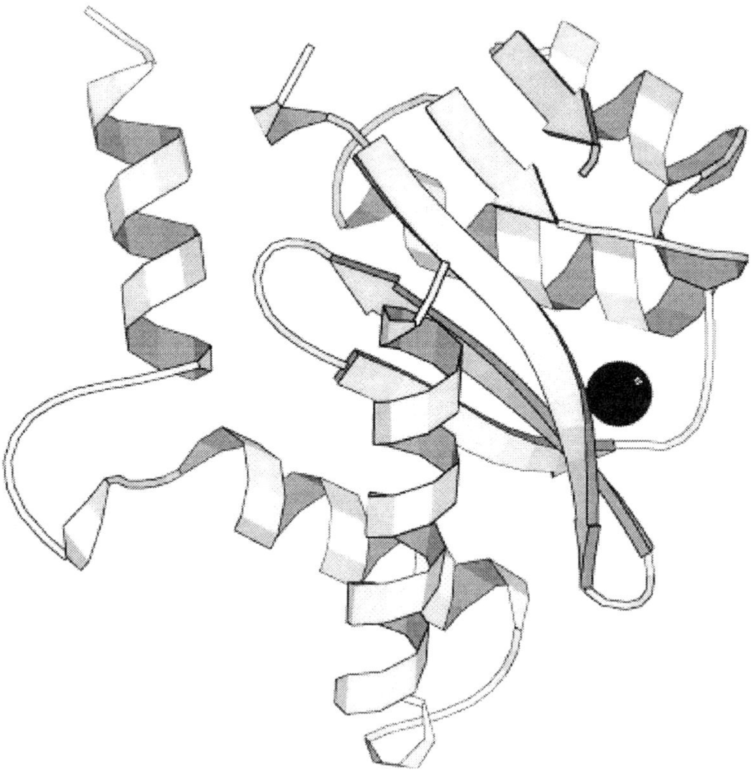

Figure 9. Ribbon structure of the central catalytic domain of integrase, based on the structure by Goldgur *et al.* (1998); PDB code 1BIU. The ball corresponds to a Mg^{2+} ion.

much better outcome, although there is no clear agreement at this time on which particular drug combinations might be the best. In particular, it is not clear whether it is better to use combinations of different drugs from the same family, or drugs belonging to different classes. On one hand, combinations such as ritonavir-saquinavir, nelfinavir-saquinavir, or ritonavir-indinavir combine two similar drugs but of distinct resistance pattern and, especially in the case of ritonavir, of different metabolism. On the other hand, combinations of indinavir or nelfinavir with nevirapine, or indinavir plus efavirenz, assure that the development of resistance will require mutations in two different enzymes, making such events less likely. In any case, starting from the initial clinical trials most therapeutic regimens utilizing protease inhibitors and non-nucleoside RT inhibitors also utilized nucleosides as backup. Since the combinations of nucleoside and non-nucleoside drugs are often effective, they are sometimes preferred in order to leave PR inhibitors as potential backup (Coleman and Holtzer, 1998) if the RT therapy fails.

In Western countries, drug treatment is reducing AIDS to a manageable and treatable long-term disease. However, even with eleven drugs already on the market, it is clear that the serious nature of the AIDS pandemic and the limitations of the therapies make it necessary to continue the process of drug development. Until a safe, effective vaccine against HIV has been developed, it will always be necessary to introduce new therapies and combinations of drugs to counteract the development of resistant variants. It is also important to consider expense, as the current drugs cost thousands of dollars a year per patient. Any drug that is to be of use in those Third World countries where AIDS is most prevalent must be cheap and easy to synthesize. It may also be necessary to develop drugs specifically targeted to HIV-2 to counteract the African pandemic. The understanding of the drug-target interactions on the molecular levels, coupled with extensive studies using the techniques of molecular biology, are of great help in achieving rapid success of this work.

ACKNOWLEDGEMENTS

We thank Jerry Alexandratos for help in preparation of the figures. A.W. would like to thank the Master and Fellows of the Sidney Sussex College and the Department of Biochemistry, University of Cambridge, for a Visiting Fellowship during which tenure this paper was written. The contents of this publication do not necessarily reflect the views or policies of the Department of Human Services, nor does mention of trade names, commercial products, or organizations imply endorsement by the U.S. Government.

REFERENCES

Baba, M., Tanaka, H., De Clercq, E., Pauwels, R., Balzarini, J., Schols, D., Nakashima, H., Perno, C. F., Walker, R. T., and Miyasaka, T. 1989. Highly specific inhibition of human immunodeficiency virus type 1 by a novel 6-substituted acyclouridine derivative. *Biochem. Biophys. Res. Commun.* **165**: 1375-1381.

Balkenkohl, F., Von Dem Bussche-Hunnefield, C., Lansky, A.and Zechel, C. 1996. Combinatorial synthesis of small molecules. *Angew. Chem. Int.* **35**:2288-2337.

Bujacz, G., Jaskólski, M., Alexandratos, J., Wlodawer, A., Merkel, G., Katz, R. A.and Skalka, A. M. 1996. The catalytic domain of avian sarcoma virus integrase: conformation of the active-site residues in the presence of divalent cations. *Structure* **4**: 89-96.

Cai, M., Zheng, R., Caffrey, M., Craigie, R., Clore, G. M. and Gronenborn, A. M. 1997. Solution structure of the N-terminal zinc binding domain of HIV- 1 integrase. *Nat. Struct. Biol.* **4**: 567-577.

Carrillo, A., Stewart, K. D., Sham, H. L., Norbeck, D. W., Kohlbrenner, W. E., Leonard, J. M. Kempf, D. J. and Molla, A. 1998. In vitro selection and characterization of human immunodeficiency virus type 1 variants with increased resistance to ABT-378, a novel protease inhibitor. *J. Virol.* **72**: 7532-7541.

Cherepanov, P., Este,J. A., Rando, R. F., Ojwang, J. O., Reekmans, G., Steinfeld, R., David, G., De Clercq, E., and Debyser, E. 1997. Mode of interaction of G-quartets with the integrase of human immunodeficiency virus type 1. *Mol. Pharmacol.* **52**: 771-780.

Coleman, R. L. and Holtzer, C. 1998. AIDS KNOWLEDGE DATABASE. 4) HIV-Related Drug Information. *http://hivinsite.ucsf.edu/akb/1997/.*

Condra, J. H., Schleif, W. A., Blahy, O. M., Gabryelski, L. J., Graham, D. J., Quintero, J. C., Rhodes, A.,

Robbins, H. L., Roth, E., Shivaprakash, M., Titus, D., Yang, T., Teppler, H., Squires, K. E., Deutsch, P. J.and Emini, E. A. 1995. *In vivo* emergence of HIV-1 variants resistant to multiple protease inhibitors. *Nature* **374**: 569-571.

De Clercq, E. 1998. The role of non-nucleoside reverse transcriptase inhibitors (NNRTIs) in the therapy of HIV-1 infection. *Antiviral Res.* **38**: 153-179.

Dyda, F., Hickman, A. B., Jenkins, T. M., Engelman, A., Craigie, R. and Davies, D. R. 1994. Crystal structure of the catalytic domain of HIV-1 integrase: Similarity to other polynucleotidyl transferases. *Science* **266**: 1981-1986.

Eijkelenboom, A. P., Lutzke, R. A., Boelens, R., Plasterk, R. H, Kaptein, R.,and Hård, K. 1995. The DNA-binding domain of HIV-1 integrase has an SH3-like fold. *Nat. Struct. Biol.* **2**: 807-810.

Eijkelenboom, A. P., van den Ent, F. M., Vos, A., Doreleijers, J. F., ard, K., Tullius, T. D., Plasterk, R. H., Kaptein, R. and Boelens, R. 1997. The solution structure of the amino-terminal HHCC domain of HIV-2 integrase: a three-helix bundle stabilized by zinc. *Curr. Biol.* **7**: 739-746.

Este, J. A., Cabrera, C., Schols, D., Cherepanov, P., Gutierrez, A., Witvrouw, M., Pannecouque, C., Debyser, Z., Rando, R. F., Clotet, B., Desmyter, J. and De Clercq, E. 1998. Human immunodeficiency virus glycoprotein gp120 as the primary target for the antiviral action of AR177 (Zintevir*). Mol. Pharmacol.* **53**: 340-345.

Goldgur, Y., Dyda, F., Hickman, A. B., Jenkins, T. M., Craigie, R. and Davies, D. R. 1998. Three new structures of the core domain of HIV-1 integrase: an active site that binds magnesium. *Proc. Natl. Acad. Sci. USA* **95**: 9150-9154.

Greenlee, W. J. 1990. Renin inhibitors. *Med. Res. Rev.* **10**: 173-236.

Huang, H., Chopra, R., Verdine, G. L., and Harrison, S. C. 1998. Structure of a covalently trapped catalytic complex of HIV-1 reverse transcriptase: implications for drug resistance. *Science* **282**: 1669-1675.

Hubbard, R. E. 1997. Can drugs be designed? *Curr. Opin. Biotechnol.* **8**: 696-700.

Jacobo-Molina, A., Ding, J., Nanni, R. G., Clark, Jr., A. D., Lu, X., Tantillo, C., Williams, R. L., Kamer, G., Ferris, A. L. and Clark, P. 1993. Crystal structure of human immunodeficiency virus type 1 reverse transcriptase complexed with double-stranded DNA at 3.0 A resolution shows bent DNA. *Proc. Natl. Acad. Sci. USA* **90**: 6320-6324.

Jager, J., Smerdon, S. J., Wang, J., Boisvert, D.C. and Steitz, T. A. 1994. Comparison of three different crystal forms shows HIV-1 reverse transcriptase displays an internal swivel motion. *Structure* **2**: 869-876.

Kohl, N. E., Emini, E. A., Schleif, W. A., Davis, L. J., Heimbach, J. C., Dixon, R. A., Scolnick, E. M. and Sigal, I. S. 1988. Active human immunodeficiency virus protease is required for viral infectivity. *Proc. Natl. Acad. Sci. USA* **85**: 4686-4690.

Kohlstaedt, L. A., Wang, J., Friedman, J. M., Rice, P. A.and Steitz, T. A. 1992. Crystal structure at 3.5 Å resolution of HIV-1 reverse transcriptase complexed with an inhibitor. *Science* **256**: 1783-1790.

Kroeger, S. M., Michejda, C. J., Hughes, S. H., Boyer, P. L., Janssen, P. A., Andries, K., Buckheit, R. W. J. and Smith, R. H. J. 1997. Molecular modeling of HIV-1 reverse transcriptase drug-resistant mutant strains: implications for the mechanism of polymerase action. *Protein Eng.* **10**: 1379-1383.

Leach, A. R. 1996. The use of molecular modelling to discover and design new molecules, In *Molecular Modelling Principles and Applications* (Leach, A. R., ed.) Longman Publishers Ltd., Singapore.

Lodi, P. J., Ernst, J., Kuszewski, J., Hickman, A. B., Engelman, A., Craigie, R., Clore, G. M. and Gronenborn, A. M. 1995. Solution structure of the DNA binding domain of HIV-1 integrase. *Biochemistry* **34**: 9826-9833.

Maignan, S., Guilloteau, J. P., Zhou-Liu, Q., Clement-Mella, C. and Mikol, V. 1998. Crystal structures of the catalytic domain of HIV-1 integrase free and complexed with its metal cofactor: high level of similarity of the active site with other viral integrases. *J. Mol. Biol.* **282**: 359-368.

Merluzzi, V. J., Hargrave, K. D. Labadia, M., Grozinger, K., Skoog, M., Wu, J. C., Shih, C. K., Eckner, K., Hattox, S. and Adams, J. 1990. Inhibition of HIV-1 replication by a nonnucleoside reverse transcriptase inhibitor. *Science* **250**: 1411-1413.

Miller, M., Jaskolski, M., Rao, J. K. M., Leis, J.and Wlodawer, A. 1989. Crystal structure of a retroviral protease proves relationship to aspartic protease family. *Nature* **337**: 576-579.

Navia, M. A., Fitzgerald, P. M., McKeever, B. M., Leu, C. T., Heimbach, J. C., Herber, W. K., Sigal, I. S., Darke, P. L. and Springer, J. P. 1989. Three-dimensional structure of aspartyl protease from human immunodeficiency virus HIV-1. *Nature* **337**: 615-620.

Pauwels, R., Andries, K., Desmyter, J., Schols, D., Kukla, M. J., Breslin, H. J., Raeymaeckers, A.,Van Gelder, J., Woestenborghs, R. and Heykants, J. 1990. Potent and selective inhibition of HIV-1 replication in vitro by a novel series of TIBO derivatives. *Nature* **343**: 470-474.

Pearl, L. H. 1987. The catalytic mechanism of aspartic proteinases. *FEBS Lett.* **214**: 8-12.

Pommier, Y., Pilon, A. A., Bajaj, K., Mazumder, A. and Neamati, N. 1997. HIV-1 integrase as a target for antiviral drugs. *Antivir. Chem. Chemother.* **8**: 463-483.

Ratner, L., Haseltine, W., Patarca, R., Livak, K. J., Starcich, B., Josephs, S. F., Doran, E. R., Rafalski, J. A., Whitehorn, E. A. and Baumeister, K. 1985. Complete nucleotide sequence of the AIDS virus, HTLV-III. *Nature* **313**: 277-284.

Ren, J., Esnouf, R., Hopkins, A., Ross, C., Jones, Y., Stammers, D. and Stuart, D. 1995. The structure of HIV-1 reverse transcriptase complexed with 9-chloro-TIBO: lessons for inhibitor design. *Structure* **15**: 915-926.

Sansom, C. 1998. Extending the boundaries of molecular modeling. *Nat. Biotechnol.* **16** : 917-918.

Schinazi, R. F., Larder, B. A. and Mellors, J. W. 1997. Mutations in retroviral genes associated with drug resistance. *Intl. Antiviral News* **5**: 129-142.

Spruance, S. L., Pavia, A. T., Mellors, J. W., Murphy, R., Gathe, J. J., Stool, E., Jemsek, J. G., Dellamonica, P., Cross, A. and Dunkle, L. 1997. Clinical efficacy of monotherapy with stavudine compared with zidovudine in HIV-infected, zidovudine-experienced patients. A randomized, double-blind, controlled trial. Bristol-Myers Squibb Stavudine/019 Study Group. *Ann. Intern. Med.* **126:** 355-363.

Swain, A. L., Miller, M. M., Green, J., Rich, D. H., Schneider, J., Kent, S. B. and Wlodawer, A. 1990. X-ray crystallographic structure of a complex between a synthetic protease of human immunodeficiency virus 1 and a substrate-based hydroxyethylamine inhibitor. *Proc. Natl. Acad. Sci. USA* **87**: 8805-8809.

Vacca, J. P. and Condra, J. H. 1997. Clinically effective HIV-1 protease inhibitors. *Drug Discov. Today* **2**: 261-272.

Volberding, P. A., Lagakos, S. W., Koch, M. A., Pettinelli, C., Myers, M. W., Booth, D. K., Balfour, H. H. J., Reichman, R. C., Bartlett, J. A. and Hirsch, M. S. 1990. Zidovudine in asymptomatic human immunodeficiency virus infection. A controlled trial in persons with fewer than 500 CD4-positive cells per cubic millimeter. The AIDS Clinical Trials Group of the National Institute of Allergy and Infectious Diseases. *N. Engl. J. Med.* **322**: 941-949.

Vondrasek, J. and Wlodawer, A. 1997. Database of HIV proteinase structures. *Trends Biochem. Sci.* **22**: 183-183.

Wlodawer, A. and Erickson, J. W. 1993. Structure-based inhibitors of HIV-1 protease. *Annu. Rev. Biochem.* **62**: 543-585.

Wlodawer, A., Miller, M., Jaskolski, M., Sathyanarayana, B. K., Baldwin, E., Weber, I. T., Selk, L. M., Clawson, L., Schneider, J. and Kent, S. B. H. 1989. Conserved folding in retroviral proteases: Crystal structure of a synthetic HIV-1 protease. *Science* **245**: 616-621.

Wlodawer, A. and Vondrasek, J. 1998. Inhibitors of HIV-1 protease: A major success of structure-assisted drug design. *Annu. Rev. Biophys. Biomol. Struct.* **27**: 249-284.

Wlodawer, A. and Vondrasek, J. 1999. Database of crystal structures of HIV protease. *http://www.ncifcrf.gov/HIVdb*

Yamazaki, T., Hinck, A. P., Wang, Y.-X., Nicholson, L. K., Torchia, D. A., Wingfield, P., Stahl, S. J., Kaufman, J. D., Chang, C.-H., Domaille, P. J. and Lam, P. Y. 1996. Three-dimensional solution structure of the HIV-1 protease complexed with DMP323, a novel cyclic urea-type inhibitor, determined by nuclear magnetic resonance spectroscopy. *Protein Science* **5**: 495-506.

INDEX